Bruno Kerl, F. Lynwood Garrison, William Theodore Brannt

The Assayer's Manual

An Abridged Treatise on the Docimastic Examination of Ores, and Furnace and

Other Artificial Products

Bruno Kerl, F. Lynwood Garrison, William Theodore Brannt

The Assayer's Manual
An Abridged Treatise on the Docimastic Examination of Ores, and Furnace and Other Artificial Products

ISBN/EAN: 9783337418250

Printed in Europe, USA, Canada, Australia, Japan

Cover: Foto ©berggeist007 / pixelio.de

More available books at **www.hansebooks.com**

THE ASSAYER'S MANUAL.

AN ABRIDGED TREATISE

ON THE

DOCIMASTIC EXAMINATION OF ORES, AND FURNACE AND OTHER ARTIFICIAL PRODUCTS.

BY

BRUNO KERL,

PROFESSOR IN THE ROYAL SCHOOL OF MINES; MEMBER OF THE ROYAL TECHNICAL COMMISSION
FOR THE INDUSTRIES, AND OF THE IMPERIAL PATENT OFFICE, BERLIN.

TRANSLATED FROM THE GERMAN
BY

WILLIAM T. BRANNT,

EDITOR OF "THE TECHNO-CHEMICAL RECEIPT BOOK," ETC.

SECOND AMERICAN EDITION.

EDITED WITH EXTENSIVE ADDITIONS
BY

F. LYNWOOD GARRISON,

MEMBER OF THE AM. INST. OF MINING ENGINEERS, FRANKLIN INSTITUTE, AND ACADEMY OF
NATURAL SCIENCES, PHILADELPHIA; FELLOW OF THE GEOLOGICAL SOCIETY
OF LONDON; MEMBER OF VEREIN DEUTSCHER EISENHÜTTENLEUTE
AND IRON AND STEEL INSTITUTE.

ILLUSTRATED BY EIGHTY-SEVEN ENGRAVINGS.

PHILADELPHIA:
HENRY CAREY BAIRD & CO.,
INDUSTRIAL PUBLISHERS, BOOKSELLERS, AND IMPORTERS,
810 WALNUT STREET.
LONDON:
E. & F. N. SPON, 125 STRAND.
1889.

COPYRIGHT BY
HENRY CAREY BAIRD & CO.,
1889.

COLLINS PRINTING HOUSE,
705 Jayne Street.

PREFACE TO THE SECOND AMERICAN EDITION.

The first American edition of KERL'S ASSAYER'S MANUAL has met with so much success that it has been determined to issue a second, containing, in addition, numerous and approved assaying methods which have, since the appearance of the first edition, been introduced. Although there have been many improvements in the apparatus used in assaying, it seems hardly the province of a work of this kind to constitute itself a catalogue of such apparatus. Only a few of the most notable of these novelties have therefore been mentioned. With the exception of the omission of the English equivalents of the weights and measures but few and unimportant changes have been made in the old text. The use of the metric system has become so universal amongst chemists and assayers in this country that the insertion of English equivalents was thought unnecessary and confusing. In the Appendix, however, there will be found tables of the relative values of the metric and English weights and measures.

While there has been but little improvement in the old *fire assay* methods the advancement in *wet* and particularly *electrolytic* methods has been notable. In this particular, the Editor has drawn largely from *Balling's Fortschritte im Probirwesen* (1887). The entire chapter on Tungsten was translated from that work, and it might be observed in this connection that it is believed that the element tungsten has never before been treated at length in any work on assaying. Although Kerl has confined the assaying of iron to a separate work on that subject, it has been deemed expedient in this case to prepare and insert in the Appendix a somewhat lengthy article on the Dry Assay of Iron Ores.

The Editor here desires to express his thanks to a number of his professional friends for assistance in preparing this work, and he feels under special obligations to Prof. Richard Smith of the Royal School of Mines, London; Dr. Edgar F. Smith of the University of Pennsylvania; and Prof. R. H. Richards of the Massachusetts Institute of Technology.

The portions of the volume contributed by the present Editor are inserted in smaller type than the text of Kerl. In some instances foot-notes to the original matter, added by the Editor, are inclosed in square brackets, thus, [ ]. These arrangements of matter will show at a glance who is responsible for any statement in the book.

F. LYNWOOD GARRISON.

"CHEPSTOW," RADNOR,
DELAWARE CO., PA., April 25, 1889.

AUTHOR'S PREFACE.

The object of the "Assayer's Manual" is to give directions for executing docimastic tests of natural and artificial products by methods taken mostly from practice, and which are of interest especially to metallurgists, but also to other technologists. To those possessing some knowledge of chemical and docimastic manipulations, the aphoristic mode of expression I have chosen will be sufficiently clear to enable them to execute the tests without further instruction. Those less skilled are referred for details to my larger work on "Assaying," Leipzig, 1866. Copious bibliographical references are given in the foot-notes, which are calculated to help both the teacher and scholar.

As the assaying of iron has been thoroughly treated in my manual of "Assaying of Iron," Leipzig, 1875, the subject has been omitted in the present treatise.

B. KERL.

Berlin, September, 1879.

CONTENTS.

GENERAL DIVISION.

PAGE

Object of the art of assaying; The dry method; The wet method; Precipitating metals by electrolysis 17
Volumetric assays; Colorimetric assays; The blowpipe . . . 18

I. MECHANICAL MANIPULATIONS.

Sampling; Non-alloys; Substances in fragments 18
Homogeneous fragments; Samples from the heap; Samples taken while the ore, etc., is being weighed; Samples by rasping; Slag samples; Heterogeneous fragments; Sampling by the cross method . . 19
Sampling by dropping the ore; Small ore and pulverized substances; Sampling while weighing; Samples of goldsmith's sweepings . . 20
Sampling before weighing; Substances in a state of fusion . . . 21
Alloys; Sampling by cutting; Samples from refined Upper Harz silver 22
Sampling by boring; Sampling by dipping; Sampling by granulation; Samples for producing coins 23
Preparation of the sample; Determination of moisture; Water-baths; Fresenius's drying disk, illustrated 24
Pulverizing the desiccated mass; Sifting 25
Washing; Schulze's washing apparatus, illustrated; Iron vanning-shovel, illustrated; Mariotte's bottle 27
The art of vanning; Weighing and measuring 28
Weighing; A pulverulent sample; Alloys; Fluxes; Weighing the button; Measuring of fluxes 29
Manner of charging the sample; The mixing scoops, illustrated . . 30

II. CHEMICAL OPERATIONS.

Classification; Working by the dry method; Ignition, carbonizing, calcining, in a neutral atmosphere, with exclusion of air, with admission of air, with reagents for decomposing 31
Roasting; Roasting dish and spatula, illustrated 32

	PAGE
Modifications	33
Fusion; Oxidizing fusion; Reducing fusion	34
Purifying fusion; Precipitating fusion; Mixing fusion; Remelting; Liquating fusion (liquation); Sublimation and distillation; Operations by the wet method; Assays by gravimetric analysis	35
Method for metallic sulphides	36
Evaporation of the solution; Precipitation of the solution; Kipp's apparatus, illustrated	37
Debray's apparatus, illustrated; Filtration; Filtering apparatus	38
Decantation; Drying precipitates; Igniting precipitates	39
Blast-lamp, illustrated; Desiccator, illustrated; Assays by volumetric analysis; The final reaction	40
Operations; Solution; Preparation of the standard solution	41
Measuring and titration of the assay liquid; Stoppered measuring flasks, illustrated; Pipettes, illustrated	42
Burettes, illustrated; Assays by colorimetric analysis	44

III. ASSAY FURNACES.

General remarks; Muffle furnaces	45
Furnaces for solid, free burning, flaming fuel	46
Pluttner's muffle furnace for coal, with the stoke hole in front, illustrated	47
Muffle furnace with the stoke hole at the back, illustrated; Charcoal and coke furnaces; Assay furnaces for charcoal, illustrated	49
Gas furnaces (coal gas); Perrot's gas muffle furnace, illustrated	50
Draught or wind furnaces	51
Placing of the assay vessels in the furnace; Firing; Taking the vessels from the furnace	52
Furnaces used in the Berlin School of Mines, illustrated	53
Wind furnaces for free-burning (flaming) coal; Wind furnaces for illuminating gas; Perrot's furnace, illustrated	55
Wiessnegg's gas furnace; Roessler's furnace for the production of high temperatures, illustrated	56
Fletcher's direct draft crucible furnace, illustrated	57
Blast furnaces	59
Sefström's furnace, illustrated; Fletcher's gas-injector furnace, illustrated	60
Furnaces for sublimation and distillation; A sublimation furnace, illustrated; A distillation furnace, illustrated	61
Retorts, illustrated	62
Organic combustion furnaces	63

IV. ASSAY VESSELS.

	PAGE
General remarks; Assay vessels for the dry method; Clay vessels	63
The principal vessels, etc.; Vessels without feet; Roasting dishes; Scorification or calcining vessels; Refining dishes; Crucibles, illustrated	64
Graphite crucibles; Soapstone crucibles; Vessels with feet; Crucibles for lead and copper smelting; Other clay vessels; Wrought-iron vessels; Crucibles for assay of lead, with or without lip	65
Vessels of bone ash; Cupels, illustrated	66
Wait's machine for making cupels	67
Vessels of other materials; Assay vessels for the wet method; For assays by gravimetric analysis; Articles of glass; Of porcelain; Of other materials; For assays by volumetric analysis; For assays by colorimetric analysis	68

V. BALANCES AND WEIGHTS.

Balances; An ore balance; Bullion or button balance; An apothecary balance; A rough scale; Weights; The gramme weight; Centner; Austrian and English assay weights	69
American assay weights	70

VI. TOOLS AND IMPLEMENTS.

General remarks; Furnace tools	70
Implements; For preparing the assay sample; Sampling; For drying the samples; For comminuting the samples; For sifting; For washing; For weighing; For charging	71
Implements for transporting the assaying vessels and for manipulating the same in the furnace, illustrated	72
Implements required for the reception and further treatment of the assay samples after they have been taken from the furnace	73

VII. ASSAY REAGENTS.

Reagents for dry assays; Reducing agents; Estimation of the reducing power	75
Oxidizing agents; Quantities of litharge required for the decomposition of the various metallic sulphides; Preparation of litharge entirely free from silver	76
Solvent agents; Acid; Basic	77
Precipitating or desulphurizing agents; Sulphurizing agents; Concentrating fluxes; Preparation of pure silver	78

	PAGE
Decomposing and volatilizing fluxes; Air-excluding fluxes	79
Reagents for wet assays; For assays by gravimetric and colorimetric analysis; Acids; Bases and salts; Metals for precipitation; For volumetric assays	80

SPECIAL DIVISION.

I. LEAD.

Lead ores; Assays of lead in the dry way; Sulphurized substances; Galena, etc., without foreign metallic sulphides; Precipitation assay	81
Rich galena (with little earths); Assay in an iron pot (Belgian assay)	82
Assay of lead matt free from copper, poorer ores, and slag, in different countries	83
Assay with potassium cyanide in clay crucibles	84
Galena with more earths; Assay with black flux (potassium carbonate and flour) and metallic iron, in clay crucibles	85
Practice in different countries	86
Upper Harz, assay with potassium carbonate; Charge	87
Galena containing large quantities of Earths; Lead monosulphide with foreign metallic sulphides; Roasting and reducing assay	88
Practice in Hungary; Assay with sulphuric acid	89
Oxidized substances; Lead oxides free from earths (litharge, minium, skimmings (Abstrich), etc.); Lead oxides with earths; Salts of lead oxide, viz., lead carbonate (cerussite), lead chromate (crocoisite), lead phosphate (pyromorphite), mimetene (lead arsenate), and yellow lead ore (wulfenite)	90
Lead sulphate (anglesite, lead fume, dross, sweepings, tailings, skimmings, etc.); Lead silicate (slags)	91
Alloys of lead; Wet assays; Assay by gravimetric analysis; Assays in Bleiberg in Carinthia, and other places	92
Mohr's process; Practice of various assayers	93
Volumetric processes; Colorimetric processes; Electrolytic processes	94

II. COPPER.

Ores; Native copper	95
Dry assays	96
German copper assay; Ores with sulphur, antimony, or arsenic; Roasting; Pyrites; Reducing and solvent fusion	97

	PAGE
Examples of charges; American charge	98
Refining; Lead; Refining on the dish with borax	99
By itself without borax and lead (Hungarian speiss assays)	100
Differences allowed in assays; Refining with lead and borax (Müsen Assay)	101
Refining by cupellation	102
Refining with the blowpipe; Oxidized substances without sulphur	103
Alloys of copper; Cornish copper assay; Wet assays; Gravimetric assays	104
Modified Swedish assay	105
Precipitation with iron	106
Correction for iron that may be contained in the precipitated copper; Color of pure precipitated copper	107
Precipitation with zinc free from lead and arsenic; Granulated zinc	108
Gravimetric method for the determination of copper recommended by Dr. F. A. Genth	109
Electrolytic assays	110
Use of Meidinger-Pinkus's battery and Clamond's thermo-electric battery; Platinum spiral and platinum foil	111
Platinum dish, illustrated and described	113
Classen's method for the electrolytic determination of metals, illustrated and described	114
Copper as recently precipitated sulphide; Determination of the copper in the form of cuprous sulphide	116
Apparatus for igniting in a current of hydrogen	117
Assay with sulpho-cyanide; Nickel coins	118
Copper alloyed with tin (bronze); Volumetric assays	119
Parkes's assay with potassium cyanide	120
Steinbeck's modified method of assaying copper; Examination of copper precipitated with iron or zinc; Fleitmann's method with ferric chloride	122
Method with sodium sulphide in an ammoniacal solution; Method with protochloride of tin	124
Haen's method recommended by Fresenius; Colorimetric methods; Heine's assay for poor ores and products (slags, etc.)	125
Jaquelin-Hubert's method for considerable percentages of copper	126
Determination of arsenic in copper; Determination of phosphorus in copper	127

III. SILVER.

Principal ores; Assays for non-alloys	128
Wet assays; Fire assays; Collecting the silver with lead; Scorification assay; The quantity of lead to be used	129

	PAGE
The quantity of borax; The number of samples to be taken	130
Table of charges	131
Practice in Hungary and the Lower Harz	133
Crucible assay	134
Charges of various countries	135
Other charges for sweepings; Results obtained in assaying sweepings and other refuse, recorded by Gorz	137
Combined lead and silver assay; Litharge; Cupellation of the argentiferous lead (assaying by the cupel or cupellation)	138
Assay of native silver ores; Examination of lead ores for traces of silver	141
Wet assays; Balling's volumetric assay	142
Assays of alloys; Dry assays; Lead bullion; Silver amalgam; Copper poor in silver	143
Cupriferous silver or fine silver (coins, refined silver, etc.)	144
Correction table for the absorption by the cupel, determined by the French Commission on Coinage and Medals; Results obtained in Freiberg; Wet assays	147
Volumetric assays; Gay-Lussac's method with sodium chloride; Preparation of the assay solution; Flasks, illustrated and described	148
Gay-Lussac's apparatus, illustrated and described	150
Calculation	151
Preparation of the normal solution; Gay Lussac's Tables for determining the fineness of silver alloys	152
Volhard's assay with sulpho-cyanide	155
Cobalt and nickel; Mercury	156
Determination of silver and copper in one solution; Assay by iodide of potassium and starch	157
Gravimetric analysis; Determination of selenium in silver	158
Electrolytic determination of silver	159
Hydrostatic assay	160

IV. GOLD.

Gold ores; Non-alloys; Mechanical assay by washing for determining the approximate percentage of gold in earthy and gravelly minerals poor in gold; Practice in Montana	161
Practice in Australia; Fire or fusion assays; Smelting the gold with lead; Scorification assay for ores of every kind; Crucible assay	162
Substances with earths and oxides (gold quartz, slag, gold sweepings); American gold ores; Rheinsand; Assay with nothing but red lead (litharge) as flux	163
Ores, etc., with combinations of sulphur, antimony, or arsenic; Pyrites poor in gold	164

	PAGE
Tailings containing a small amount of auriferous pyrites; Hungarian smelting works; Gold ores containing tellurides	165
Cupellation of the auriferous lead; When the laminated button breaks up; When the flattened button does not break up	166
Wet assay (Plattner's chlorination process); Alloys of gold; Gold amalgam; Auriferous lead and bismuth	167
Auriferous iron, steel, etc.; Alloys of gold and silver, with or without copper; Preliminary test for alloys free from copper; The color of the alloy	168
An examination of the touchstone; Preliminary assay of cupriferous alloys by cupellation; With lead alone; The quantity of lead required for alloys of gold with silver and copper	169
With an addition of lead and silver	170
Coins, the standard of which is known; Preparation of chemically pure gold; Roberts's process	171
Roll assay for argentiferous gold; Preliminary assay; Weighing the assay sample; Charging; Cupelling	172
Flattening (laminating) the button	173
Boiling in nitric acid	174
Washing (rinsing off the rolls)	175
Drying and annealing of the rolls; Weighing of the rolls; Loss of gold in cupelling	176
Action of platinum, rhodium, and iridium	177
D'Hennin's process for separating iridium; Palladium; To obtain good results in cupelling gold	178
Pulverulent assay (Strubprobe) of auriferous silver	180
Different modes of determining the gold; Separation of auriferous silver grains from samples of ores; Alloys of gold with copper; Quartation of gold with cadmium	181
Advantages offered by this method of quartation; Juptner's volumetric assay of gold	182
Preparation of the ferrous ammonium sulphate solution; Preparation of the permanganate solution; Separation of gold from platinum by electrolysis	183
Toughening brittle gold	184

V. PLATINUM.

Ores; Native platinum	184
Assay of platiniferous ores; Fire assays; Percentage of sand; Percentage of gold; Percentage of platinum; Wet assay	185
Alloys of platinum; Gold with platinum; Silver with platinum; Silver and gold with platinum	186
Electrolytic assay	187

VI. NICKEL.

	PAGE
Ores	187
Fire assay (Plattner's assay); Compounds free from copper; Compounds containing metallic sulphides; Arsenizing	188
Compounds free from sulphur and rich in arsenic; Reducing and solvent fusion; Modifications which may occur; Addition of iron filings	189
Arsenizing and fusion in one operation; Slagging off of the arsenical iron	190
Modifications; Dearsenizing; Slagging off the cobalt arsenide; Cupriferous compounds; Plattner's method	191
When the percentage of copper is small; Wet method partially used when the percentage of copper is large; The processes	192
Nickeliferous pyrrhotine; Compounds soluble with difficulty, as, for instance, slags	193
Compounds containing antimony; Wet assay; Gravimetric assay; Electrolytic assay	194
Processes when iron is present in small and large quantities	195
Process when zinc is present	197
Determination of nickel in pyrites and matt; assay of New Caledonia ores	198
Electrolytic determination of copper, nickel, and cobalt in speiss	199
Other assays	200
The separation of nickel and cobalt; separation of the iron; Nickeliferous solution from the assay with sulpho-cyanide for determining copper in nickel coins	201
Volumetric assay with sodium sulphide	202
Separation of cobalt; Volumetric assay of nickel and cobalt according to Donath	203
Colorimetric assay	204

VII. COBALT.

Ores; Assays of cobalt; Object of the assays; Determination of cobalt by the dry or wet method; Determination of the blue coloring powder (density), and the beauty of the colors; Smalt assay	204
Assay to determine the quality of color; assay to determine the intensity	206

VIII. ZINC.

Ores; Dry assays; Assay by distillation	207
Indirect assay	208
Wet assays; Gravimetric assays; Determination of zinc as zinc sulphide	209

	PAGE
Determination of zinc as zinc oxide; Electrolytic assay	210
When copper is present; Parodi and Mascazzini on precipitation of zinc from its sulphate solution	211
Zinc assay according to Rheinhardt and Ihle	212
Millot on the electrolytic determination of zinc in ores; According to Classen	213
Volumetric assays; Shaffner's assay with sodium sulphide	214
Indicators for recognizing the final reaction	215
Points to be observed; The quantity of the hydrated ferric oxide; Shade of coloration of the hydrated ferric oxide; The quantity of fluid; Measuring the volume of liquid	216
Removal of admixtures having a disturbing effect—copper, manganese, lead, etc.	217
Zinc used for fixing the standard solution; Arrangement of the light; Assay with potassium ferrocyanide	218
Galetti on determining zinc with potassium ferrocyanide; Determination of zinc by decomposing the sulphide of zinc with chloride of silver, and determining the zinc from the equivalent content of chlorine, according to Mann	219
Preparation of the titrating solutions; Determination of zinc from its combinations with sulphur by decomposition with nitrate of silver, Balling's method	220
Determination of lead, copper, and zinc in one solution, according to this method; Schober's volumetric assay	221

IX. CADMIUM.

Ores; Electrolytic assay; Processes of Clark and Yver	222
Processes of Classen, Beilstein, and Jawein	223

X. TIN.

Ores; Determination of tinstone by washing; Saxon assay of tin; Determination of tin by washing in Cornwall	223
Fire assays; German assay	224
Modifications which are necessary—When the ore contains many earthy admixtures; When the ore contains foreign metallic sulphides, arsenides, and antimonides	225
Modification necessary on account of the ease with which tin oxide is slagged off; When separate grains of tin are found; When tin oxide is combined with silicate (as, for instance, in tin ore slags)	226
Cornish assay of tin; Levol's assay with potassium cyanide	227
Wet assays; Gravimetric assays	228

xvi   CONTENTS.

                                                            PAGE

Determination of tin in tin slags 229
Volumetric assays; Determination of tin by means of iodine . . 230
Determination of tin by means of potassium permanganate . . 231
Determination of lead in tin; Detection of tin in presence of antimony;
    Electrolytic determination of tin 232

XI. BISMUTH.

Ores 233
Fire assays; Ores and compounds free from sulphur (native bismuth;
    tetradymite, bismuthic cupel ash, etc.); Sulphurized bismuth ores;
    Various processes 234
Wet assays; Assay of ore; Separation of bismuth from lead; According to Patera 235
Determination of the percentage of lead and silver; according to Ullgren; Tin oxide; Lead and bismuth; Electrolytic assay . . 236

XII. MERCURY.

Ores 236
Fire assays; Assays yielding free mercury; Different processes . . 237
Combustion furnace, illustrated and described 238
Assays in which the mercury is determined in combination with gold;
    Eschka's process 239
Prevailing working assay in Idria (Austria); Table of compensating
    differences allowed in the results: Teuber's process for testing mercury 240
Küstel's assay; Assay of cinnabar; Wet assays; Gravimetric assay;
    Volumetric assays 241
Electrolytic determination of mercury; Processes of Classen, Smith,
    and Knerr 242

XIII. ANTIMONY.

Ores; Fire assays; Liquation process for determining antimonium crudum; Determination of antimony in antimony sulphide; Assay by
    precipitation 243
Roasting and reducing assay 244
Wet assays; Gravimetric assay; Becker's method . . . 245
Volumetric method; Weil's method of determining antimony with protochloride of tin 246
Table of cubic centimeters of protochloride of tin solution used for
    every 25 cubic centimeters of assay solution (2 grammes=250 cubic
    centimeters) correspond to per cent. of antimony 248

Determination of the simultaneous presence of iron, copper, and antimony; Volumetric estimation of antimony in the presence of tin; Electrolytic determination of antimony; Processes of Parodi, Mascazzini, Luckow, and Classen 249

XIV. ARSENIC.

Ores; Fire assays; Native arsenic; Arsenious acid 251
Realgar (red orpiment); Assays for the determination of realgar; Assays for the determination of yellow orpiment; Wet assay; Gravimetric assays 252
Wet method combined with the dry 253
Volumetric assays; T. Mohr's estimation of arsenious acid; Pearce's method of determining arsenic 254
Canby's modification with oxide of zinc 255
Arsenic in glass 255

XV. URANIUM.

Ores; Wet assays; Gravimetric assays; More accurate analytical process 255
Patera's technical test; Alibegoff's process 256
Volumetric assay; Zimmermann's process 257
Method of determining uranium with potassium dichromate and iodine (Zimmermann's process) 258

XVI. TUNGSTEN.

Wolfram 258
Fire assay; Wet assays; Gravimetric assays; Sheele's method; Berzelius's method; Margueritte's method 259
Cabenzl's method; separation of tungsten from tin; Talbott's method 260
Volumetric assay; Zettner's method; Preparation of the titrating solution 261

XVII. CHROMIUM.

Ores; Wet assays 261
Gravimetric assays; Direct assay; Pourcel's method 262
Best methods for decomposing chrome iron ores; Hager's method; Fels's method; F. Clarke's method; Indirect assay; Volumetric assay 263

B

XVIII. MANGANESE.

	PAGE
Ores; Assays of pyrolusite	264
Gravimetric assays	266
Method of Fresenius-Will, illustrated	267
Method of Fikentscher-Nolte	268
Volumetric assays	269
Bunsen's method with iodine; Apparatus	270
Levol's method with iron	272
Volhard's method	273
Execution of the assay; Determining the strength of the permanganate; Hampe's method	274
Belani's method; Electrolytic assays	275

XIX. SULPHUR.

Ores	276
Assays by distillation for the determination of the amount of sulphur which an ore may yield; Sulphur earths; Gerlach's method; Iron pyrites; Assays of sulphur for the determination of the quantity of sulphur contained in a substance; Dry assay (raw-matt assay)	277
Hungarian assay	278
Wet assays; Gravimetric assays; Gravimetric determination of sulphur in pyrites according to Lunge	279
Fresenius's observations; Deutocom's method; Bodewig's method	280
Determination of sulphur in pyrites waste; Working test for determining the residue of sulphur in the roasting charge	281
Volumetric assays; Wildenstein's assay	282
Volumetric determination of sulphur in ores which contain either sulphur alone or also sulphates	283
Estimation of sulphur in metallic lead	284

XX. FUELS.

Assays of fuels; Determination of the amount of hydroscopic water; Yield of carbon	285
Determination of the coking quality of coal according to Richter; Volatile products; Determination of the ash	286
Amount of ash in different kinds of fuel; Determining the amount of sulphur in a coal or its ash; Determination of heating power	287
Berthier's method of determining the absolute heating power; Welter's law	288
Determination of the sulphur	290

CONTENTS. xix

	PAGE
Physical and chemical behavior; Examination of furnace gases; Orsat's apparatus, illustrated and described	291
Bunte's burette, illustrated and described	294

APPENDIX.

Tabular synopses; Atomic weights	297
Fusing points of metals and furnace products; Glowing temperatures	298
Plattner's method; Erhard and Schertel's experiments on the fusing points of metals, alloys, furnace products, rocks, and silicates	299
Erhard and Schertel's tables	300
Table of furnace products and gangues; Lower Harz working assays; lead	301
Copper; Iron; Zinc	302
Silver; Gold; Sulphur; Schaffner's assay of zinc as modified by Brunnlechner	303
Experiments on a heat regulator at the United States Assay Office, New York; Automatic heat regulator, illustrated and described	307
Dry assay of iron ores	310
Proportion of fluxes	312
Fluxes best adapted for ores or metallurgical products	313
Fluxes; Silica; Glass; China clay; Lime; Assay in large unlined crucibles	314
Reducing agent; Results	315

THE METRIC SYSTEM OF WEIGHTS AND MEASURES.

The metric system	317
Weights and measures; Apothecaries' weight, U. S.; Avoirdupois weight; Relative value of troy and avoirdupois weights; Wine measure, U. S.; Imperial measure, adopted by all the British colleges; Relative value of apothecaries' and imperial measures	319
Relative value of weights and measures in distilled water at 60° Fahr.; Value of apothecaries' weight in apothecaries' measure; Value of apothecaries' measure in apothecaries' weight; Value of avoirdupois weight in apothecaries' measure; Value of apothecaries' measure in avoirdupois weight; Value of imperial measure in apothecaries' and avoirdupois weights	320

TABLES SHOWING THE RELATIVE VALUES OF FRENCH AND ENGLISH WEIGHTS AND MEASURES.

	PAGE
Measures of length	321
Superficial measures; Measures of capacity	322
American measures; British imperial measures; Weights	323
Different values for the gramme; Avoirdupois; Troy (precious metals); Apothecaries' (pharmacy); Carat weight for diamonds	324
Proposed symbols for abbreviations	325
Ready-made calculations	326

HYDROMETERS AND THERMOMETERS.

Hydrometer (areometer)	329
Thermometers	331
Centigrade and Fahrenheit	332
Comparison of Centigrade and Fahrenheit scales and approximate steam pressure in pounds and atmospheres per square inch due to the temperature	334

INDEX 335

ASSAYING.

GENERAL DIVISION.

1. OBJECT OF THE ART OF ASSAYING.[1]

THE art of assaying (*docimacy*, from δοχιμαζειν, to test) is a branch of analytical chemistry. Its object is the quantitative determination, in the shortest possible time, of the products of mining and metallurgical operations, as well as the quantitative examination of many natural and artificial products derived from other sources, such as coins, fuels, etc. Formerly, in order to reach the result with the greatest expedition, the dry method (*dry assay*) was chosen for producing the chemical reaction; but, as this was frequently done at the expense of accuracy in the result of the assay, the *wet method* (wet assay, *gravimetric analysis*, *analysis by measure* or *volumetric analysis*, and *colorimetric analysis*) is also used in modern times. But it has by no means entirely displaced the dry method, for the latter is employed in all cases where results sufficiently accurate are more quickly reached, or where suitable, simple wet assay methods (as in assaying lead, cobalt, nickel, gold, and silver) cannot be substituted for it.

Sometimes a combination of both is employed (assays of lead, gold, etc.). Recently the method of *precipitating* metals *by electrolysis*[2] has been employed to great advantage.

---

[1] Kerl, Eisenprobirkunst, Leipzig, 1875. Balling, Probirkunde, Braunschweig, 1879. Mitchell, Manual of Practical Assaying, London, 1888. Ricketts, Notes on Assaying and Assay Schemes, New York, 1887.

[2] B. u. h. Ztg., 1869, p. 181 (Luckow); 1875, p. 155; 1877, p. 5 (Schweder). Grothe, polyt. Ztschr., 1877, p. 11 (Bertrand). Quantitative Chemical Analysis by Electrolysis (Classen), Trans., New York, 1887.

Volumetric assays can mostly be performed in a shorter time, which is an important item where much assaying has to be done. The results they yield are either very accurate, or at least sufficiently exact[1] for metallurgico-technical purposes; they are less expensive, but require greater experience and more chemical knowledge on the part of the operator, and special apparatus of accurate construction. While by the dry method the metal assayed, or one of its alloys having a known composition, is weighed directly, in volumetric assays it is calculated from certain reactions of the reagents employed, and the result may possibly be vitiated on account of the presence of foreign substances, of whose presence there is not always an indication.

Colorimetric assays are chiefly employed for determining very small quantities of metals which either could not be detected by other methods, or, if so, then only by very tedious processes (copper, lead); but recently they have been developed so as to adapt them for substances rich in metal (copper).

The *blowpipe* is frequently used for a preliminary assay.[2]

I. Mechanical Manipulations.

2. SAMPLING.

It is absolutely necessary that the small quantity of sample with which the assay is made should represent the average composition of the ore-heap, etc., from which it is taken. The manner of taking samples varies according to the character of the substances to be assayed, viz:—

A. Non-alloys (ores, matt, speiss, slag, etc.).

1. *Substances in fragments*, either homogeneous or heterogeneous in composition.

---

[1] B. u. h. Ztg., 1869, p. 330. (Compare gravimetric and volumetric assays of Cu, Fe, Zn, Sb.)

[2] Berzelius, Anwendung des Löthrohrs, Nürnberg, 1828. Scherer, Löthrohrbuch, Braunschweig, 1857. Birnbaum, Löthrohrbuch, Braunschweig, 1872. Simmler, Löthrohrchemie, Zürich, 1873. Hirschwald, Löthrohrtabellen, Leipzig, 1875. Landauer, Löthrohranalyse, Braunschweig, 1876. Kerl, Löthrohrprobirkunst, Clausthal, 1877. Landauer, systematischer Gang der Löthrohranalyse, Wiesbaden, 1878. Plattner-Richter, Probirkunst mit dem Löthrohr, 4 Aufl., Leipzig, 1878.

a. *Homogeneous fragments,* many iron ores, lead and copper ores, etc.

α. *Samples from the heap.*[1]—Pieces are taken at random (it is best to do so with bandaged eyes) with the hand or a shovel, from different places on the circumference of the heap, and also from the interior, after the upper layer which has been dried by the atmosphere has been removed. The collected lumps (about 100 kilogrammes, 220.54 lbs.; in Freiberg, for certain ores, one-tenth of the heap) are comminuted to pieces of the size of a bean, either by means of rollers or stamps, or with a sledge-hammer. A square or conical heap is then formed of the pieces, and this is divided into four parts. One of these is taken, the pieces forming it are still further comminuted, and then again formed into a heap, which is divided as before. The comminution and reduction are repeated several times, finally upon an iron plate provided with a rim (reducing-board), until at last from $\frac{1}{2}$ to 1 kilogramme (1 to 2 lbs.) of the sample remains, in such comminuted form that it will pass through a sieve having 30 by 30 meshes to the square centimeter (about 75 meshes to the square inch).

β. *Samples taken while the ore, etc., is being weighed.*—Pieces are taken at random from every lot weighed, and the collected pieces comminuted and reduced according to paragraph α (Upper Harz copper pyrites).

γ. *Samples by rasping.*—Fuels, etc., which cannot be pulverized are comminuted by a rasp, and a reduced sample made from this.

δ. *Slag samples.*—A piece of slag is taken every time the slag is tapped or run off, or a piece is broken off from every cone formed. The pieces of one charge are comminuted and reduced in the manner above stated.

b. *Heterogeneous fragments.*—(Gold and silver ores, many copper ores, coal with slate and pyrites, etc.)

α. *Sampling by the cross method.*[2]—When the grains are too dissimilar and too coarse, the entire heap is broken up. The broken fragments (as large as a walnut for less valuable ores, and about the size of a hazelnut or bean for the more valuable)

---

[1] B. u. h. Ztg., 1868, p. 26; 1872, p. 59.
[2] Preuss. Ztschr. xvii. 137 (Mansfeld); xviii. 223, 224 (Swansea).

are passed through screens or cylindrical sieves. An oblong or square heap, 30 to 40 centimeters high, is then formed. Ditches about 20 to 30 centimeters wide, and crossing each other, are dug out with a shovel, and the samples are then taken by digging them out from the top down to the bottom of the squares, which remain standing between the ditches. These samples are comminuted to the size of millet-seed, thoroughly mixed, and formed into a new rectangular heap, which is again crossed, and samples taken from it in the manner stated. This operation is repeated, finally by using the spoon, until the samples are reduced to a powder. (Method in the great ore markets of Swansea and Liverpool, for American silver ores in the Upper Harz and in Freiberg, etc.)

β. *Sampling by dropping the ore.*[1] (Sturzprobe.)—The ore is dropped through a funnel standing over a pyramid of sheet-iron, which is divided into four divisions by partitions projecting over the edge, into which the ore is distributed. The ore from one of the divisions is comminuted and dropped through a funnel into a similar but smaller pyramid. This operation is repeated, smaller pyramids being used every time, until a sufficiently reduced sample has been obtained (method in Chili and Colorado).

2. *Small ore and pulverized substances.*

a. *Sampling while weighing.*—The small ore must be carefully mixed, and, in case it should be rich, it is best to have it in such a condition that not over 15 per cent. of coarse material will remain behind in passing it through a sieve having 10 meshes to the square centimeter (25 meshes to the square inch), but, otherwise, it may be coarser. It is generally weighed in quantities of 50 to 100 kilogrammes. Three spoonfuls are taken from different places of every lot weighed, and placed in wooden troughs standing near the scale. All the samples taken from one lot are mixed together and a heap is formed from them. This is reduced by quartering, or by the crossing method.[2]

Samples of goldsmiths' *sweepings* (scrapings, fragments of

---

[1] Preuss. Ztschr. xxiv. 49. B. u. h. Ztg. 1872, p. 232 (Chili). Ann. d. Mines, 1878, XIII. 606 (Colorado).

[2] Kerl, Oberharzer, Hüttenprocesse, 1860, p. 195.

crucibles, rags, etc.) are taken by incinerating the entire mass in order to destroy organic matter. The mass is then comminuted by stamping or trituration, and passing it through a sieve having meshes the size of a grain of sand (less than 1 millimeter, 0.039 inch). The iron in the coarse mass remaining in the sieve is extracted by using a magnet, and the residue fused with soda and borax, cast into a bar, weighed, and sampled by chipping from top and bottom. The portion that has passed through the sieve, freed from iron with the magnet, is then weighed, sampled from every weighing, then united, quartered, and triturated until everything passes through a very fine sieve. It is then assayed, and the yield of metal obtained from both coarse and fine material is calculated.

b. Sampling before weighing.—This is done by passing with a hollow sheet-iron cylinder in several places through the heap down to the bottom. The lower end of the cylinder is provided with a valve which when closed retains the charge, and with a handle. The samples are then mixed and reduced as above described (Freiberg).

In some of the German smelting works, the ore,[1] when it is bought, is sampled by weighing—

to within 10 lbs. when it contains up to 0.5 per cent. Ag, or 0.01 per cent. Au.

" 1 lb. when it contains over 0.5 to 5 per cent. Ag, or 0.0105 to 0.1 per cent. Au.

" 0.1 lb. when it contains over 5 to 50 per cent. Ag, or 0.1005 to 1 per cent. Au.

" 0.02 lb. if it contains more.

" 10 lbs. if the ores contain no gold or silver.

3. *Substances in a state of fusion.*—An average sample may be obtained in the following manner: A dry tapping-bar, previously heated, is held in the fluid mass (slag, etc.). When the bar has become cold, the adhering mass is broken off, comminuted, and

---

[1] Bezahlungstarife für den Einkauf von fremden Erzen u. Gekrätzen auf den fiscalischen Hüttenwerken bei Freiberg. Clausthal, 1870. Oestr. Ztschr. 1869, No. 44. Engl. Standard f. Zinnerze, B. u. h. Ztg. 1862, p. 261; f. Kupfererze, 1862, p. 316. Spanische Tarife, B. u. h. Ztg. 1868, p. 26. Tarnowitzer Erztaxe in Preuss. Ztschr. xiv. 217.

then mixed and reduced. In this operation, however, the iron must not decompose the fused material (a possible separation of lead may occur, for example, from lead matt and lead speiss).

B. Alloys.—These are homogeneous when in a fluid state, especially after they have been stirred. But when solidified, they show a different composition in different places (edge, centre, top, bottom).[1] In sampling, this must be taken into consideration.

1. *Sampling by cutting.*—The sample (2.5 grammes from every ingot of silver, 1.5 grammes from those of gold) is cut from the upper and lower sides on opposite ends of the ingot by means of a hollow chisel and hammer. (In England and the United States, the opposite edges are chipped off.) The samples are hammered or rolled out, the resulting sheets cut into shreds, and each sample is assayed by itself (0.5 to 1 gramme of each is weighed off for the purpose). The yield is calculated, and either the average is given (gold assays), or the lowest yield (sometimes in silver assays). This method is best adapted for alloys of tolerably uniform composition, but is also employed for those showing a considerable difference in the lower and upper sampling.

For instance, the lower sample from refined Upper Harz silver is from $\frac{3}{1000}$ to $\frac{4}{1000}$ richer than the upper, the percentage of gold increasing towards the bottom. The centre, as a rule, contains more silver than the edge. In the "*five-mark piece*" the centre is $\frac{2}{1000}$ richer than the edge, and the same is the case with the "*thaler,*" as they are stamped from a bar poorer on the edge than in the centre. For this reason, when taking samples from such coins, it is best to cut out a quadrant, cut off the corners, and assay them. In this way the assay samples represent the composition of both periphery and centre of the bar from which the coins have been stamped. Fewer differences occur in gold than in silver coins.

---

[1] Kerl, Grundr. d. allgemeinen Hüttenkunde, 2 Aufl., 1879, p. 15. Dingler, cciii. 106; ccxv. 431. Ber. d. deutsch. chem. Ges. 1874, p. 1548, B. u. h. Ztg. 1874, p. 63; 1875, p. 251.

2. *Sampling by boring.*[1]—This is done by boring through the edge and centre of the ingot. By this means a sample is obtained from the centre of the ingot, which is not the case in chipping a sample, but the ingot is made unsightly. It is very difficult to mix the borings uniformly, and it is therefore better to fuse them under a covering of charcoal powder. It is best to use a mechanical contrivance for boring through thick pieces. This consists of a lever weighted at one end, and a drill, operated by the hand in the centre.

3. *Sampling by dipping.*—This is obtained in the same manner as mentioned on p. 21, for instance, from refined copper. Another method of taking samples is, by dipping the curved, bright end of a pair of pincers, or of an iron rod, into the metal bath. When the pincers or iron rod has become cold, the crust adhering to the end is broken off. A sample is generally taken from the surface of silver while it is being refined, and one from the under side of the congealed refined silver.

4. *Sampling by granulation.*—This sample indicates in the most reliable manner the average value of the metal. It is obtained in the following manner : the bars of precious alloys are smelted in a black-lead crucible. The mass, while in fusion, is stirred with a rod of iron or clay, and a sample is scooped up with a small ladle from the bottom of the crucible. It is then poured in a thin stream into a copper vessel filled with warm water, to which a gentle rotary motion is given by means of a broom; or the sample is directly poured through a birch broom. The resulting granulated metal is then carefully dried. Alloys of base metal (as granulated lead[2]) are fused under coal-dust and then directly poured upon an iron plate.

The following samples are taken for producing coins : *ingot sample* from the metals to be alloyed; *granulated* or *crucible sample* from the fused alloys; *stock sample* from sectors of the finished coins; and a *sample* from defective coins which have been thrown out and fused together, for instance inside of four weeks.

[1] Mitchell, Prac. Assaying, 1888, p. 364.
[2] B. u. h. Ztg., 1869, p. 278 (Brixlegg).

3. PREPARATION OF THE SAMPLE.

Alloys are prepared by rolling out and cutting up the resulting sheets; or the granulated metal is used without further preparation. The following operations may be required for non-alloys:—

1. *Determination of moisture.*—The sample is divided, by weighing with reduced weights, into as many centners or kilogrammes or pounds as are actually contained in the lot (for instance, in Freiberg, 1 centner weight = 75 grammes). The weighed portion is heated in an iron or copper pan or directly in the removable scale-pan of the balance, by holding it over a heated stove or a brazier of charcoal, and constantly stirring it until a cold plate of glass or slate, when held over it, shows no deposition of moisture, and two successive weighings agree. The heating should be carefully conducted, so that, with sulphur compounds for example, no odor of sulphurous acid shall be developed, and, with organic substances, no carbonization shall take place (this may be guarded against by holding a piece of paper in the mass). Water-baths are used for drying the sample at 100° C. (212° F.). These are copper boxes, with double walls, the intermediate space (Fig. 1) containing water; or, they consist of hemispherical copper or enamelled iron vessels, placed one within the other, leaving an intermediate space for water; or of two cylinders, one placed within the other (*Scheibler's*[1] steam apparatus). Where a determinate temperature, high or moderate, is required, it is best to use *Fresenius's drying disk* of cast iron, which is heated from below (Fig. 2). It is 25 centimeters in diameter, and 4 centimeters thick; *b* is the handle, 36 centimeters high; *c* are small brass dishes with numbered handles fitting into suitable recesses; *d*, a case filled with copper-filings for the reception of the thermometer *e*. Besides these, *air-baths*,[2] commonly called thermostats, are also used.

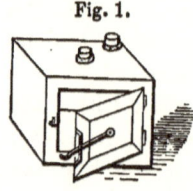

Fig. 1.

---

[1] Dingler, ccxxiii. 312. Muspratt's Chemie, v. 1635.

[2] Bunsen's Luftbad mit Temperaturregulator in Kerl's Grundr. der Eisenprobirkunst, 1875, p. 3. Raulin's Wärmeregulator in Dingler, ccxxvii. 263.

2. *Pulverizing the desiccated mass.*—This should be done carefully, to avoid loss of dust, either in a covered mortar, or on a

Fig. 2.

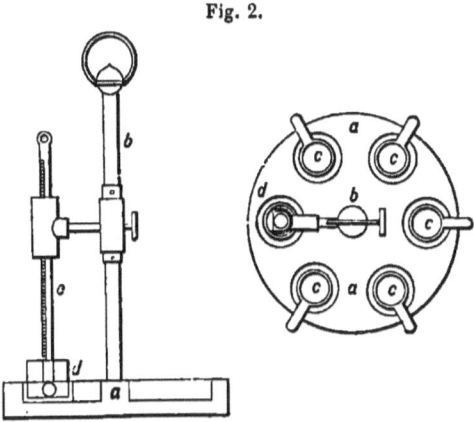

flat disk of cast iron as is shown in Fig. 3. The iron plate should be perfectly true and have a smooth surface. The rub-

Fig. 3.

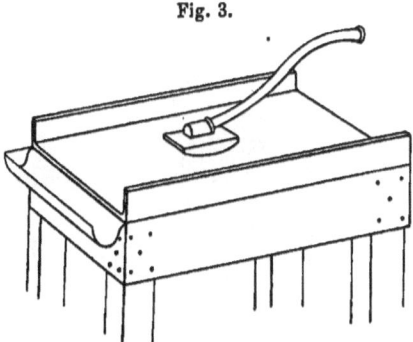

bers are made of iron and weigh from 15 to 60 pounds (6.8 to 27.2 kilogrammes).

3. *Sifting.*—Brass wire sieves are generally used in preference to hair sieves. For less valuable ores the sieves have from 14 to 20 meshes to the square centimeter (about 35–50 to the square inch), and more valuable ores from 28 to 32 meshes to the square

centimeter (from 70–80 to the square inch). Brittle substances will pass through the sieves without difficulty, but those with malleable admixtures will leave a flattened residue in the sieve; as, for example, ores carrying native silver and copper, silver glance, granules of lead in slag and thin matt, sweepings containing gold and silver (p. 20), etc. In case *hard* gangue (quartz) is to be sifted, the fine mass which has passed through the sieve is several times rubbed together with the coarse residue remaining upon the sieve, until everything has passed through it. The residue of *soft* gangue is weighed and at once assayed by itself, and the fine siftings separately also, after they have been mixed upon glazed paper and passed several times through a coarse sieve. The entire yield is then calculated by adding the product

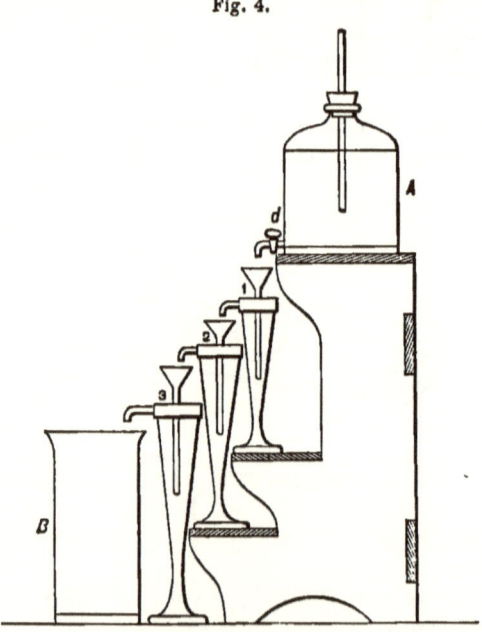

Fig. 4.

of both together. The material ready for assaying is packed in wooden boxes, glass bottles, or small linen bags. Dishes of ordinary potter's clay (Mehlscherben) may serve for the reception of the stamped ore and lots of samples while they are weighed out for assaying; they have an outer diameter of from 80 to 100 millimeters on the bottom, 100 to 200 millimeters on the top, and are from 40 to 45 millimeters high (Hungary).

4. *Washing.*—This is done to separate specifically heavier substances from lighter ones (roasted tin ore, gold gravel, etc.), or to obtain a uniform grain (smalt assays). The mass, comminuted as fine as possible, is placed in a beaker-glass and water poured upon it. It is then thoroughly stirred up, when it is allowed to settle, and the turbid liquid poured off; or it is washed in Schulze's washing apparatus (Fig. 4) by allowing water to flow

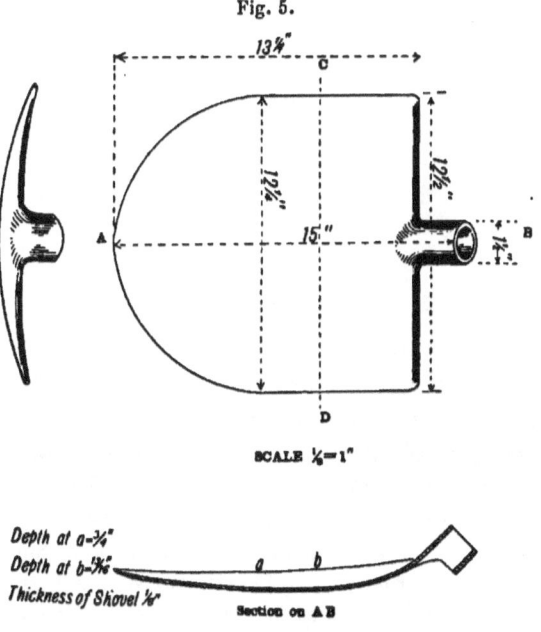

Fig. 5.

from a *Mariotte's* bottle, A, through d, upon the mass contained in the glass 1; the different sizes of grain will settle in the glasses 2 and 3, the finest grains in the beaker-glass B. Another mode of washing the substances is by using an iron vanning shovel (Fig. 5). The washed mass should be occasionally examined with a magnifying glass or blowpipe as to its physical characteristics (for instance, silver ore for the presence of metallic sulphides, antimony and arsenic in order to regulate the addition of lead before smelting).

The art of vanning is apt to be one of the most useful accomplishments to an assayer or miner. It is, however, somewhat difficult to learn, and requires considerable practice before reliable results can be obtained. Perhaps the best way to learn vanning is to thoroughly mix known quantities of finely powdered tin ore and sand together, and see, by repeated trials, how nearly they can be separated. Thus, for example, thoroughly mix 20 grains (1.30 grammes) of tin ore with 500 grains (32.4 grammes) of finely powdered sand. Place on shovel or trough with just sufficient water to cover the mixture, and as the sand is separated out from time to time, wash it in a dish. When, as far as possible, all the sand has been separated out, dry the residue (tin ore) remaining on the shovel and weigh. The sand which was washed out into the dish is then taken and revanned in the same manner in order to separate out any tin ore yet remaining; dry, weigh, and add to weight first determined.

4. WEIGHING AND MEASURING.

Before every weighing, the balance must be tested as to its equilibrium. The substance to be weighed should be cold, and must not be placed directly upon the scale-pan, but upon suitable smaller pans, watch-glasses, etc.; hygroscopic substances in closed tubes. The balance beam should be raised from the knife edge every time before a weight is put into or removed from the pan. The weights must not be put in the pan at random, but systematically, as this is the only way of saving time. The highest probable weight should be added first, then the next lowest, and so on until the equilibrium has been established. The pans should then be changed in order to test the correctness of the weight. Perfect equilibrium of the balance is, however, not absolutely essential, as a correct weighing may be obtained by placing the substance to be weighed in one pan, and in the other pan any convenient material as a make-weight, such as tinfoil, shot, granules of lead, etc., until equilibrium is established. The balance is then raised, the weighed substance is removed from the pan, and sufficient weights to counterpoise the balance are put in its place. The sum of these will give the correct weight of the sample.

1. *Weighing:* a. *A pulverulent sample.*—The dried sample is poured upon glazed paper and spread out in spirals with the spatula. It is then drawn together towards the centre by radial bands. Some of it is now taken with the spatula from the bottom to the top of the heap, the weights are placed upon the left pan of the balance, and the sample is continuously poured into a counterpoised saucer or watch-glass placed upon the other pan, by gently tapping on the handle of the spoon, until equilibrium has been established. In case too much has been poured in, some of it is removed, but the balance, while this is done, must be arrested. Large quantities are weighed upon a *watch-crystal* counterpoised by granules of lead or shot. Hygroscopic substances are conveniently weighed by filling a *stoppered glass tube*, 12 to 14 centimeters long, and 8 to 10 millimeters wide, with them, noting the weight, and, after pouring out the requisite amount of sample, again weighing the tube; the difference between the two weighings will give the weight of the quantity abstracted.

b. *Alloys.*—They are converted either into granulated form or into small strips or splinters. These are collected in a glass or copper saucer and placed by means of the forceps in the right pan of the balance while the weights are placed in the left.

c. *Fluxes.*—In weighing these, very great accuracy is not of so much importance. They are placed either directly upon the pan of the balance, provided it is not attacked by them, or otherwise upon a tared watch-crystal.

2. *Weighing the button.*—The button is taken hold of with the forceps and placed upon the left pan of the balance, and the weights are then put upon the right; but, as has been stated, the balance must always be arrested before the weights are put on or removed.

3. *Measuring of fluxes.*—Granulated lead (test lead) free from silver is measured with large, gauged, iron spoons numbered on

the handles, or with a glass-tube, one end of which is closed with a stopper, while in the other is a wooden cylinder provided with a scale.

5. MANNER OF CHARGING THE SAMPLE.

The sample is poured either directly into the crucible without any fluxes (as in roasting), or the fluxes are added in such a manner that—

1. The sample lies on the bottom of the crucible, and the fluxes are placed upon it in consecutive order without stirring the mass up. When this is done, the mass will not puff up as easily when it is heated (charges with carbonaceous mixtures, for instance, assay of lead with carbonate of potassa, flour and iron).

2. The sample is added to the fluxes already in the crucible, in cases where the puffing up of the charge on heating is not feared, and is intimately mixed together (for instance, assay of lead with potassium carbonate).

3. The sample is mixed with the fluxes before it is placed in the crucible. The mixing is done in a mixing-scoop of copper (Fig. 6) by means of a spatula. The scoop is about 140 millimeters long and 40 millimeters wide.

Fig. 6.   Fig. 7.[1]

The mixture is poured into the crucible through the spout of the scoop, about 20 millimeters wide, a brush being used to brush out the last traces of the mixture; or, in case a very vigorous chemical reaction is desired, the sample and fluxes are first intimately rubbed together in a mortar of stone (porcelain, serpentine, agate), or of metal (steel, cast-iron, brass). Open mixing-scoops of copper provided with a handle (Fig. 7) are used for mixing the charge, or for receiving, in consecutive order,

---

[1] [Care should be taken that they are bright and smooth.—G.]

the substances constituting it, or for pouring them into the glowing crucible standing in the heated furnace (assays of lead, English assay of copper).

II. Chemical Operations.

6. CLASSIFICATION.

These operations are divided into those by the *dry* and those by the *wet method*, and are either preliminary (roasting, etc.), or capital operations (smelting, etc.).

7. WORKING BY THE DRY METHOD.

These operations are carried on, either below the fusing point (*ignition, carbonizing, calcining, roasting*), with or without admittance of air; or at a fusing heat (smelting); or volatile substances are to be expelled by heat, and their vapors condensed to the liquid state (*distillation*), or to the solid state (*sublimation*).

1. *Ignition, carbonizing, calcining.*[1]—Heating without fusing—

a. In a *neutral atmosphere,* to drive out volatile substances (for instance, water and carbonic acid from iron ores), or to change their molecular condition (for instance, annealing gold and silver alloys before rolling them out, etc.).

b. With exclusion of air in covered pots or crucibles, to decompose metallic sulphides and arsenides (iron and arsenical pyrites in the dry assay of blende), to effect reduction (ignition of tin ore with charcoal), or to arsenize or dearsenize the substances (assay of nickel and cobalt).

c. With admission of air in roasting dishes (determination of ash in fuel, combustion of bitumen in copper schist and blackband iron ore, oxidation of cement copper, etc.).

d. With reagents for decomposing substances insoluble in acids (for instance, silicates with four times their weight of a mixture

---

[1] Gaslampen in Dingler, ccxxiv. 617; ccxxv. 83 (Müncke). Fresenius's Ztschr. 1879, p. 257 (Ebell). Bunsenbrenner von Glas in Dingler, ccxxvii. 85, 398.

of 13 parts of potassium carbonate and 10 parts of anhydrous sodium carbonate in a platinum crucible).

2. *Roasting.*—Metallic sulphides, arsenides, and antimonides are heated in presence of air to a temperature insufficient for fusion, but which permits of their oxidation; metallic oxides are, therefore, produced, while sulphurous, arsenious, and (sometimes) antimonious acids are volatilized.

The process is as follows: The powdered sample is spread out in a shallow, smooth roasting dish. This is about 50 to 52 millimeters wide in the clear, and 8 to 10 millimeters deep, made of not too refractory clay, and has rather thin sides. It is lined with reddle, chalk, or oxide of iron, and, if necessary, in order to increase the surface, the lined sides are marked with a spatula in such a manner that radial furrows, running from the centre towards the edges, are formed. It is then placed in the muffle of a muffle-furnace (Fig. 27) and heated at a gradually increasing temperature until it glows; the heating must be the more gradual the more fusible the sample. (Antimonial and arsenical metals are more easily fused than metallic sulphides, antimony glance, lead sulphate, and "fahlerz" containing mercury). The mouth of the muffle is left open, with the exception of a low layer of pieces of wood charcoal touching each other, and continued in a forward direction. These pieces in a glowing state heat the oxidizing air current. The roasting dish must occasionally be turned around during the operation. It is taken from the muffle when the mass has ceased to burn, and oxidation is complete. This is indicated by the heated mass ceasing to fume and no longer emitting odors of sulphurous or arsenious acid, and the metallic lustre having been replaced by an earthy appearance. If this should not be the case, the roasting dish must be placed back into the muffle until these signs make their appearance. The now roasted sample may be somewhat sintered together. It is then rubbed with the iron knob, *b*, of a wooden-handled spatula, after being loosened from the edges of the roasting dish with the knife's edge, *a*, of the rod. This tool (Fig. 9) is about 195 millimeters long, and consists

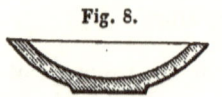

Fig. 8.

of the iron head b, about 16 millimeters diameter, and the steel knife-blade a, set in a wooden handle. The roasting is repeated once or several times, but the mass must be rubbed up previous to each roasting. It is then mixed with 1 or 2 volumes of powdered wood charcoal, or 20 to 25 per cent. of graphite, and the roasting dish with its contents is again placed in the muffle and brought to a glow. By this process the sulphates, antimoniates, and arseniates formed during the oxidizing period are reduced to metallic sulphides, antimonides, and arsenides, while the volatile products of oxidation escape (*reducing roasting*). These compounds when all the carbon has been consumed (which may be readily recognized by the manner of glowing) will be again converted into oxides; sulphurous, arsenious, and antimonious acids being evolved in the operation. But new sulphates, antimoniates, and arseniates will constantly be formed, and these, if the sample is to be roasted as completely as possible (for instance, copper ores, but lead ores in a less degree), can only be removed by repeating the rubbing up of the assay sample twice or three times, mixing it with charcoal powder, and glowing until the coal is *completely* consumed, although even after this small quantities of sulphates will nevertheless remain. When the roasted sample has become sufficiently cold, it is placed in an iron mortar and mixed with 20 to 50 per cent. of ammonium carbonate. A small conical heap of the mixture is formed in the roasting dish; this is covered with an empty roasting dish and quickly ignited until the odor of ammonia can no longer be detected. When this is the case, the last traces of sulphuric acid in the roasted sample will have been volatilized in the form of ammonium sulphate. (Lead and bismuth sulphates are only incompletely decomposed by ammonium carbonate.) The roasting dish is now taken from the muffle and allowed to cool off. The sample is then placed in a mortar and rubbed up.

Fig. 9.

Modifications.—When the ores are refractory (for instance, copper pyrites), powered charcoal or graphite is added to the sample before roasting, in order to shorten the time required for the operation. Very fusible substances which evolve vapors (such as "*fahlerz*" containing mercury) must be heated very gradually. To diminish the loss of metal (for instance, of silver and gold) the temperature must not be raised higher than is absolutely necessary. The loss from this cause is greatest with ores containing antimony, arsenic, zinc, etc.

3. *Fusion.*—The sample is brought into a liquid state, either by itself, or with fluxes. During this process the resulting products (metal button or regulus, speiss, matt, slags) arrange themselves in layers according to their specific gravities, and are separated from each other, either by breaking to pieces the clay assay-vessels in which they have been fused, after they have become cold, or they are poured out while still in a fluid state, into iron or bronze moulds, where the separation then takes place. Sometimes the fluid, oxidized substances are absorbed by the porous sides of the assay-vessel, leaving the metal button behind (cupellation of lead refining copper on the cupel). The following distinctions are made according to the object of the fusion:—

a. Oxidizing fusion.—In this process the following may serve as oxidizing agents: the oxygen of the air, demanding open vessels for the operation (cupels, calcining and roasting dishes), which must be heated in the muffle-furnace (for instance, cupellation of lead, refining copper, assay of cobalt and nickel); or fluxes yielding oxygen, and then open or covered assay-vessels (pots, crucibles), and muffle, wind, and blast furnaces may be used (saltpetre in the Cornish assay of copper and in the assay of chromium, lead oxide in the assays of fuel, silver and gold); or both at the same time (refining of black copper). The resulting oxides are more frequently slagged off by themselves or by solvent agents added as a flux (borax, glass, etc.), than absorbed by the porous vessel used for fusing (cupels).

b. Reducing fusion.—This operation is seldom executed by itself with reducing agents (coal, flour, colophony, potassium cyanide), but generally in connection with fluxes (potassium or sodium carbonate), in order to allow of a better collection of the particles of metal (as from litharge, white-lead ore); or in connection with reducing, fluxing, and solvent agents (borax, glass, phosphorus salt). A definite *low* temperature must then be used to reduce one metallic oxide, while the metallic oxides with more difficulty reducible, are slagged off with the earths which may be present (assays of lead, copper, and tin ores). Muffle, wind, and blast furnaces are used. The vessels used for this process (crucibles, pots) should be roomy, as the mass puffs up. This is caused by the formation of carbonic oxide which ignites above the vessels. This phenomenon is called "*flaming*," the end of the operation being generally indicated by its cessation.

c. Purifying fusion.—This is more frequently used in connection with oxidizing fusion (p. 34) and reducing fusion (p. 34) than by itself (assay of smalt, assay of thin matt).

d. Precipitating fusion.—By this process metallic sulphides (in assays of lead, bismuth, and antimony) or arsenical metals (in the assay of lead ores and nickel and cobalt ores containing bismuth) are decomposed by iron. The desulphuration of the metals is promoted by suitable fluxes (potassium or sodium carbonate, black flux), or the slagging off of earthy and other admixtures is effected (borax, glass, alkalies).

e. Mixing fusion, to prepare alloys by fusing different metals together (gold and silver in quartation).

f. Remelting, in order to produce the sample in another form (as, for instance, by granulation, p. 23).

g. Liquating fusion (liquation).—Liquation of easily fusible substances from more refractory substances (assay of antimony glance).

4. *Sublimation and distillation.*—The sample is placed, either by itself or with fluxes, in crucibles, tubes, or retorts, and heated until the substances volatilize, and the vapors are then condensed as sublimates (flaky arsenic, flowers of sulphur, realgar), or as distillates (mercury, zinc) in suitable condensers.

8. OPERATIONS BY THE WET METHOD.

These may be—
1. *Assays by gravimetric analysis.*[1]
   a. The sample is dissolved in acids, in a porcelain dish cov-

---

[1] Rammelsberg, quant. Analyse, Berlin, 1863. Wöhler, Mineralanalyse, Göttingen, 1861. Sonnenschein, quant. Analyse, Berlin, 1864. Rose-Finkener, Mineralchemie, Leipzig, 1865. Fresenius's quant. Analyse, 6 Aufl. 1871.

ered with a watch-crystal. Or a bellied flask is used for the purpose (Fig. 10). This either stands upright and is provided with a funnel, or is placed in a slanting position to prevent the liquid, in case it effervesces, from being thrown out of the mouth of the flask. The vessel may be heated on a sand-bath, or upon a wire gauze over a lamp, until the solution is complete, or a residue showing no trace of ore, etc., remains.

The following method is used for metallic sulphides, which, when they are dissolved with acids, separate sulphur which incloses some of the ore. The solution is evaporated to dryness in a porcelain evaporating dish. The dry mass is heated over a lamp until the sulphur is burned. The

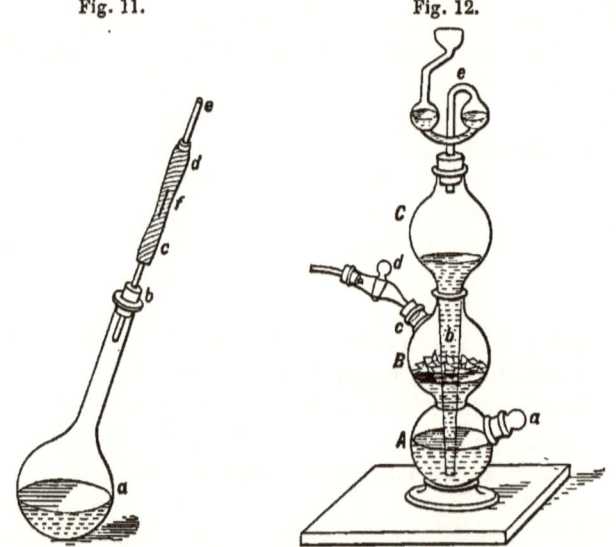

residue is digested with a small quantity of acid, water is added to this, and the fluid then partly filtered; but in doing this great care must be observed. The residue is again treated with acid,

Classen, quant. Analyse, Stuttgart, 1857. Menschutkin, Analyt. Chemie, Leipzig, 1878. Bolley, techn.-chem. Unters., Leipzig, 1879. Muspratt's tech. Chemie, 3 Aufl.

evaporated to dryness, the sulphur burned, etc. If it is necessary to exclude the air, the apparatus in Fig. 11 is used. It consists of a flask, *a*, with a rubber cork, *b*, and provided with a rubber tube, *c d*, having a slit at *f* and closed at *e* by a small glass rod.

b. Evaporation of the solution in a glass flask (Swedish assay of copper, assay of lead sulphate), or in a covered porcelain dish by heating it in the sand-bath, over a lamp, or on the water-bath.

c. Precipitation of the filtered or unfiltered solution; or where a mass evaporated to dryness is to be treated, it is moistened with a little acid, allowed to stand for a few minutes, and then boiled with the addition of a small quantity of water. It is then filtered, etc.

Kipp's apparatus (Fig. 12) is well adapted for precipitation with sulphuretted hydrogen.

C, a glass bulb, receives the diluted sulphuric acid from the funnel tube *e;* the acid enters the glass bulb *A* through the tube *b*. It rises in this and comes in contact with ferrous sul-

Fig. 13.

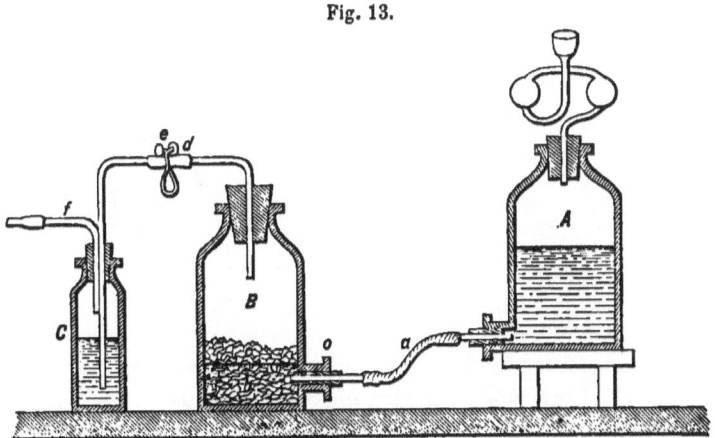

phide or calcium mono-sulphide contained in the bulb *B*, and the gas generated escapes by the tubulure *c* through the lateral tube controlled by the cock *d;* the tubulure *c* also serves for filling the bulb with ferrous sulphide; *a* is the tubulure for emptying *A*.

Debray's apparatus (Fig. 13) is arranged in the following manner: A is the vessel for the diluted sulphuric acid, provided with a safety tube and a rubber tube a. B is a vessel containing a layer of glass splinters, piled up so high that the ferrous sulphide, lying upon it, is above the opening o. d is a glass tube provided with a clip (compression stop-clock) e; C is a wash-bottle, f the pipe for conducting away the gas. By opening the cock e the acid flows from A into B, and sulphuretted hydrogen is disengaged. By closing the cock the fluid is forced back from B to A; the pressure of the gas may be increased by placing A higher up. Instead of sulphuretted hydrogen, sodium hyposulphite may be used as a precipitating agent.

d. Filtration.—A funnel, the sides of which have a slope of 60°, is generally used for this process; a filter of paper is folded into it, and, if necessary, covered with a watch-glass. If the filtration is to be done quickly, the filter is connected with an air-pump,[1] or compressed air is used.[2] Fig. 14 shows a *filtering apparatus* connected at a with a water air-pump. The mouth of the flask is furnished with a rubber stopper perforated for the

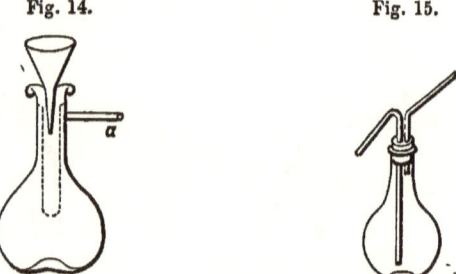

Fig. 14. Fig. 15.

reception of the funnel. The dotted lines below this represent a perforated test-tube, through which the liquid is drawn into the flask. The precipitate on the filter is washed[3] by means of a wash-bottle (Fig. 15).

---

[1] Fresenius, quant. Analyse, 1871, p. 97. Fresenius's Ztschr. f. analyt. Chem. ii. 359; iv. 46; 1875, p. 308.

[2] Fresenius's Ztschr. xvi. 92. Dingler, ccxxv. 81, 105.

[3] Bunsen's Auswaschen der Niederschläge in Fresenius's Ztschr. viii. 174.

e. Decantation.—When a precipitate thoroughly settles, the clear supernatant liquid may be poured off. The precipitate is then repeatedly washed with water and decanted. To dry the precipitate, the contents of the flask (a glass vessel (Fig. 16) with straight sides), are washed into a crucible or evaporating dish, with as little wash-water as possible (precipitated copper and gold, tin stone purified by boiling with acid); or the precipitate is filtered off, and, if necessary, also the sediment remaining, after the water used for decanting has been poured off.

Fig. 16.

f. Drying precipitates.[1]—The filter,[2] without being taken from the funnel, is covered with paper, to protect it against dust, etc., and dried in an air-bath, or a water-bath. Or, it is removed from the funnel, folded up and dried first between blotting paper, and then in a covered roasting-dish, in the muffle-furnace (assay of lead with sulphuric acid).

g. Igniting precipitates.[3]—If the substance is not to be weighed upon the dried filter, it is highly heated with the filter in the roasting-dish after the lid has been removed. Or, the precipitate is carefully detached from the filter; the latter is folded up,

Fig. 17.   Fig. 18.

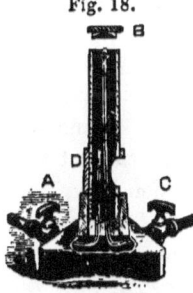

wrapped around with platinum wire, and is then burned over a flame or directly upon the cover of a platinum crucible, or in a

---

[1] Mürrle, of Pforzheim, furnishes distilling apparatus and sand-bath very suitably combined, with steam and air drying closets (these are in use in the Berlin School of Mines). Bestimmung der Niederschläge ohne Filtriren, Auswaschen und Trocknen in Fresenius's Ztschr. 1877, 157; 1879, p. 14.

[2] Filtrirpapiere in Fresenius's Ztschr. xvi. 59; xviii. 246, 260.

[3] Fresenius's Ztschr. 1875, p. 328.

roasting dish. The residue, together with the ashes of the filter, is placed on a roasting dish and ignited in a muffle-furnace, or in a platinum and porcelain crucible over a Bunsen burner, an ordinary spirit-lamp (Fig. 17), or a blast-lamp.

The best form of blast-lamp is shown in Fig. 18; it consists substantially of a Bunsen burner with a blast attachment. The blast flame, when confined by the loose cap, B, is compact and extremely powerful owing to the fact that the air mixture is partially made before the blast begins to act. The taps A and C, respectively, admit the gas and air.

If necessary the crucible is allowed to become cold by placing it in the desiccator (Fig. 19), in which are fragments of caustic alkali, calcium chloride, or sulphuric acid (caustic soda attracts water with the greatest avidity, next follow caustic potassa and calcium chloride in the order named). In the figure, a is a smoothly ground glass plate, with which b, a bell-glass, makes an air-tight joint; c, cup with concentrated sulphuric acid, and d, the dish or crucible with the precipitate to be dried, resting upon it.

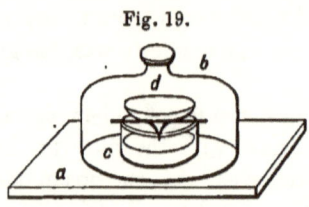
Fig. 19.

2. *Assays by volumetric analysis*[1] (p. 17).—By this method, the quantity of a substance in solution is determined from the volume of a solution of another body, which produces with the first a definite reaction, and the strength of which per unit of volume of its solution is known (called a standard solution or normal solution). The result is then found by calculation from the quantity of normal solution employed. The *final reaction*, which can sometimes be recognized only by a change occurring in another substance, especially added to the fluid (*indicator*), may be known—

a. In saturating a base or an acid with the normal solution

---

[1] Schwarz, Maassanalyse, Braunschweig, 1853 und 1873. Schwertfeger, Maassanalyse, Regensburg, 1857. Gräger, Maassanalyse, Weimar, 1866. Fleischer, Maassanalyse, Leipzig, 1867. Fleischer, Titrirmethode, Leipzig, 1871. Rieth, Volumetrie, Bonn, 1871. Mohr, Titrirmethode, Braunschweig, 1874. Muspratt's Chemie, vii. 167.

(analysis by saturation), by a *change of color*, or by *decolorization* of a colored solution (assay of copper with potassium cyanide), or by an indicator such as litmus, which is added for the purpose, as in the estimation of acids or alkaline carbonates.

b. In precipitating the body to be determined, with a standard solution, *when precipitation ceases* (Gay Lussac's silver assay), or by some *change in an added indicator* (Schaffner's zinc assay; Pelouze's copper assay); and frequently also by the *drop-test*, that is, a drop of the assay fluid and of the indicating fluid are brought in contact upon a porcelain plate by means of a glass-rod, or alongside of each other upon filtering paper, in such a manner that the edges of the drops run together, or by allowing a drop of the assay fluid to flow down over paper saturated with the indicating substance, etc.

c. In oxidizing or reducing the substance to be determined by means of a standard solution without adding an indicator, the final reaction will be recognized by the appearance or disappearance of certain colors (chameleon assay), or by adding an indicator (starch in the assay of copper, assays of manganese, etc.).

The *operations* which may occur are as follows:—

a. Solution, that is to say, bringing the substance to be tested into a state of solution as in 1, *a* (p. 35).

b. Preparation of the standard solution, namely—

α. By dissolving a *weighed* quantity of a chemically pure solid substance, and diluting the solution to a definite volume, so that the chemical power of a unit of volume of the solution is known. These liquids are called *normal solutions* when as many grammes of the substance have been dissolved and diluted to 1 liter as are equal to the atomic weight of the substance, and *decinormal solutions* when a quantity of substance corresponding to $\frac{1}{10}$ of the atomic weight has been used for the solution.

β. By dissolving an *unweighed* quantity of the solid substance and making an *empirical solution* by diluting it in a corresponding manner, that is to 1 liter. The titer of this is determined by allowing it to act upon a measured volume of a solution containing a known quantity of the body to be determined, until the reaction takes place. The titer is then found from the volume of the empirical solution consumed.

42   ASSAYING.

The standard (titer) of normal solutions subject to chemical alterations must be verified from time to time.

c. *Measuring and titration of the assay liquid.*—For this are required—

α. *For measuring, stoppered measuring flasks* (Fig. 20) divided up to a mark on the neck into 1, ½, ¼ liter and into small divisions (200, 100 cubic centimeters, etc.); a *stoppered mixing cylinder* (commonly called a test-mixer) (Fig. 21), having a capacity of

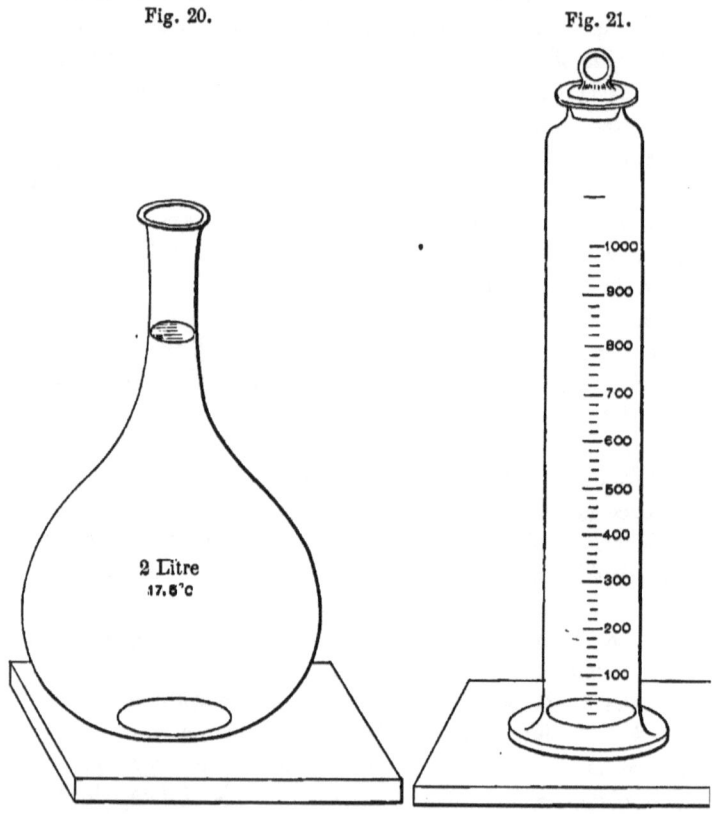

Fig. 20.   Fig. 21.

from one to two liters also divided into cubic centimeters. By means of this, fluids can be measured off diluted, and mixed in definite proportions. *Pipettes* (*measuring-pipettes*) (Fig. 22) divided into whole and $\frac{1}{10}$ cubic centimeters; and *whole pipettes*, capable of holding a certain number of cubic centimeters up to a mark. The latter are used for transferring a certain quantity of assay fluid to a beaker glass, flask, etc. In doing this the lower end of the pipette is either held against the side of the

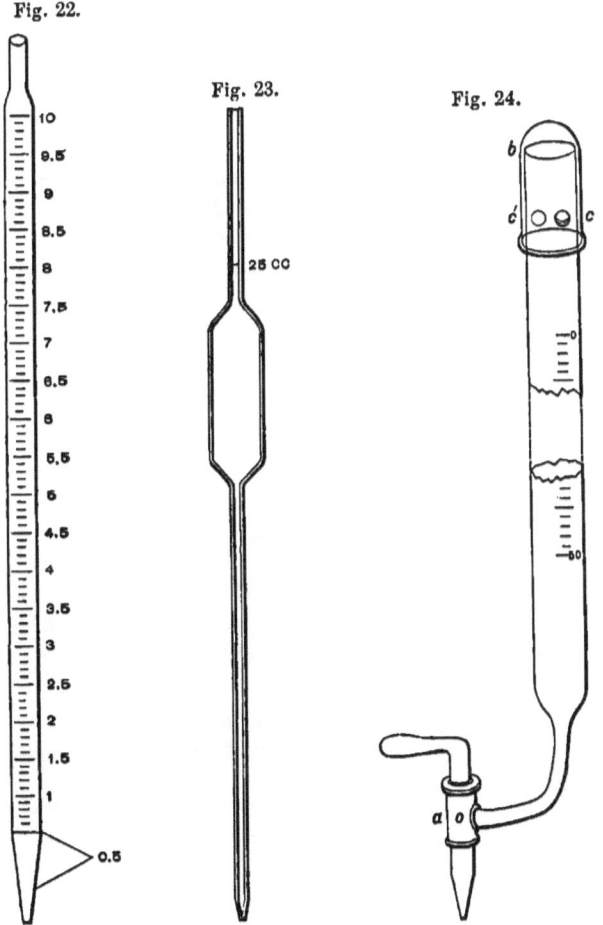

Fig. 22. Fig. 23. Fig. 24.

vessel and the fluid allowed to run down on it, or it is held free. *Stohmann's siphon-pipette* is used for removing the clear supernatant liquid from precipitates, or poisonous, bad-smelling liquids, etc.

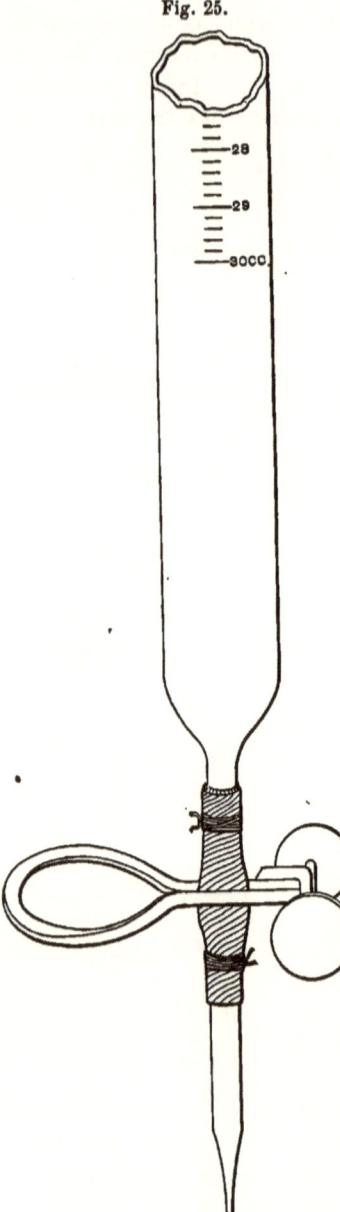

Fig. 25.

β. *For titrating.*—Burettes[1] for measuring the number of cubic centimeters of standard solution which have been allowed to run into the assay fluid until the final reaction is reached. For measuring assay and normal solutions, it is a very good plan to place two burettes in the same stand side by side. The burette represented by Fig. 24 (p. 43) is well adapted for all uses. It is provided with a glass-cock a; b is a glass-cap to protect the liquid from dust; $c\, c'$ are openings in it for the admission of air. *Mohr's* burette is the simplest form of the apparatus, and has the preference over all others for general purposes. It is, however, not to be recommended in cases where the rubber of the pinch-cock will be liable to act chemically on the liquid employed. (Fig. 25.)

3. *Assays by colorimetric analysis.*— This method is based upon the principle that

---

[1] Stender's glass manufactory in Lampspringe furnishes graduated glass vessels with graduation in red burned in with enamel. König's Ventilburette in Dingler, ccxvii. 134. Kleinert's Chameleon-burette in Fresenius's Ztschr. 1878, p. 183. Bürettenstative in Dingler, ccxxii. 465; ccxxix. 366. Fresenius's Ztschr. 1877, p. 82, 228.

equal volumes of solutions of an equally intense color contain also equal quantities of coloring matter. By comparing fluids of an equal intensity of color, and taking the volume into consideration, a conclusion is formed as to the percentage of the coloring body which is contained in the one to be determined. The same manipulations occur here as in assays by gravimetric analysis, namely, solutions, precipitation, etc., and in addition comparison of the colored assay solution with standard colored solutions contained in tubes or tapering glasses of known cross-sections, measuring the solutions in calibrated cylinders, etc.

III. Assay Furnaces.

9. GENERAL REMARKS.

The choice of an assay furnace will depend chiefly on the degree of heat to be obtained, and whether the substances are to be oxidized or reduced, or only calcined, fused, sublimed, or distilled. Furnaces, accordingly, are divided into *muffle furnaces*, *draught* or *wind furnaces*, *blast furnaces*, *sublimation furnaces*, and *distillation furnaces*. In regard to their construction, they vary chiefly according to the fuel to be used (flaming or glowing fuel).

10. MUFFLE-FURNACES.[1]

The principal part of this is the muffle (Fig. 26). It is usually made of refractory clay, sometimes, though rarely, of iron. It is open in front, and closed at the rear; and the semi-cylindrical body is often provided along the sides with draft orifices, as shown. It is either connected with the bottom, or stands loose upon it.

Fig. 26.

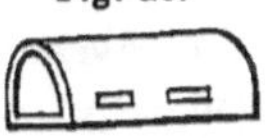

It serves for the reception of the assay charge, and is heated from the outside by a glowing or flaming fire. These furnaces are absolutely necessary for oxidizing processes (calcining, cupellation, refining), but they are also adapted for operations requiring

---

[1] Engin. and Min. Journ. 1878, No. 26, p. 443. Silliman, Double Muffle-Furnace, 1876, vol. xxii. No. 17.

only the production of a high temperature (glowing, reducing, and purifying fusion, etc.), that is to say, when only temperatures not exceeding the fusing point of gold and copper (about 1200° C., 2192° F.) are required (they are, therefore, not available for assays of cast-iron). In the latter cases the fuel is not completely utilized, and besides, they are more difficult to attend than the wind and blast furnaces, where the crucibles, etc., are placed directly in the glowing fire, or come in direct contact with the flame.

The furnaces are either *bricked in* (for instance, large muffle-furnaces for burning coal), or they are *portable*. In the latter case, the furnace for receiving the muffle is constructed of fire-clay which is sometimes surrounded with a casing of sheet-iron (mint furnaces). The work connected with the muffle-furnace consists chiefly in heating it, regulating the temperature (by reducing or urging the fire, regulating the admission of air, opening or closing the mouth of the muffle, by removing or piling up fuel, etc.), in stirring the fire regularly (in doing this the fuel must be piled chiefly upon the front part of the grate and only a thin layer upon the back part), in ventilating the grate frequently, in repairing (that is, lining defective places in the walls of the furnace, filling in of cracks in the bottom of the muffle with fire-clay, or scraping the bottom and lining it by strewing it with powdered fire-clay, cupel ashes, chalk, pounded assay vessels, etc.), introducing and removing the assay vessels in the muffle, cleansing the furnace after the work is finished by drawing the glowing cinders from the grate and allowing the fire-door to remain open, etc.

According to the kind of fuel used, we may divide them into—

1. *Furnaces for solid, free-burning, flaming fuel.*—These are generally used with large muffles, and with such fuel the heat can be better regulated than in furnaces heated by a glowing fire, but they require more care in attending them. Stoking is done from the front (*Plattner's* furnace[1]), or from the back (Schemnitz, Pribram[2]). With the latter arrangement the opera-

---

[1] Freiberger, Jahrb. 1842, p. 1. Ztschr. des Ver. deutsch. Ingen. 1877, Plate 12, Figs. 3 to 5.

[2] Rittinger's Erfahr. 1857, p. 29. B. u. h. Ztg. 1876, p. 353; 1876, p. 61.

tor, working in front of the furnace, is not exposed to the direct heat, but it also prevents him from giving immediate attention to the firing should the assay require it.

Fig. 27.

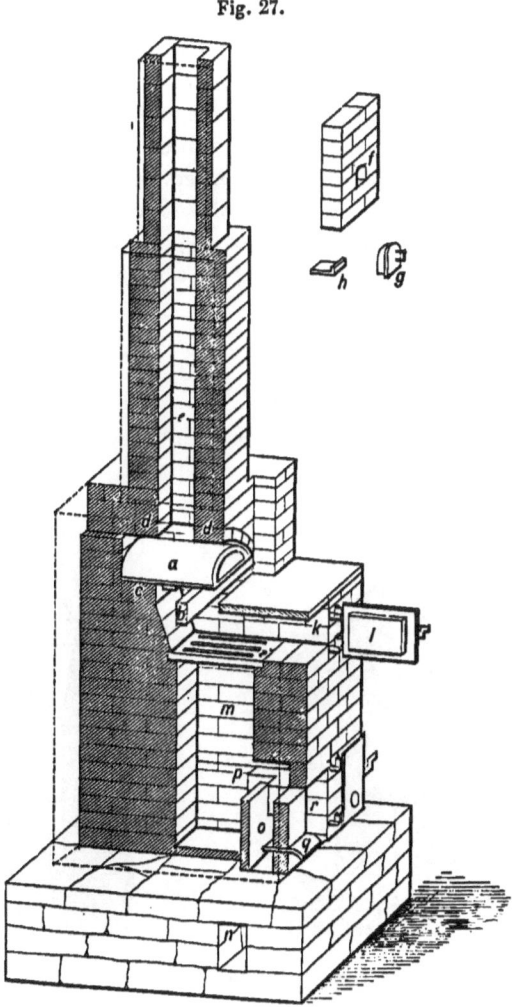

Fig. 27 represents *Plattner's muffle-furnace for coal, with the stoke-hole in front.* *a*, muffle of fire-clay, 36.6 centimeters long,

17.6 centimeters high, and 34.2 centimeters wide, with an ascending slope of 2.4 centimeters. It rests upon the support b, and three legs c; d is the vault. There is a space of 4.9 centimeters between it and the walls of the furnace. e, the chimney, 14.7 centimeters wide, and 3 to 4 meters high. f, mouth of the muffle, 12 centimeters wide, and 14.6 centimeters high, which can be closed by the fire-clay door g. Another door h is used for covering a slit sometimes provided over the muffle (for heating plates of metal, etc.), but it is usually omitted; i is the grate, 26.8 centimeters wide, and 51.4 centimeters long, 28.1 centimeters below the muffle; k, the stoke-hole, 22 centimeters high, and 26.8 centimeters wide; l, fire-door; m, ash-pit, 76.8 centimeters long, and 26.8 centimeters wide; n, a channel, 22 centimeters

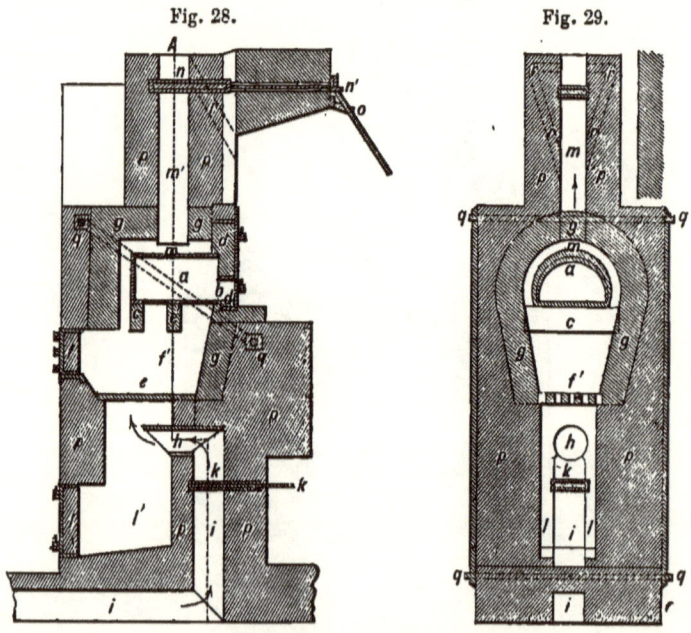

Fig. 28. Fig. 29.

wide, communicating with the open air for conducting air under the grate through the flue o, p, 9.8 centimeters wide, which is provided with a damper q; r, ash-pit door, 26.8 centimeters wide, and 34.2 centimeters high.

Figs. 28 and 29 represent a *muffle-furnace with the stoke-hole at the back*. a, muffle, resting upon the supports c and c'; b, mouth of the muffle; d, front wall; e, grate; f, fire-door; f', fire-box; g, refractory lining; h, i, channel for conducting the external air beneath the grate; k, damper; l, ash-pit door; l', ash-pit; m, fire-space surrounding the muffle; m', chimney (it is better to place it nearer d), with damper n, n', and lever o, for regulating the same; p, brickwork of the chimney, with flues, r, for carrying off the fumes coming from the mouth of the muffle; q, hooping.

2. *Charcoal and coke furnaces.*—Coke, as a general rule, requires a grate under the muffle, and a strong draught. With charcoal this arrangement is not so essential, though in order to secure a more uniform supply of air a grate is usually provided. The ashes from coke are more difficult to remove and attack the walls of the furnace more than wood ashes. Smaller furnaces of this kind are much used for assaying gold and silver; and also larger ones, in which the heat can be better regulated (the Schemnitz charcoal furnaces are of this construction), and where the stoke-hole is in the rear, or the firing is done through two channels on the sides.

Assay furnaces for charcoal.—Fig. 30 shows such a furnace. b, muffle of fire-clay, 14 centimeters long, 7.5 centimeters high, 9 centimeters wide, with walls 8 millimeters thick, and resting upon two rails passing through openings in the iron casing. The inside of the casing is lined with fire-clay from 15 to 20 millimeters thick. In front of the muffle is a shelf of sheet-iron resting upon the rails supporting the muffle; c is the mouth of the furnace through which the charcoal is fed, and the products of combustion escape into a hood or through a sheet-iron smoke-stack. The mouth of the muffle and the flues above and below it can be closed by dampers; a, the cupel.

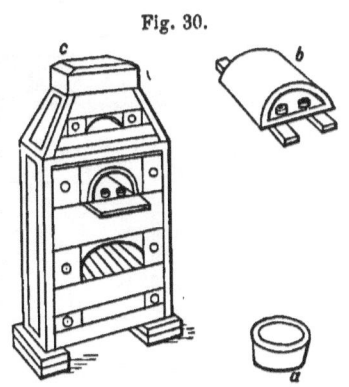

Fig. 30.

3. *Gas furnaces (coal-gas).*—By using these furnaces the work can be carried on in a very cleanly manner, and the temperature can be very perfectly regulated. The gas is introduced either by means of burners placed at the rear wall (*Perrot's* furnace, used in the Berlin School of Mines), or from below through four straight burners standing alongside each other beneath a slit in the bottom (furnaces of *Lenoir* and *Forster* of Vienna, used in the laboratory of the Schemnitz School of Mines, etc.), or through curved burners arranged in the form of a circle beneath the furnace (furnace of the Société genevoise pour la construction d'instruments de physique à Genève, used in the Berlin School of Mines). The oil furnaces of *Andouin-Deville* of Paris (using the vapors of crude petroleum) are said to be cheaper in operation than the gas furnaces just described. The oil trickles from funnels upon the hot grate-bars set obliquely and channelled. There it is instantly vaporized and burns.[1]

Fig. 31 shows *Perrot's gas muffle-furnace.* a, muffle of fireclay, with refractory coating and movable cover b; e, f, g, furnace

Fig. 31.

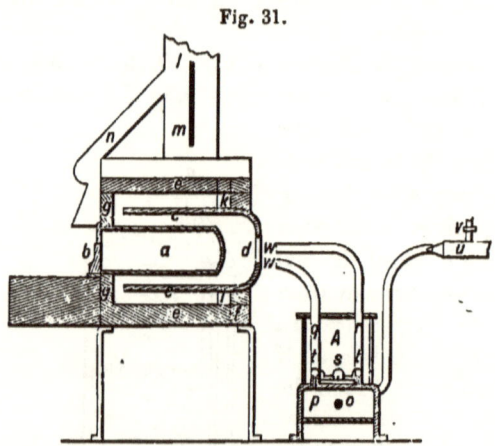

walls of sheet-iron with refractory lining; A, burner, with chamber p, into which coal-gas enters at o, from the pipe u, provided with manometer v. From here it passes through narrow chan-

[1] Ztschr. des Ver. deutsch. Ingen. xxi. 225.

nels into the burner-tubes q and r, which are provided below with openings furnished with valves t for regulating the admission of air; w, w, the nozzles from which the flame passes through d into the space around the muffle, and escapes through the flue k into the chimney l, in which is a damper m. The chimney also receives, through the pipe n, the fumes which may escape from the mouth of the muffle.

11. DRAUGHT OR WIND FURNACES.

These consist—

1. In case *carbonized* fuel (coke, charcoal) is used, of a round, rectangular or oblong fire-place, separated from the ash-pit at the base by a grate, and provided with a fire-clay or cast-iron cover or top-plate. A lateral flue connects the fire-place with the chimney. The furnace is either bricked in (Fig. 32) or is portable. In the latter case, the body of the furnace is made of a sheet-iron cylinder lined with refractory material. It is also a very good plan to set a furnace of this kind into brick-work, leaving an intermediate space between the two, in which case the usual binding with strap-iron may be dispensed with. The cover or top-plate over the fire-place consists of two fire-tiles provided with some convenience for easily removing and replacing them. It is best to place a small carriage in the ash-pit (Fig. 33) for receiving and removing the ashes (Berlin School of Mines).

Fig. 32.

The degree of temperature possible to attain depends on the height of the shaft between the grate and the flue, the height of the chimney, and the quality of the fuel used (coke will give a higher temperature than charcoal). The temperature can be increased with the aid of a flue leading from the ash-pit into the open air, or by an under-grate blast, and is regulated by a damper fixed in the door of the ash-pit, or in the flue or chimney. The highest temperature is found at about 4 to 6 centimeters above the grate, which should be taken into consideration in placing the crucibles in the furnace.

The labor attending these furnaces consists of—

a. The placing of the assay vessels in the furnace by hand. If it is necessary to look into them during the operation (assay of lead in iron crucibles, Cornish roasting assay of copper), they are placed in a hollow made in the fuel, generally coke; or, if this is not required (assays for lead, copper, tin, iron, etc., in clay crucibles), the assay vessels are placed immediately upon the grate, leaving sufficient space between them for the necessary fuel, and in such a manner that the part of the vessel which is to be heated the strongest stands about 4 to 6 centimeters above the grate. If, therefore, vessels with feet are used, they must be placed directly upon the grate, while those without feet (crucibles) are supported on a block or stand of fire-clay.

b. Firing.—This, as a general rule, is done from *below* by putting glowing coals between the assaying vessels, filling up the shaft with fuel, and then gradually closing the top-plate of the furnace. But if the heating must take place very slowly, the firing is done from *above*, by placing the glowing coals on top of the fuel with which the shaft is filled. The fire, when the mouth of the furnace is closed, will then gradually work down. (In the Schemnitz laboratory, the lateral flue is placed below the grate and the air required for combustion is introduced from above.) The temperature is regulated in the manner indicated on p. 51, and, if necessary, fuel is added from time to time, but, before this is done, the glowing coal must be poked down to do away with empty spaces.

c. Taking the vessels from the furnace.—This is done by lifting the vessels out at the top of the furnace, by means of crucible tongs (Fig. 60), either out of the coke, or from the grate, after the fuel has burned down, or, in the latter case, it may be more convenient to remove them through an opening in the side (*t* in Fig. 32), but this must be closed up during the operation of the furnace. Either the contents of the crucibles are poured out and the crucibles while still glowing placed back in the furnace and again charged from the mixing capsules (Fig. 7) (assay of lead in an iron crucible), or the clay crucibles are allowed to cool off and are then broken up.

Fig. 33.

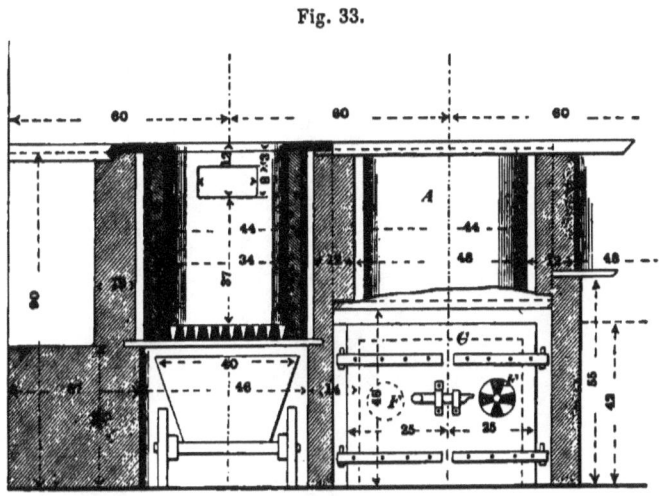

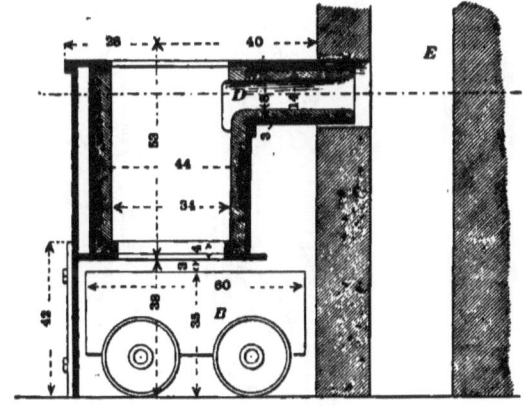

Fig. 34.

The furnaces used in the Berlin School of Mines are shown in Figs. 33, 34, and 35. They consist of an iron cylinder A set into brick-work and lined with refractory material. The grate

ASSAYING.

Fig. 35.

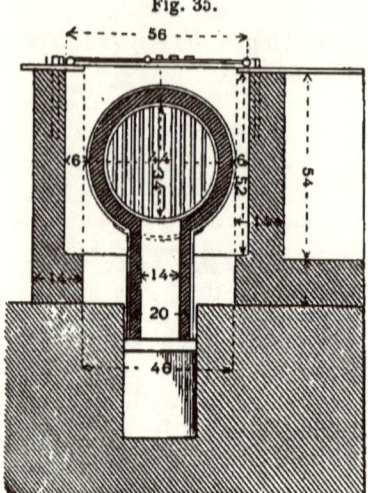

Fig. 36.

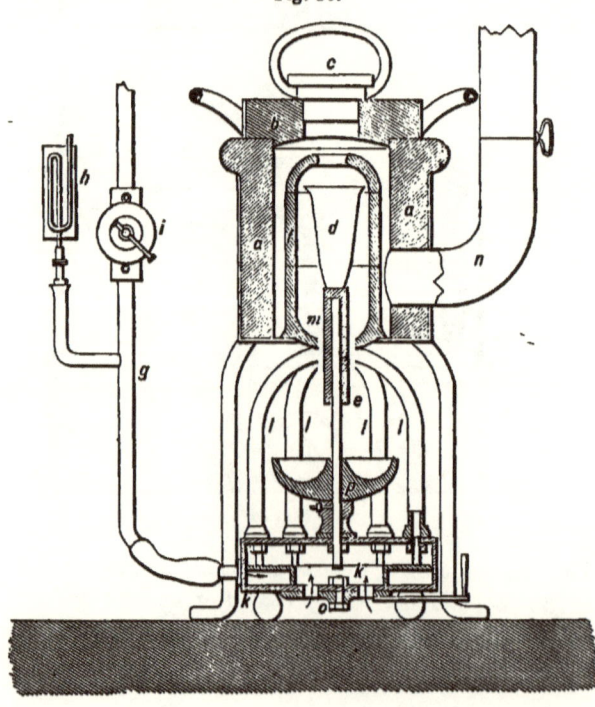

forms the bottom, beneath which is placed a truck B for the ashes. The door of the ash-pit C is provided with register F. D represents the flue and E the chimney. The height of the furnaces varies according to the kind of assays to be made in them. The refractory lining is usually about 5 centimeters in thickness, giving the furnace a clear interior diameter of about 34 centimeters. For lead assays they are 20 centimeters high, for copper 26 centimeters, and for iron 35.5 centimeters. The chimney is 10 meters high, provided with a damper for regulating the draught.[1]

2. *Wind furnaces for free-burning (flaming) coal.*—The assay vessels stand over the grate upon a tile of fire-clay in the same manner as in *Plattner's* furnace, except that there is no muffle. (Freiberg.)

3. *Wind furnaces for illuminating gas.*[2]—These furnaces are easily attended, the work can be carried on in a very cleanly manner, and at the same time with the greatest accuracy, as the assayer can conveniently look into the crucible during the operation.

Fig. 36 represents *Perrot's furnace*. a, the outer shell, with cover b, and sight-hole c; d, the crucible, upon a movable stand, e; f, inner shell; g, pipe, with manometer h, and cock i, for conveying gas into the annular chamber k, from which it passes through the pipes l, through the annular opening m, into the inner space of the furnace, where flame plays around the crucible d, and finally escapes through the upper opening of the inner chamber into the exterior annular space, and is carried off below through the pipe n, leading to the chimney; o, openings for admitting the air required for combustion, which is mixed in the pipes l with the gas. The admission of air is regulated by a cut-off. p is a cup for the reception of any metal which may overflow and escape from the crucible.

---

[1] B. u. h. Ztg., 1880, p. 2.
[2] B. u. h. Ztg. 1873, p. 284 (Perrot); Dingler, ccvi. 360 (Wiessnegg); Dingler, clxxx. 220; clxxxix. 376; Oestr. Jahrb. v. Hauer, 1878, p. 123 (Schlösing). Mitchell, Pract. Assaying, 1888, pp. 74 to 105. Hempel's Gasofen mit Oxydationsvorrichtung, z. B. zum Abtreiben, in Fresenius's Ztschr. xvi. 454; xviii. 404 (may be had of Desaga in Heidelberg).

56   ASSAYING.

Wiessnegg's gas furnaces are of simpler construction, and consume less gas. The flame plays around the crucible in the form

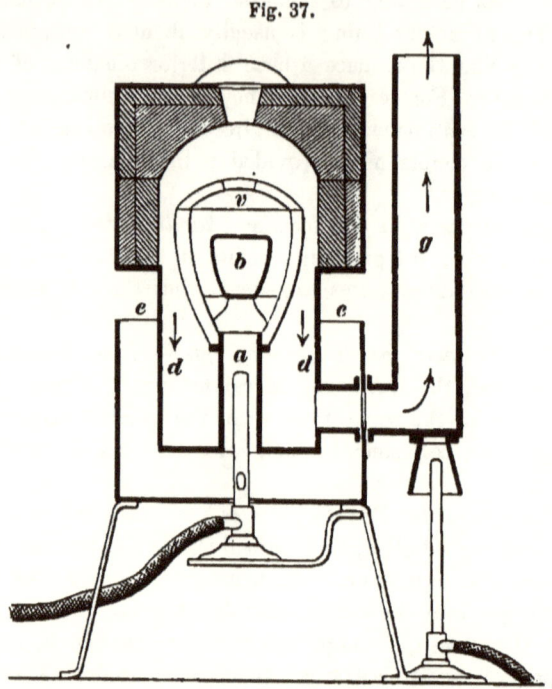

Fig. 37.

of a spiral. By this means air and gas are more intimately mixed, and higher temperatures can be obtained. The same result is attained in Schlösing's furnace. The attainable heat of *Perrot's* furnace is about 1560° C. (2840° F.).

Roessler[1] describes a small furnace for the production of high temperatures (Fig. 37) which is heated by a Bunsen burner. A considerable quantity of pure gold, silver, etc., can be melted in it in from fifteen to twenty minutes. The cold air enters at *e* and is heated by the hot walls of the sheet-iron jacket *d*. It then passes

---

[1] Polytechn. Notizblatt, 1884, p. 308. Chemikerztg, 1884, p. 1220.

to the burner and thence towards the top at *a* into the flame and together with the latter under the crucible *b* where combustion takes place. The products of combustion pass out through the aperture *v*, through the iron jacket, and after heating its inner walls, are carried off through the chimney *g*. Beneath the latter is placed a second burner whose flame has to be so regulated that only sufficient air for complete combustion is drawn in.

For melting with a coke-fire the same principle has been modified as shown in Fig. 38. The cold air enters at *a* and after being heated passes through *c* into the closed ash-pit under the

Fig. 38.

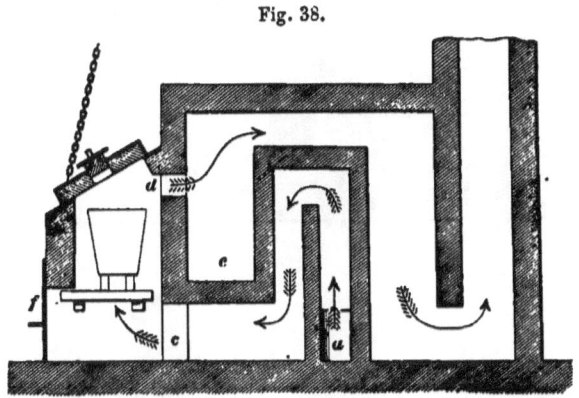

grate. The products of combustion escape through *d*, deposit at *e* any dust or ashes they may contain, and after heating the air for combustion passes out through the chimney. By this preparatory heating the air acquires a temperature of nearly 300° C. (572° F.) and a combustion temperature of about 1400° C. (2552° F.).[1] The tightly closed ash-pit door is shown at *f*.

Fig. 39 shows Fletcher's direct draft crucible furnace. It consists of a fire clay body held together by sheet-iron bands. The heat and flame pass through the body of the furnace to the chimney. It can be used either for scorifying or cupelling, and by removing the top cover the heat has full play upon a roasting dish placed upon it.

[1] Dingler's Journ. Bd. 257, p. 153. Chemikerztg, 1885, p. 1359.

58 ASSAYING.

Fig. 40 is a new form of gas furnace suggested by Brown.[1] In this furnace the flame from the burner, shown in the left of the

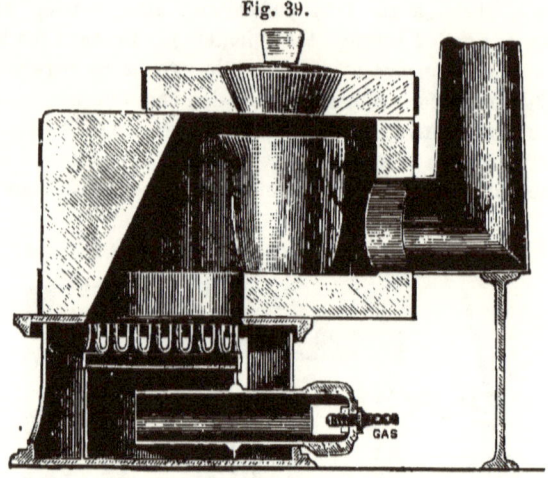

Fig. 39.

Fig. 40.

engraving, rises and passes horizontally through the body of the furnace to the chimney. Its form and operation are somewhat similar

[1] Manual of Assaying. Walter Lee Brown. Chicago, 1886.

to those of a reverberatory furnace, the movable bricks when in place forming the roof. The exterior dimensions are 20 inches (50.6 centimeters) long, 7 inches (17.71 centimeters) wide, and 5½ inches (13.91 centimeters) deep. In the interior, upon the bottom, are four little wedge-shaped bridges of fire-clay which are movable; and upon these rests a false bottom, also movable. The latter corresponds to the muffle-bottom of an ordinary furnace, and upon it is done all the work.

12. BLAST FURNACES.

These are low cylindrical shaft furnaces, constructed of fire-resisting material, or of a sheet-iron cylinder lined with refractory clay. At some distance above the hearth of the furnace are one or several tuyeres symmetrically arranged. The mouth of the furnace is provided with a movable sheet-iron chimney. The furnace, in order to increase the temperature, is well supplied with air heated in the space between the two iron shells surrounding the shaft (*Sefström's* furnace) or in a reservoir below the perforated hearth of the furnace (*Deville's* furnace).[1] The hearth in *Welch's* furnace may be easily separated from the furnace body.[2] The fuel used in these furnaces should be in lumps about the size of a walnut and uniform in size. A very high temperature can be produced in a shorter time, and with the consumption of less glowing fuel than in wind furnaces, but a certain amount of power is required for operating the blast (bellows, fan, Root's blower),[3] and the fire must frequently be stirred and fuel added. If only one crucible is used, it is placed in the centre of the furnace. But if more are placed in the furnace at one time, each is placed at the same distance from one of the tuyeres. *Raschette's* furnace furnishes very high temperatures; it has an oblong cross-section, and the tuyeres are arranged alternately in rows. *Munscheid's*[4] *gas blast furnace* gives also very high temperatures. Gas and air mixed are drawn into it by means of an exhaust fan.

[1] Kerl, Thonwaarenindustrie, 1879, p. 76.   [2] Dingler, ccxxix. 159.
[3] Root's blowers are well adapted for this purpose.
[4] B. u. h. Ztg. 1878, p. 361.

Fig. 41 shows *Sefström's furnace.* *b* is the space between two sheet-iron cylinders closed on top by *a*. The inner cylinder is lined with a fire-resisting material, *c* (1 part clay and 3 to 4 parts quartz sand). The air enters at *d*, and, after having been heated in the intermediate space, is carried through the tuyeres *o*. The dimensions of a furnace for six small iron crucibles are as follows: 18 centimeters in diameter; total height 15 centimeters; a collar 7 centimeters high, upon which sits the sheet-iron chimney; width and height in the clear, 10.5 centimeters; thickness of the refractory lining 2.5 centimeters; distance between the two cylinders, on the sides 1.2 centimeters, and on the bottom 2.5 centimeters. The manometer is placed on the exterior shell. *Lang's* blast-furnace[1] for larger masses has an annular air-conduit.

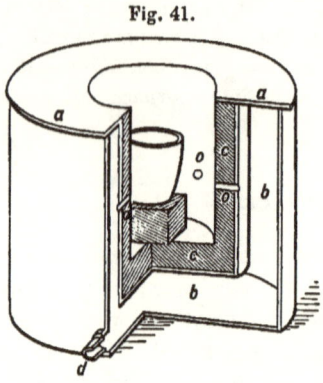

Fig. 41.

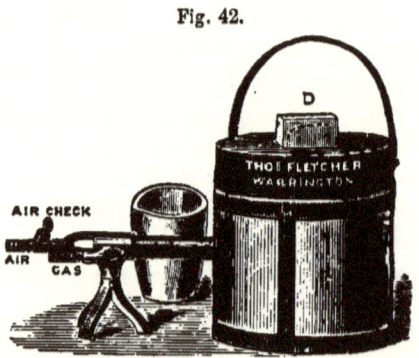

Fig. 42.

Fig. 42 shows Fletcher's injector gas furnace. It consists of a hollow cylinder of fire-clay inclosed in a sheet-iron casing and containing an interior space sufficiently large to hold a crucible about 3

---

[1] Kärnthn. Ztschr. 1879, No. 8, p. 287.

FURNACES FOR SUBLIMATION AND DISTILLATION. 61

inches (7.58 centimeters) in diameter. Extremely high temperatures can be obtained in this little furnace, when a free supply of gas is introduced through the blast burner.

13. FURNACES FOR SUBLIMATION AND DISTILLATION.

These consist of a fire-space to be heated, for the reception of variously shaped vessels (tubes, retorts, boilers, etc.) of clay, porcelain, glass, or iron, in which the substances are heated without addition (as, arsenical pyrites, iron pyrites, amalgam), or with fluxes (preparation of realgar, separation of mercury from cinnabar, etc.). They are usually provided with a receiver, cooled off, for condensing the volatile products into solid bodies (sublimation), or to fluids (distillation).

Fig. 43 shows a *sublimation furnace*. a are clay tubes for the reception of the assay sample (as, for example, arsenical pyrites or iron and arsenical pyrites for the production of realgar); b, receiver for the sublimate (arsenic, realgar); c, the grate; d, flues.

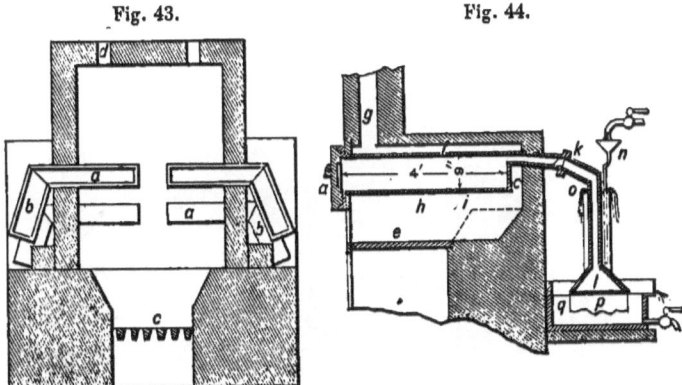

Fig. 43.   Fig. 44.

Fig. 44 represents a *distillation furnace*. a, tube for the reception of the assay sample (for instance, gold or silver amalgam); e, grate; h, combustion chamber; g, chimney; c, pipe for carrying off vapors (of mercury) into the condensing pipe k, provided with a funnel l, the edge of which is covered with a cloth q, upon which cold water flows from n, and overflows at o.—Or Fig. 45.

a, a retort, with tubulure *b*; *c*, receiver, covered with paper or cloth kept cool by water which is allowed to trickle upon it.—

Fig. 45.

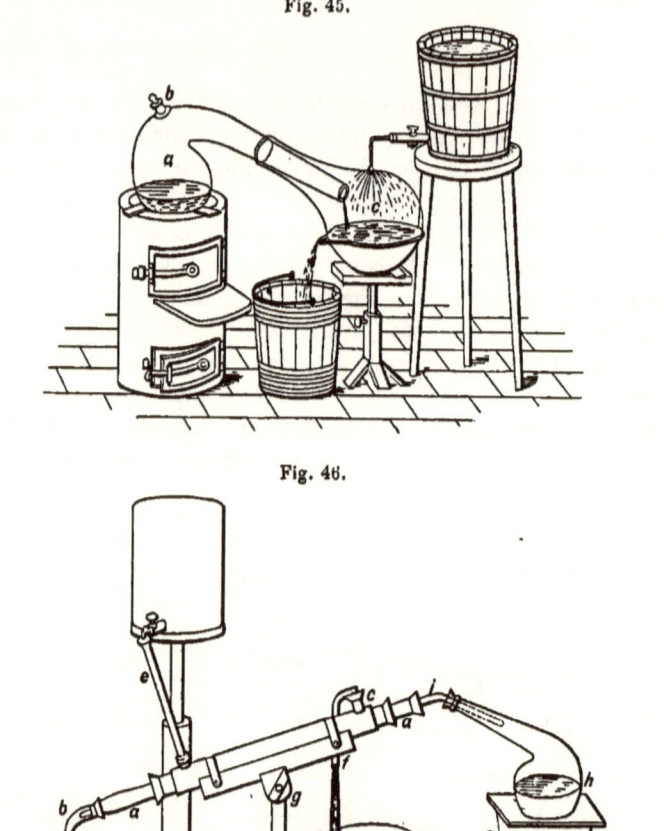

Fig. 46.

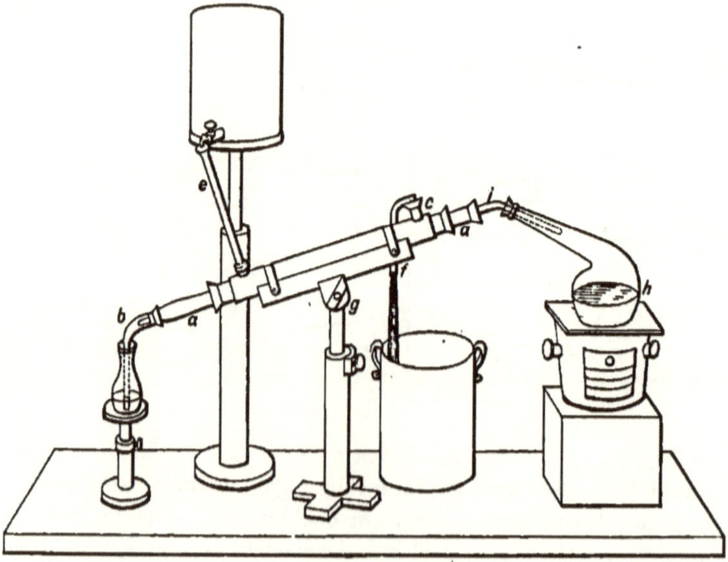

Or Fig. 46, if a more thorough cooling off is required. *h*, a retort, from which the vapors pass through *i* into the cooling-pipe *a*.

ASSAY VESSELS FOR THE DRY METHOD.

This rests upon a stand g, and is surrounded by a sheet-iron cylinder, into which cold water passes through e, and is discharged at c, f, while the liquid formed from the condensed vapors is collected in b.

Organic-combustion furnaces (Fig. 77) may also be used for heating tubes.

IV. Assay Vessels.

14. GENERAL REMARKS.

The form of the assay vessels, as well as the materials from which they are made, varies according to the object for which they are to be used. The principal distinction is, whether they are to be employed in the *dry* or *wet* method.

15. ASSAY VESSELS FOR THE DRY METHOD.

A. Clay vessels.[1]—These are required to be more or less refractory according to the heat to which they are to be exposed (their refractory quality depends on the proportion between silicic acid and alumina and the quantity of fluxing agents—ferric oxide, lime, alkalies, magnesia—which may be present). They must allow of being suddenly heated and cooled without cracking (fat, contracting clay requires to be mixed with quartz, chamotte,[2] graphite). The vessels should be corroded as little as possible by the substances heated in them, but, as a general rule, this can never be entirely prevented. (It may be done to some extent by making the sides of the crucible thicker, or by giving a finer grain to the stuff of which they are made. This should be made as compact as possible, by mixing the clay with chamotte instead of quartz. The interior of the crucible should be made very smooth, and it should be fired in the kiln as strongly as possible.) The vessels should further be very compact. This can be

---

[1] Kerl, Thonwaarenindustrie, 1879, p. 491. Percy, Metallurgy, vol. I. 1875, p. 111.

[2] [Chamotte is a mixture of unburnt fire-clay and dust of fire-bricks, glass pots, or seggars.—TRANSLATOR.]

accomplished by giving a suitable grain to the mass, exercising great care in moulding and firing them strongly. The compactness of the vessels is tested by fusing metallic sulphides, such as galena, several times in them. They are made either by a *plug* and *mould* (roasting dishes and scorifiers, Upper Harz crucibles for lead smelting) or they are turned upon the potter's wheel (crucibles and larger melting pots).

The principal vessels, etc., are—

1. *Vessels without feet.*

   a. Roasting dishes (Fig. 8).—They are flat, smooth inside, not very refractory, 8 to 10 millimeters deep and 50 to 80 millimeters wide. They are used in the manner indicated on p. 32.

   b. Scorification or calcining vessels (Fig. 47).—They have a thick bottom and sides, very smooth interior, and are very compact. To avoid being corroded by lead oxide, they should be made of clay mixed with chamotte. They are 40 to 50 millimeters wide in the clear, 15 to 20 millimeters deep, with a bottom 10 millimeters or more thick.

Fig. 47.

   c. Refining dishes.—They have either the same form as the flat roasting dishes, but are fire resistant and one edge is somewhat ground down, or they are made from fragments (Fig. 48) of crucibles (Fig. 52), and are then 70 to 80 millimeters long; or they are shaped like a flat saucer with feet. These are 30 millimeters wide, with a total height of 25 millimeters (Hungary).

   d. Crucibles.—These are of various forms and sizes, large and small (Figs. 49 to 51). They are respectively 32 and 45 milli-

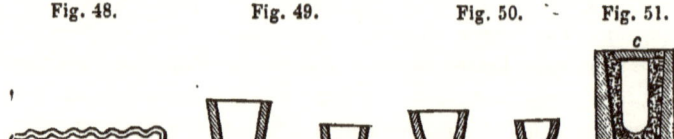

Fig. 48. Fig. 49. Fig. 50. Fig. 51.

meters high in the clear, and have a total height of 39 and 52 millimeters, a clear width of 33 and 43 millimeters, and are not very refractory. Fig. 50 shows a large and a small Cornish crucible for the assay of copper. They are very refractory. Their respective dimensions are: diameter on the top 80 and 68 millimeters, total height 84 and 60 millimeters. Fig. 51 represents a crucible for iron. This is lined by means of a wooden plug with charcoal powder, *b* (this is first moistened with starch paste, molasses, or clay), or it is lined with a mixture of 90 to 95 per cent. retort graphite, 5 per cent. rosin, and some petroleum, and burned with exclusion of air. They are covered with the perforated lid *c*. They are 37 millimeters high and 25 millimeters wide, and, after they have been lined, respectively 22 millimeters, and 10 millimeters. They are very refractory (Hessian pots). The *French* pots are especially refractory and smooth inside.

Graphite crucibles are made of graphite mixed with clay. They are very smooth inside, and very refractory. Those used in Cornwall for assay of tin are 80 millimeters wide on the top, and 50 millimeters on the bottom, have a clear height of 74 millimeters, and a total height of 90 millimeters.

Soapstone crucibles, if gradually heated, are adapted for all smelting purposes. They are infusible, not affected by alkalies, and become harder by burning.

2. *Vessels with feet. Crucibles for lead and copper smelting* (*a*, Fig. 52).—The latter are more refractory than the first. They are 25 to 32 millimeters wide on the top, 40 to 50 millimeters in the centre, 83 to 85 millimeters high in the clear, with a total height of 110 to 120 millimeters. Sometimes there is a depression in the bottom for the reception of the regulus, and the foot, when broken off, may serve as a cover.

Fig. 52.

Fig. 53 shows a crucible for smelting iron, lined with powdered charcoal (see above). These crucibles are 45 millimeters wide, and 55 millimeters high in the clear, with a total height of 90 millimeters.

Fig. 53.

3. *Other clay vessels.*—To this category belong *muffles* (p. 45), *retorts*, and *tubes*.

B. *Wrought-iron vessels. Crucibles for assay of lead, with or without lip.*—These are from 8 to 12 centimeters high, and 5 to 8 centimeters wide. The sides are from 10 to 12 millimeters thick, and the bottom from 2 to 3 centimeters. Other iron vessels used are, *tubes* and *retorts*, and cast-iron *muffles*.

C. *Vessels of bone-ash: Cupels* (Fig. 54).—They are made either of bone-ash alone, or with an addition of a little wood-ash, or pearl-ash, to the water used for moistening the bone-ash, which addition decreases their power of conducting heat. The bones are burnt white throughout, are then powdered and washed. The dried powder, which should be about as fine as coarse wheat flour, is used for the principal mass, while a finer flour is reserved for a final coating. The cupels are formed by filling and driving the prepared bone-ash into a mould made for the purpose (Fig. 55), *B*, the pestle; *A*, the mould; or they are pressed.[1] Ordinary Freiberg cupels for ores consist of 3 volumes of soap-boilers' ash, and 1 volume bone-ash. Their outer diameter is 35 millimeters, diameter in the clear 24 to 25 millimeters. They are 10 to 12 millimeters high in the clear, with a total height of 18 millimeters. Fine or mint cupels consist of 2 volumes soap boilers' ash, and 3 volumes bone-ash. Their total diameter is 26 millimeters, with a clear diameter of 18 millimeters; their total height is 14 millimeters. Cupels in order to be perfect should dry very slowly, and be thoroughly ignited before they are used. They should be white, and, besides a certain degree of solidity, should possess the requisite porosity to absorb litharge (when taken up with the tongs they must not crumble, but it should be possible to crush them with the fingers). They must neither undergo any perceptible change nor crack, when exposed to a white heat, and should develop no gases and form no chemical combinations with the substance fused in them. When too solid they crack easily, absorb the litharge too slowly, cupellation being thereby prolonged, which causes a loss of silver.

Fig. 54.

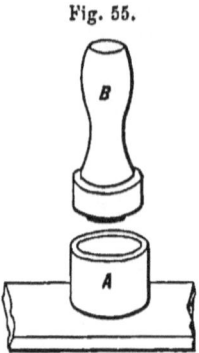
Fig. 55.

[1] B. u. h. Ztg. 1868, p. 154.

ASSAY VESSELS FOR THE DRY METHOD. 67

If, on the other hand, they are too porous, they absorb too much silver and gold with the lead oxides. (A loss of metal can never be entirely avoided in the assay.)

Wait[1] suggests the very simple machine for making cupels as is shown in Fig. 56. It consists of a common letter-press, the mov-

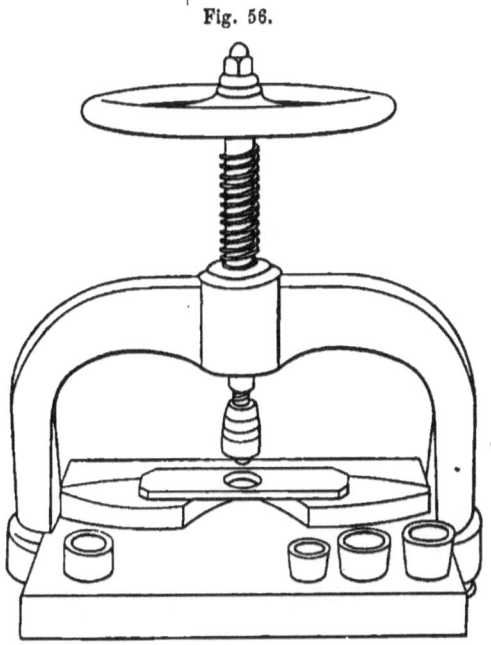

Fig. 56.

able platen of which has been removed, and replaced with a plunger, which turns with the screw, moving up and down, and making the depression in the cupel. The ring holding the bone-ash is held in place directly underneath the plunger by a wooden guide fastened to the base of the press. There is also fastened to the top of the guide for the ring, at a distance above the base of the press equal to the height of the cupel-ring, a piece of wood $\frac{3}{8}$ inch thick, under which the ring will just slide into place, and through which there is a circular hole into which the plunger exactly fits; this last

[1] Note on a cupel-machine. Prof. Charles E. Wait, Trans. Amer. Institute of Mining Engineers, vol. xiv. p 767.

guide not only directs the plunger, but also prevents the lifting of the ring by the plunger. There is also a small spring (not shown in the figure) which pushes forward the ring after the ash has been compressed. To make a cupel, fill the ring with slightly moistened ash, the plunger being raised just high enough to allow the ring to be put in place with the left hand, with the right hand give the wheel about one turn, depending upon the pressure required, reverse the wheel until the plunger is raised, and remove the thumb suddenly, at which moment the spring will push the ring to the front. Cupels of almost any size may be made by having plunger, ring, and guides to suit.

D. Vessels of other materials.—Cupels in the form of a prism, about 2.5 centimeters high and 4 centimeters thick, are chiselled out of charcoal, or turned from hard wood and then carbonized. *Coke cupels* are made of powdered and sifted coke, which is kneaded with liquid pitch. The stiff mixture formed after the mass has become cold, is pulverized, and some more powdered coke added to it (4 parts of coke to 1 part of pitch). The entire mass is then passed through a hair sieve, heated, and stamped in a cupel-mould about 2.5 centimeters high, with a diameter of 3.7 centimeters on the top and 3 centimeters on the bottom. The finished cupels are then ignited, the air being excluded during the operation.

16. ASSAY VESSELS FOR THE WET METHOD.

1. *For assays by gravimetric analysis.*—*Articles of glass:* digesting-flasks or matrasses, beaker-glasses, funnels, watch-glasses, wash-bottles, stirring-rods, retorts, tubes, apparatus for generating sulphuretted hydrogen, etc. *Of porcelain:* crucibles, evaporating dishes, tubes, etc. *Of other materials:* forceps, crucible tongs, wire triangles, wire gauze, etc.[1]

2. *For assays by volumetric analysis,* see p. 40.

3. *For assays by colorimetric analysis.*—Tapering glasses or

---

[1] Muencke's Klemmvorrichtung in Dingler, ccxxv. 387. Dreiecke und Tiegelzangen mit Porzellanarmirung in Fresenius's Ztschr. 1879, p. 259. Doppelaspirator, Dingler, ccxxv. 619.

tubes of a uniform diameter for comparing colors; graduated measuring vessels of glass or porcelain, divided into centimeters, ounces, etc.; dissolving vessels, etc.

V. Balances and Weights.

17. BALANCES.

Of these will be required—

1. *An ore balance,* for weighing ores and the regulus of base metals. This should be capable of carrying from 30 to 50 grammes, and, with 5 grammes in each pan, must be distinctly sensitive to an addition of 1 milligramme.

2. *Bullion or button balance,* inclosed in a case, for weighing gold and silver beads and alloys of precious metals. It should be capable of carrying 5 grammes at the utmost, and must distinctly turn with $\frac{1}{10}$ to $\frac{1}{20}$ milligramme when both pans are loaded with 1 gramme.

3. *An apothecary balance,* for weighing larger quantities. It should be sensitive to 5 milligrammes.

4. *A rough scale,* for weighing approximately larger quantities (fluxes, etc.). A grocer's scale will answer the purpose.

18. WEIGHTS.

The following are used:—

1. *The gramme weight,* from 50 to 0.001 gramme; for silver coins from 1 gramme = 1000 parts to $\frac{1}{1000}$ part; for assays of gold from $\frac{1}{2}$ gramme as the unit = 1000 parts to $\frac{1}{1000}$ part.

2. *Centner.*—1 assay centner = 5 grammes (Upper Harz) or = 3.75 grammes (Freiberg) = 100 pounds, which is divided into 100 parts of pounds, or the *quint,* the smallest weight being $\frac{1}{2}$ of the quint.

In *Austrian* smelting works, etc., 1 assay centner = 10 grammes, which is divided into 100 parts, called pounds; the pound is divided into 32 loth, the loth into 4 quentchen, and this into 4 denär, the smallest weight being 1 denär = 0.195 milligramme.—*English grain weight.* The unit is usually 1000 grains. The smallest weight for gold and silver buttons is 0.001 grain. For an assay of ore, generally a sample is taken weighing 400 grains. The divisions of this system are as follows: 1 ounce = 480 grains = 20 pennyweights (24 grains to the pennyweight).—*American assay weight.* 1 assay ton = 29.166 grammes (450.26 grains); 1 pound avoirdupois (commercial weight) = 7000 troy grains (apothecaries' weight); 1 ton = 2000 pounds avoirdupois; 2000 × 7000 = 14,000,000 grains troy in 1 ton avoirdupois; 480 grains troy = 1 ounce troy; 14,000,000 ÷ 480 = 29,166 troy ounces in 2000 pounds avoirdupois; one assay ton contains 29,166 milligrammes, therefore, 2000 pounds are to 1 assay ton as 1 ounce troy weight to 1 milligramme. If, for instance, an assay ton yields 1 milligramme of gold or silver the result will be 1 ounce troy in 2000 pounds avoirdupois without further calculation.[1]

VI. Tools and Implements.

19. GENERAL REMARKS.

We will only consider the tools and implements required for *the dry method*, as those for the *wet method*[2] do not differ from those used in analytical chemistry (stands, forceps, crucible tongs, cork drill, etc.).

20. FURNACE TOOLS.

The following tools are used in attending the furnaces. *Shovels* with perforated blades for handling the fuel; large and small *iron pokers* and *scrapers* for cleansing the grate and muffle, poking the coal, etc.; *coal* and *ash sieves* with meshes respectively 1 centimeter and 3 millimeters wide; *iron boxes* filled with water for cooling the tools, etc.

---

[1] For general practice it is far preferable to use the French metric system of weights, instead of the arbitrary and varying German systems. The simplicity and convenience of the American assay ton leave nothing to be desired.—TRANSLATOR.

[2] Neuere Geräthschaften in Fresenius's Ztschr. f. anlyt. Chemie.

21. IMPLEMENTS.

The following are required :—

A. For preparing the assay sample.

1. *Sampling.*—*Iron spoons* having a diameter of 4 centimeters; *shovels; troughs, wooden boxes,* etc., for the reception of the assay samples; *files* and *cold chisels; hollow chisels; drills; hollow cylinders of sheet iron* for small ore; *magnifying glass,* etc.

2. *For drying the samples* (p. 24).—*Drying pans* of sheet iron or copper; *drying frames; iron spatulas; drying disk* (Fig. 2, p. 25); *water-baths* (Fig. 1, p. 24); *air-baths; desiccators* (Fig. 19, p. 40), etc.

3. *For comminuting the samples* (p. 25).—*Grinding plate* and *rubber; mortars; hammers; anvils; rolls; common scissors* and *plate shears; files; rasps; pliers; vise,* etc.

4. *For sifting.*—A series of sieves of from 20 to 100 meshes to the inch, for sifting ores, fluxes, etc. A box sieve, consisting of a round tin box, into which a sieve can be snugly fitted, is very useful, as in sieving the pulverized ore there is no dust. If desired, a tin cover can be made to inclose the whole.

5. *For washing* (p. 27).—*Beaker glasses;* glass cylinders; iron vanning shovels (Fig. 5, p. 27), etc.

6. *For weighing.*—*Brass pincettes* with fine ivory points for taking up small weights, metal buttons, etc., and others with blunt or broad points for lifting larger weights and heavier buttons of precious and base metals; *brass mixing spoons,* 18 centimeters long and 2 centimeters wide, having on one end a *spatula* 1.2 centimeters wide; *camel's hair* and *bristle brushes; watch-crystals; small glass* or *porcelain vessels; glass tubes,* one end fused shut and the other closed with a cork or glass stopper; *glazed* paper; artistically closed *cornets* of fine letter paper of different colors. They are used for the reception of shavings, granules, etc., of alloys, etc.

7. *For charging.*—*Mixing scoops* (Figs. 6 and 7, p. 30); mixing *spatula* of brass or horn; *bristle brushes; measuring spoon* for granulated lead: *touch stones* and *touch needles,* etc.

72 ASSAYING.

B. Implements for transporting the assaying vessels and for manipulating the same in the furnace.—*Iron tongs* (Fig. 57) for catching hold of the vessels. For large muffle-furnaces they are

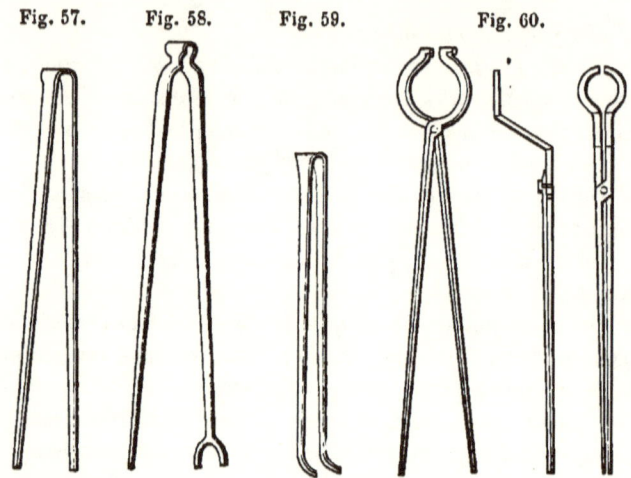

Fig. 57. Fig. 58. Fig. 59. Fig. 60.

80 to 100 centimeters long, and, for smaller furnaces, 50 to 60 centimeters. Scorification tongs with one arm forked, as shown in Fig. 58, the horse-shoe part just large enough to fit the scorifier, 60 millimeters long and 45 millimeters wide; *crucible tongs* (Fig. 60) for wind and blast furnaces. Small *assay plates* of sheet-iron, with handles. They are about 14 centimeters square, and are provided with 9 depressions, each 28 millimeters wide, in which cornets, simple weights of lead (Bleischweren), etc., are kept. The following implements are required for manipulations in the furnace during roasting, fusing, etc.: *curved stirring rods* and *spatulas* of iron; *iron ladles; tongs* with curved tips (Fig. 59) for taking hold of cornets, buttons, etc.; *cooling iron* (Fig. 61). The blade of this for large muffle-furnaces is 9 centimeters long, 7 centimeters wide, and 1 centimeter thick. It is provided with a handle 85 centimeters long. For small furnaces the respective dimensions are: 5 centimeters, 4 centimeters, 0.7 centimeter, and 70 centimeters.

Fig. 61.

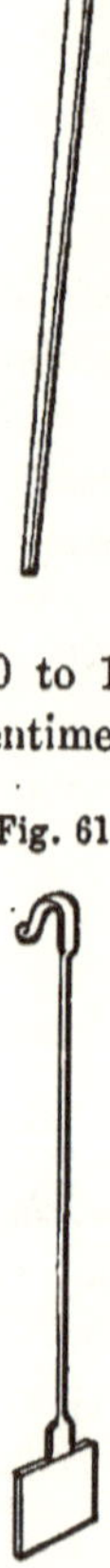

IMPLEMENTS.   73

Perhaps the best tongs for all around work are the kind shown in Fig. 62. They are about from 1½ to 2 feet (4.5 to 6 decimeters) long and have the joint about 3 inches (7.6 centimeters) from the end.

Fig. 62.

Figs. 63 and 64 show respectively Judson's patent steel scorification and cupel tongs, designed to enable the operator to remove the scorifiers or cupels from the rear of muffle without disturbing those in front.

C. Implements required for the reception and further treatment of the assay samples after they have been taken from the furnace.

1. *For the reception of the assay samples* are required: *sheet-iron plates* with handles. They are divided into squares by strips of sheet iron crossing each other, or have depressions, each about 4 centimeters wide, in which the assay vessels are placed; open and closed *ingot moulds* for casting lead and silver bars, ingots, etc.; *small iron or leaden plates* (Kornbleche), about 10 centimeters long and

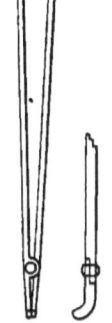

Fig. 63.

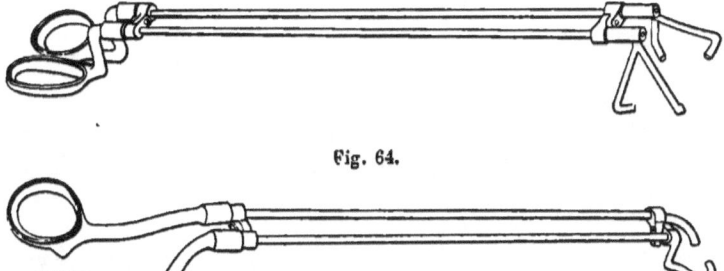

Fig. 64.

60 millimeters wide, with depressions 3 centimeters wide for the reception of gold and silver buttons from the cupel; and *boards of wood* with larger depressions for the reception of buttons of base metals.

2. *For the further treatment of the samples after they have become cold.*— *Hammers* for breaking the clay crucibles, etc., and removing the slag from the buttons. The body of these is about

9 centimeters long, the head square, the other end running into a point (also smaller hammers); an *anvil* with a plate beneath it about 6 centimeters square; *cupel tongs* (*pliers*), 160 centimeters long, for taking the buttons of silver and gold from the cupels; a *button brush* consisting of a brass holder with bristles at both ends; a *bar magnet* for extracting particles of iron from the slag, etc.

VII. Assay Reagents.

22. REAGENTS[1] FOR DRY ASSAYS.

According to their action they may be divided into—

1. *Reducing agents.*—*Charcoal* in the form of powder, or of small pieces placed on top of the charge (assays of lead, copper, etc.), or of crucible lining (assays of iron); when it is generally mixed with other reducing agents (potassium carbonate, sodium carbonate, etc.), because the presence of a large percentage of charcoal without additions in smelting processes renders the charge more difficult to fuse.

Powdered coke, anthracite, and graphite[2] may also be used instead of powdered charcoal, but they are less combustible. Rosin, fat oils, tallow, sugar, etc., were formerly also employed. Bitartrate[3] of potassium (argol) ($KC_4H_5O_6$), either crude or refined, yields considerable carbon in becoming carbonized, and in consequence exerts a vigorous reducing effect, but causes refractoriness. For this reason its percentage of carbon is reduced, if necessary, by mixing it in different proportions with saltpetre. The mixture is poured into a red-hot crucible, placed under a well-drawing chimney. The mixture deflagrates and emits empyreumatic vapors, when, by reason of a partial oxidation of the carbon, a mixture of potassium carbonate and carbon is formed. This is known as *black flux*. For vigorous reduction it is made from 1 part of saltpetre and 3 parts of argol; for less vigorous, either of 1 and $2\frac{1}{2}$, or 1 and 2 parts respectively. Another flux

---

[1] Muspratt's techn. Chemie, 3d Aufl. Bolley, Handb. der techn. chem. Untersuchungen, 1879.

[2] Werthbestimmung in Fresenius's Ztschr. 1875. p. 394.

[3] Werthbestimmung in Fresenius's Ztschr. vii. 149.

containing potassium carbonate (without carbon), mostly with undecomposed saltpetre, and known as *white flux*, consists of 1 to 2 parts of saltpetre, and 1 part of argol; *gray flux* has 3 of argol to 2 of saltpetre. A mixture of argol and saltpetre, before it is deflagrated, is called *raw flux*. As black flux, on account of its hygroscopic properties, must be frequently prepared fresh, and this work is unpleasant by reason of the evolution of bad odors, a mixture of potassium carbonate (or soda) and flour (starch $= C_6H_{10}O_5$), which is also cheaper, is used in preference. Usually 20 to 25 per cent. of wheat flour is taken, but for more vigorous reductions 30 to 35 per cent. (for instance in assays of copper), and even as much as 50 per cent. (assay of tin). When this mixture is used, a separation of carbon, in a fine state of division, takes place, with the exhibition of a yellow flame of carburetted hydrogen, caused by the carbonization of the flour during the fusing of the assay sample. The flame of the carburetted hydrogen must not be confounded with the *blue flame* of carbonic oxide. Mixtures of potassium carbonate (or of sodium carbonate) and coal-dust are less intimate, and their action is consequently less energetic than that of mixtures in which the carbon has been separated from organic substances in a very finely divided condition (tartaric acid, flour). *Potassium cyanide*, with 64.1 K and 35.9 Cy, is an energetic reducing (also desulphurizing) agent, even at a low temperature (assays of tin). *Potassium ferrocyanide*, $K_4Fe(CN)_6$, yields, on heating, a mixture of iron carbide, ferrous and ferric oxides, free carbon, and a small quantity of potassium cyanide. It has also a vigorous desulphurizing action.

The reducing power is estimated from the quantity of lead which is yielded by fusing 1 to 2 grammes of the reagent with 60 grammes of litharge and 15 grammes of sodium carbonate or potassium carbonate.

According to *Berthier*, the reducing power of the various agents is as follows: 1 part hydrogen, 104; pure carbon, 34.31; well-glowed wood charcoal, 31.81; ordinary wood charcoal, 28.00; tallow, 15; sugar, 14.5; kiln-dried starch, 13; ordinary starch, 11; tartaric acid, 6; oxalic acid, 0.90; black flux with 2 parts of argol, 1.40; black flux with $2\frac{1}{2}$ parts of argol, 1.90; black flux with 3 parts of argol, 3.80; 94 parts soda and 6 wood charcoal, 1.80; 88 parts soda with 12 charcoal, 3.60; 90 parts of soda with 10 of sugar, 1.40; 90 parts of soda with 10 of starch, 1.15; 80 parts of soda with 20 parts of starch, 2.30; crude argol, 5.60; purified argol, 4.50; pure argol (carbonized), 3.10; potassium binoxalate (salt of sorrel), 0.90; white soda-soap, 16 parts.

2. *Oxidizing agents.*—*Saltpetre*, KNO_3, with 46.56 K_2O, and 53.44 N_2O_5. It should be as free from sulphates as possible.—*Litharge*, PbO, with 92.83 Pb, and 7.17 O, exerts an oxidizing effect upon metals and metallic sulphides (scorification), as well as upon organic substances (assay of fuel). As generally used it is in the form of red litharge free from particles of metallic lead.[1] It should completely dissolve in acetic acid, and be as free from gold and silver as possible. When it contains silver, *white lead* not adulterated with heavy spar, $2\ (PbCO_3) + Pb(HO)_2$, with 86.27 PbO, may be substituted for it. It is best to prepare this by the wet method (for instance, by that of *Dietel*, of Eisenach, which yields a product from gold, silver, antimony, and copper).

According to *Berthier*, 1 part of the various metallic sulphides requires the following quantities of litharge for its decomposition:—

	Parts.
PbS	1.87
HgS	10 to 12
BiS	20
Sb_2S_3	25
ZnS	25
FeS	30
SnS_2	25 to 30
Copper pyrites	30 to 35
FeS_2	50
As_2S_3	50 to 60

Litharge entirely free from silver can be prepared by oxidizing the purest *Pattison* or *Villach* lead by cupellation, or bringing such lead, after it has been granulated, into fused saltpetre; or by gradually strewing charcoal powder upon litharge fused in refractory crucible, whereby some lead will be reduced which, in sinking down, withdraws the silver from the litharge.

[1] Dingler, cxciv. 84.

3. *Solvent agents.*

a. Acid; such as powdered quartz; powdered glass[1] free from arsenic and lead. It should contain 60 to 70 per cent. or more of SiO_2, 5 to 22 per cent. of alkalies, 6 to 25 per. cent. of lime, 0.5 to 5 per cent. alumina, and its fusing point should be between that of borax and fluor-spar, or about 1200° C. (2192° F.). *Borax,* $Na_2B_4O_7 + 10\ H_2O$, with 16.37 Na_2O, 36.53 BO_3, and 47.10 H_2O. It should be completely dehydrated, or in the condition known as *borax-glass.* This is produced by fusing borax in a clay crucible, and then pouring it upon a bright metallic surface. It is more fusible than glass, and the boracic acid forms combinations with nearly all the bases as well as with silicic acid. *Salt of phosphorus* (*sodium-ammonium-hydrogen phosphate*), or *microcosmic salt,* $NaH(NH_4)P + 4H_2O$, with 14.90 Na_2O, 12.46 NH_4O, 4.29 basic water, and 34.32 water of crystallization, and 34.03 P_2O_5. In the anhydrous condition it is a more energetic solvent agent than borax (assays of cupreous nickel). *Clay,* such as kaolin, $Al_2Si_2O_7$, with 46.40 SiO_2, 39.68 Al_2O_3, 6.96 H_2O and 6.96 *aq.;* burned China clay contains 53 Si_2O_2 and 47Al_2O_3. Most varieties of clay contain over 50 per cent. of silicic acid.

b. Basic; such as *potassium carbonate,* K_2CO_3, with 68.09 K_2O and 31.91 CO_2. It should be as free as possible from sulphates, and when mixed with carbon as *black flux* (p. 74), and with flour (p. 75), is a vigorous reducing, fluxing, and desulphurizing agent. *Sodium carbonate,* Na_2CO_3, with 58.58 Na_2O and 41.42 CO_2, acts somewhat less energetically than potassium carbonate, and consequently a larger quantity of it must be used. It is less deliquescent, more fusible, and cheaper. A mixture of 13 parts of dry potassium carbonate and 10 parts of calcined sodium carbonate, furnishes a flux very easily fusible.—*Caustic alkalies* act more energetically than carbonates, but exert a very injurious effect upon the crucibles.—*Calcium carbonate,* $CaCO_3$, with 56.29 CaO, in the form of chalk (or calcite), may be used for smelting operations at higher temperatures (for instance, in assays of iron, Cornish assays of copper).—*Fluor-spar* (*calcium fluoride*), $CaFl_2$,

---

[1] Analyses of Glass, in Poggendorf's Ann. 1879, vol. 6, p. 431. Dingler, ccxxxii. 348 (Weber).

with 51.54 Ca, is more easily fusible than calcium carbonate, and is especially effective for removing silicic acid. It readily fuses down with calcium phosphate, heavy spar, and gypsum.—*Lead oxide* (litharge, minium, white lead) readily fuses with silicic acid, the alkalies, and with most of the heavy metallic oxides, but less so with the earths and alkaline earths.—*Ferric oxide*, F_2O_3, with 69.3 Fe (in assays of copper).

4. *Precipitating or desulphurizing agents.*—*Iron* in the form of iron filings (assays of zinc blende, antimony, and mercury), and as pieces of wire 4 to 5 millimeters thick, 6 to 9 millimeters long, and weighing from 0.5 to 2 grammes (assays of lead and bismuth).—*Potassium cyanide* (p. 75) and *potassium ferrocyanide* (p. 75). The *caustic alkalies* and *carbonates* (p. 77) decompose metallic sulphides. The metal is separated, and sulphites, hyposulphites, and sulphates of the alkalies together with alkaline sulphide are formed. The latter forms a sulpho-salt (for instance, with FeS, PbS, and Cu_2S) with one part of the metallic sulphide, which, as a general rule, can be decomposed with iron. Carbon promotes desulphuration (*black flux, potassium carbonate*, and *flour*, pp. 74, 75).—*Lead oxide* (p. 76).—*Saltpetre* oxidizes metallic sulphides, while the metals (silver, copper, lead) are separated.

5. *Sulphurizing agents.*—*Sulphur* in the form of flowers of sulphur; or of *iron pyrites* (Cornish assay of copper).

6. *Concentrating fluxes.*—*Lead* in a granulated condition (assay lead). It is prepared by rocking lead in the form of thick liquid paste in a trough well covered with chalk and then sifting the mass. It is also used in the form of sticks of the purest *Pattison* lead for alloying with gold and silver. (Where the assayer is obliged to make his own granulated lead, as in sections where pure lead free from silver cannot be obtained, it will be necessary for him, after granulation in the manner above described, to sample well, and test about 30 to 50 grammes for silver by the scorification assay. In using this lead, the amount of silver contained in it must, in all cases, be deducted from the results obtained in making an assay.)—*Silver*[1] for alloying with gold

---

Preparation of pure silver: Cupriferous silver or fine silver is dissolved in pure nitric acid. This is diluted with distilled water, allowed to stand for some time, and then filtered. The filtrate is strongly diluted, and chloride of silver (quartation).—*Gold* for collecting copper (assay of nickel and cobalt).—*Antimony (antimony oxide)* and *arsenic* for copper (refining).—*Copper oxide* for tin.—*Iron pyrites* as a collecting agent for copper (assay of matt).

7. *Decomposing and volatilizing fluxes.*—*Charcoal* and *graphite* for decomposing sulphates, arseniates, and antimoniates by roasting.—*Ammonium carbonate* $(NH_4)_2CO_3$ for decomposing sulphates, especially copper sulphate, at red heat (p. 33), but less completely lead and bismuth sulphates.—*Common salt (sodium chloride)* NaCl, with 39.66 Na and 60.34 Cl, for the volatilization of antimony and arsenic in refining black copper according to the Cornish method.

8. *Air-excluding fluxes* (covering agents).—*Decrepitated common salt*, as free from sulphates as possible, fuses easily, and, becoming very thin fluid, washes down particles of metal adhering to the sides of the assay vessels. It volatilizes at a red heat.—*Glass* (p. 76) in *Berthier's* assay of fuel and assay of matt.—*Refined slag* from charcoal iron blast-furnaces (in assays of zinc).

is precipitated by the addition of pure hydrochloric acid. This is washed by decantation and then boiled several times with diluted hydrochloric acid, but after each boiling it should be washed with distilled water. The moist chloride of silver (3 parts) is mixed with dry sodium carbonate (1½ parts) and with pure saltpetre (¼ part of the whole). The mixture is then dried in a porcelain dish and fused in a porcelain crucible; or 100 parts of chloride of silver are fused with 70 parts of chalk, and 4 parts of wood charcoal. (Mohr, Titrirmethode, 1874, p. 425. Fresenius's Ztschr. xiii. 179.)

[Star recommends the following process as yielding a metal which comes nearer ideal purity: Slightly cupriferous silver is converted into dry nitrate, and fused to reduce any platinum nitrate present to metal. The fused mass is taken up with dilute ammonia and then diluted to about fifty times the weight of silver it contains. The filtered solution is now mixed with an excess of a sulphide of ammonia solution, $SO_3(NH_4)_2$, and allowed to stand. After twenty-four hours, about half of the silver has separated out in crystals; from the mother liquor the rest comes down promptly on the application of a water-bath heat. The rationale of the process is that the sulphide hardly acts upon the dissolved oxide of silver, but it reduces some of the oxide of copper, $2CuO$, to Cu_2O, with the formation of sulphate $SO_4(NH_4)_2$. This Cu_2O dioxidizes its equivalent of Ag_2O, forming $Ag + Cu_2O_2$, which latter is reduced by the stock sulphide and reconverted into Cu_2O, which now acts upon a fresh equivalent of Ag_2O; and so on to the end.—G.]

23. REAGENTS FOR WET ASSAYS.

The following are principally used:—

1. *For assays by gravimetric and colorimetric analysis.—Acids:* hydrochloric, sulphuric, nitric, and acetic acid, aqua regia (nitromuriatic acid). *Bases and salts:* caustic alkalies, alkaline carbonates, potassium chlorate, ferrous sulphate, sodium sulphide, etc. —*Metals* for precipitation: *iron* in the form of wire-pins 30 to 35 millimeters long and 2 to 3 millimeters thick, or in a pulverulent condition, for copper; *zinc* in the form of such pins or of granules,[1] or in a pulverulent state,[2] as a reducing agent for iron solutions, etc.; *copper* (electrolytic copper is the purest); *bromine*[3] for decomposing sulphurets, compounds of gold, etc.

2. For volumetric assays: *Potassium permanganate* (chameleon), $KMnO_4$, with 29.8 K_2O and 70.2 Mn_2O_7; *sodium sulphide; potassium cyanide; barium chloride; potassium iodide;* free and with dissolved iodine; *sodium hyposulphite; ferric chloride; sodium chloride; potassium sulphocyanide,* etc. As indicators: *litmus tincture, tincture of Brazil wood,*[4] etc., for acids and alkalies; for sulphur: *the salts of iron, nickel, and lead,* and *sodium nitroprusside;* for iodine: *starch paste;* for iron oxide: *potassium sulphocyanide,* etc.

---

[1] [Zinc is very easily granulated by pouring the molten metal into a bucket of warm water.—G.]

[2] Dingler, ccxxviii. 378.   [3] Dingler, ccxix. 544.

[4] Fresenius's Ztschr. 1875, p. 324 (Rothholz). Bericht der deutsch. chem. Ges. 1877, p. 1572 (Fluorescein).

SPECIAL DIVISION.

I. LEAD.

24. LEAD ORES.

Galena (lead monosulphide) PbS, with 86.6 Pb; *cerusite (lead carbonate)* $PbCO_3$, with 76.6 Pb; *anglesite (lead sulphate)* $PbSO_4$, with 68.3 Pb; *pyromorphite* $3Pb_3P_2O_8 + PbCl$, with 76.3 Pb; *crocoisite (lead chromate)* $PbCrO_4$, with 63.2 Pb; *wulfenite (molybdate of lead)* $PbMoO_4$, with 57 Pb.

25. ASSAYS OF LEAD IN THE DRY WAY.

The results of these assays are inaccurate, as the lead is liable to slag off and to volatilize, which is promoted by the presence of other volatile substances (arsenic, zinc, antimony), and also by reason of a possible contamination of the lead by other metals (copper, arsenic, antimony, bismuth). The highest yield which can be obtained from pure galena is 85.25 per cent., but in poorer ores the loss of lead may be 5 per cent. greater. The results are sometimes calculated to whole per cents., but more frequently from 5 to 5 per cent.

The condition of the ore, whether the lead is sulphurized or oxidized, and whether the substance is pure or contains more or less of earths and foreign metallic sulphide, will decide the choice of the assay method.

I. *Sulphurized Substances.*

A. Galena, etc., without foreign metallic sulphides (ZnS, FeS, Cu_2S, Sb_2S_3, As_2S_3).—*Precipitation assay:* the assay sample is either decomposed by alkalies alone (Upper Harz assay with potassium carbonate, assay with potassium cyanide), or together with metallic iron (in an iron pot or clay crucible), with the following *reactions :—*

The lead sulphide is decomposed by the alkalies at a comparatively low temperature (7 $PbS + 4K_2CO_3 = 4Pb + 3(K_2PbS_2) + K_2SO_4 + 4CO_2$.) The sulpho-salt ($K_2PbS_2$), which otherwise would pass into the slag, is either decomposed by iron at a high temperature, $3(K_2PbS_2) + 3Fe = 3Pb + 3(K_2FeS_2)$ or, as is the case in the Upper Harz assay with potash, by a suitable admission of air at a lower temperature. By this process the K_2S of the sulpho-salt is completely converted into K_2SO_4, but the PbS only partly into $PbSO_4$, so that, by increasing the heat, the still remaining PbS is decomposed by the $PbSO_4$, as follows: ($PbS + PbSO_4 = 2Pb + 2SO_2$.) In the first case, the presence of carbon (black flux, flour, tartaric acid, etc.) promotes desulphurization.

1. *Rich galena (with little earths).*

a. Assay in an iron pot (Belgian assay).—This is the best method of assaying lead, as pure galena with 86.6 Pb yields 84.25 to 85.25 per cent. of lead, therefore, with a loss of but 1 to 2 per cent. and sometimes only 0.5 per cent. of lead. It also permits the use of a larger charge, and, the iron pot being a good conductor of heat, the operation can be more quickly executed and at a comparatively low temperature. When large quantities of earth are present, more slag will be formed, and consequently the resulting loss of lead will increase, according to *Percy*,[1] at the following rate: 1.80 to 7.90 per cent. when the ore contains from 10 to 90 per cent. of calcium carbonate, and 1.18 to 35 per cent. when 10 to 90 per cent. of silicic acid is present. The sample is less frequently fused without any fluxes (Flintshire) than with alkaline fluxes. An addition of carbon (black flux, flour, argol) checks the oxidation of the lead, promotes the reduction of the lead carbonate and lead sulphate which may be present, and prevents the oxidation of the iron sulphide. The latter (oxide of iron) vigorously attacks the walls of the pot and retains particles

---

[1] [Metallurgy of Lead. By John Percy. London: John Murray, 1870, pp. 113–117.—G.]

of lead when the contents of the pot are poured out. Fluorspar is a good flux for heavy spar. *Silver* and *gold* pass entirely into the lead, but only traces of *zinc* and *iron*. *Copper* divides itself between lead and slag. A large part of the *antimony* passes into the lead, and while a part of the arsenic volatilizes, very little of it passes into the lead, and the largest part forms spices with iron.

Fifty grammes of ore are placed in an iron pot previously heated to dark redness, in a coke fire in the wind-furnace, or in a gas-furnace. From 50 to 100 grammes of black flux or potassium carbonate with 15 to 20 per cent. of flour are added, then 2 to 3 grammes of borax, and finally a covering of common salt 5 millimeters thick. The charged crucible is then placed between the darkly glowing coal in the furnace. The latter is closed, and the temperature *gradually* raised in about five minutes to complete redness, and this is kept up until the contents of the crucible are in quiet fusion without foaming (three to five minutes). The granules of lead floating on the top should be submerged by means of an iron spatula or wooden rod. The furnace is then closed for a few minutes, after which the pot is taken out and allowed to cool off somewhat. Its contents are poured into a mould, which should have been previously coated with graphite and heated. If the contents are poured out while too hot, a film of lead may remain adhering to the iron, and, if too cold, the lead will partly spread over the slag. The mould, after having been allowed to cool off, is turned over, and the hard, black slag is quickly broken off from the lead button to prevent it from becoming moist, as it would then not separate quite as well. The lead button is then washed with hot water or diluted sulphuric acid, dried, and weighed. The slag is again smelted with some potassium carbonate and flour or black flux for about 10 to 12 minutes, and then poured out. The time required for the fusion, counted from placing the ore into the pot to the first pouring out, is from 10 to 15 minutes. The iron pots will bear from 40 to 50 operations. In many works *lead matt* free from copper, poorer ores, and slag are assayed according to this method.

England[1] (Flintshire): 500 grains (32.4 grammes) of rich

---

[1] [Percy, Metallurgy of Lead, pp. 106–108.—G.]

galena, 500 grains sodium carbonate, and 50 grains argol; for poorer ores: 350 grains of sodium carbonate, 150 grains borax, and 50 grains argol. The ore is mixed in a mixing scoop with a long spout (Fig. 7, p. 30), with three-fourths to four-fifths of the quantity of the flux. The mixture is pushed to the front of the scoop, next to this the remainder of the flux is placed, and behind this the borax. The whole is then carefully poured into the dark glowing pot and subjected to the fusing operation mentioned above. In pouring out the contents the slag is kept back in the pot by means of a wooden rod, and is again fused with 20 to 30 grains of sodium carbonate and 5 to 10 grains of argol. The yield of lead from pure galena is generally from 84.25 to 85.25 per cent. The difference in the results of the assay is nearly the same for the richest ores and those yielding up to 50 per cent., but is greater in poorer ores. In *Wales* and *Flintshire* a yield of 75 to 82 per cent. Pb is obtained from pure galena by placing 10 ounces troy of the ore in a covered iron dish and fusing it in an open forge fire.—*Bleiberg* in *Carinthia:* 50 grammes of ore, 2 tablespoonfuls of flux (3 parts of argol, 2 saltpetre, 1 borax), a cover of powdered glass (or common salt), smelting for 12 to 15 minutes, etc.—*Belgium:* 10 grammes of galena with 28 grammes of sodium carbonate and 5 grammes borax, or 10 grammes sodium carbonate and 10 grammes argol.—*Tarnowitz:* 50 grammes of ore, with black flux, borax, and a covering of salt resting upon a layer of a little black flux. The difference between the separate assays is not more than 2 per cent.—*Mechernich:* 25 grammes of ore with 150 grammes of borax and 100 grammes sodium carbonate and argol in equal parts. For slag more borax, for lead matt more soda.

In Friedrichshutte, near Tarnowitz, two lots of ore of 50 grammes each are melted down in a wind furnace together with 20 grammes of a flux consisting of 8 parts of potash, 1 of flour, and 10 grammes of borax. The buttons are accurately weighed to centigrammes, and the content of lead must not vary more than 1.5 per cent.[1]

b. Assay with potassium cyanide in clay crucibles.—This can be executed at comparatively low temperatures, and gives a good

---

[1] Ztschft. f. d. Bg., Httn. u. Salinenwesen im preuss. Staate. Bd. 22, p. 92.

yield, but is more expensive than the foregoing. Besides, the potassium cyanide is poisonous, and adheres to the porous mass of the crucible which may uncover the lead button and effect its oxidation.

The charge, according to *Levol*, is as follows: 100 parts galena, 100 potassium ferrocyanide, and 50 potassium cyanide with some sodium carbonate; according to *Ricketts:* 10 grammes of ore, 20 to 25 grammes of potassium cyanide, and a covering of common salt. The charge is fused for 12 to 15 minutes at a low temperature; the yield is 78.5 to 79.1 per cent. of lead.

2. *Galena with more earths.*

a. Assay with black flux (potassium carbonate and flour) and metallic iron, in clay crucibles.—When the ore contains large quantities of earth, more slag is formed. This, if the contents of the crucible were to be poured out, would retain considerable lead, which will settle if the charge is allowed to cool off in the crucible. The loss of lead is from 2 to 3 per cent. Deep crucibles (Fig. 49, p. 64) are used for this purpose, and the charges fused in a muffle or wind furnace. The work can be done more conveniently in the latter, and fuel will also be saved. Should small quantities of metallic sulphides be present, it is well to roast the ore somewhat in a covered crucible to volatilize the arsenical sulphides, the sulphur from the iron pyrites, etc. Charge: 5 grammes of galena are placed on the bottom of the crucible, upon this is put a piece of iron wire 4 to 5 millimeters thick, and up to 9 millimeters long (it should be longer or shorter according to the percentage of lead, that is to say, about 25 to 30 per cent. of the weight of the ore). Upon this are placed 15 grammes of black flux (or potassium carbonate with 15 to 20 per cent. of flour) and, in case of basic gangues, 2 to 3 grammes of borax. Upon this comes a covering of common salt 5 millimeters thick, and on top of all a piece of charcoal the size of a hazel-nut, for maintaining a reducing atmosphere. The contents of the crucible are slowly heated in the *muffle* furnace until the yellow flame caused by the carbonization of the flour is no longer visible. The heat is then raised, and tongues of bluish flames arising from the carbonic oxide will make their appearance. The contents of the crucible should not froth too strongly, and for this reason the firing must be done very carefully, especially when low crucibles are used. When the "flaming" and frothing have ceased, the heat is still kept up for ½ to ¾ of an hour to allow the sulpho-salt (p. 82) to become decomposed by the metallic iron. 25 to 30 minutes are required for fusing the charge. The sample, fuming strongly from the vapors of the common salt, is then taken out, allowed to cool off, and freed from slag. By hammering the lead flat, the iron adhering to it will fly off. The lead button, which is covered with iron sulphide, is then brushed and weighed. The success of the assay is indicated by the iron still adhering to the lead without this being wrapped around it (to prevent this, the iron wire should not be too fine), by thoroughly fused slag and a malleable lead button. If brittle, it contains sulphur. The various assays must agree within 1 to 3 per cent. according to the richness of the ore.

Freiberg: 3.75 grammes of ore, 0.92 to 1.13 grammes of iron wire; 7.5 to 9.4 grammes of black flux or potassium carbonate with flour, 1.13 to 1.5 grammes of borax, and for basic gangue 2.25 to 2.63 grammes of glass, and a covering of common salt, 5 millimeters thick. The charge is heated from ¾ to 1 hour in the wind-furnace.—*Pribram:* 0.5 gramme of crude argol is placed in the bottom of the crucible, upon this iron wire, then 5 grammes of galena, and 12 grammes of black flux, and finally a covering of common salt. The charge is heated from 20 to 25 minutes in a gently glowing wind-furnace until the fusing mass subsides. The fire is then urged on, when the assay will emit gas (boil) vigorously, and, when this is the case, the firing is continued for 5 minutes longer. A difference of 2 per cent. is allowed in the assay of ores with 0 to 50 per cent. of Pb, and 3 per cent. in those with over 50 per cent.—*England:*[1] The same quantities of ore and flux are used as for assays in the iron pot (p. 82). The ore is placed in a Hessian crucible together with ¾ to ⅘ of the flux, and a strip of wrought iron in the shape of a horse-shoe is pushed into the mass. The crucible is gradually heated, requiring from 20 to 25 minutes, and during this time the iron is moved about several times. When the flux is thin

---

[1] [Percy, Mettallurgy of Lead, 110–112.—G.]

fluid the crucible is taken from the furnace, the iron, which should be free from globules of lead, is removed, and the crucible allowed to cool. The contents are then poured out, and the lead button is freed from slag. If the heat has not been strong enough, the lead button will be hard, and will have a lustre like galena, and the slag will also be covered with a lustrous film. The yield of lead from pure galena is from 82 to 83 per cent. of Pb.—*New York:* 10 grammes of ore, 25 grammes of black flux, three loops of iron wire, which are taken out after the fusion is complete, and a covering of common salt. The yield from pure galena is 78.4 to 78.6 per cent., with a difference of 1 to 2 per cent. in the various assays.—*Upper Harz:* The assay was formerly conducted in the same manner as in *Freiberg*, but now iron pots are used.

b. Upper Harz, assay with potassium carbonate.—A muffle-furnace is required for this method of assaying. Low crucibles (Fig. 49) may be used, as the charge contains no carbon, and several crucibles can be placed in the muffle at one time. The result of this assay is not as accurate, the yield being somewhat less than with the methods described above, as the success of the operation depends on the proper "*cooling of the assay,*" for which there is no guide but experience. This method is therefore chiefly available for uniform ores only, the approximate yield of which is known. It has been almost abandoned at the present time.

Charge: 12.5 to 15 grammes of potassium carbonate are placed in a small crucible (Fig. 49, p. 64). To this are added 5 grammes of galena, and both are thoroughly stirred together with the mixing spatula. In case basic earths are present, 1 assay spoonful of borax is placed upon the mixture, and upon this a covering of common salt 5 millimeters thick. The charge is then placed in the thoroughly heated muffle-furnace, where it remains, with the mouth of the muffle closed, until it has come into perfect fusion (that is, when no more deposits are perceptible on the edges of the crucible). To decompose the sulpho-salt by oxidation, the mouth of the muffle is then opened for about 10 to 15 minutes, until the crucible appears dark and the vapors above it have greatly diminished or entirely disappeared (this is called *cooling the assay*). Thereupon the furnace is brought back to its first temperature, completely closing the muffle, in order to decompose the still remaining sulphurized lead by the sulphate which has been formed. The crucible is then taken out and allowed to cool off, and the lead buttons are freed from adhering slag. If the assay has been successful, the slag is completely fused, and the lead button has a pure lead color, but not much metallic lustre, as, if this is the case, the heat has been too strong.—For ore containing *antimony:* 10 grammes of ore, 35 grammes of potassium carbonate, 1 gramme of saltpetre, and a covering of common salt. 30 minutes are required for fusion, 10 minutes for cooling, and 10 minutes for the final heating of the assay.

3. *Galena containing large quantities of earths.*—The English method (p. 83) is employed with a strip of sheet-iron in the form of a horse-shoe, but stronger fluxes (caustic alkalies) are used, which, to be sure, attack the crucibles more energetically, and larger charges, as for instance: 100 grammes of assay sample, 100 to 150 grammes of caustic soda, 150 to 250 grammes of potassium carbonate or sodium carbonate, strip of iron in the form of a horse-shoe, 25 millimeters wide, and 4 millimeters thick. From 1 to $1\frac{1}{2}$ hours are required for perfect fusion, and until the iron is free from lead globules.

B. Lead monosulphide with foreign metallic sulphides (galena with zinc blende, pyrites, etc.; lead matt, etc.).

1. *Roasting and reducing assay.*—The result of this assay is inaccurate, as the lead oxide is liable to slag off and foreign metallic oxides to be reduced, the metal of which contaminates the lead. For this reason the assay with sulphuric acid is frequently used instead (*Rammelsberg* smelting works in the Lower Harz).

In the Lower Hungarian works and in the Klausenburg district this assay is used as a checking assay. (Einlosungsprobe.) No iron is added, but the ore is roasted as completely as possible with the addition of coal dust. In Schemnitz the following differences are allowed:—

Amount of lead in the ore.	Difference allowed.
Up to 30 per cent.	2 per cent.
Up to 40 "	4 "
Over 40 "	6 "

The assays are repeated three times.

The allowances used in the Klausenburg mining district[1] differ but little from the above.

Amount in the ores.	Differences allowed.
0 to 25 per cent.	2 per cent.
25.25 to 50 "	4 "
50.25 per cent. to the maximum content	6 "

Five grammes of ore are roasted in a roasting dish. This is mixed in the assay vessel with 7.5 to 15 grammes of potassium carbonate and flour, or black flux. Upon this is placed 1.25 to 1.5 grammes of glass, then 0.25 to 0.5 grammes of thick iron wire, upon this 1.25 to 1.5 grammes of borax, then a covering of common salt, and on top of all a piece of coal. By fusing the roasted charge in the muffle- or wind-furnace at not too high a temperature, the lead oxide and lead sulphate are reduced (if the temperature is too high, many other metallic oxides are also reduced), and the foreign oxides and earths contained in the sample are slagged off by the aid of the potassium carbonate in the black flux, as well as of the borax and glass. The heating should be done with the greatest care, as the contents swell up very much. 20 to 30 minutes are required for smelting in the muffle-furnace after the "*flaming*" in the muffle has ceased, and from 15 to 20 minutes in the wind-furnace after the flames get under way.

Hungary: 10 grammes of roasted ore are charged with 11.5 grammes of black flux, and this is covered with a layer of 15 to 20 grammes of common salt. It is fused by keeping up a strong fire under the muffle for half an hour. The yield is from 10 to 12 per cent. Pb less than from an assay with metallic iron. It will be larger if charcoal dust is added in roasting, but the button will be less pure.

2. *Assay with sulphuric acid* (combined dry and wet method).—This gives more accurate results than the foregoing processes, as the foreign metals are removed before fusion, but there will be always a loss of lead in the last-named operation.

Five to 10 grammes of the ore are ground as fine as possible. It is decomposed by digesting it with nitro-muriatic acid (*aqua regia*) in a glass flask with straight walls (Fig. 16, p. 39). A

---

[1] Fortschritte im Probirwesen. Balling. Berlin, 1887, p. 118.

few drops of sulphuric acid are added, and it is then evaporated to dryness. The dry mass is digested with diluted sulphuric acid, filtered, and washed. The filter is freed from water between blotting-paper, and, with the residuum (lead sulphate and insoluble earths, clay, etc.), is dried in a roasting dish under the muffle, and finally incinerated at as low a temperature as possible. The mass is then ground up, and charged in a high crucible with 15 grammes of black flux (1 part saltpetre and 3 parts argol), and then with 1 to 1.5 grammes of iron. The charge is slowly heated, and then strongly heated after the "flaming" in the muffle has ceased (15 to 20 minutes).—If the assay sample should contain *antimony*, which partly remains as lead antimoniate with the lead sulphate, the process is as follows: The assay sample is decomposed by nitric acid, to which some tartaric acid has been added. It is neutralized with sodium carbonate, and the antimony is extracted by digesting the mass for about half an hour in a solution of sodium sulphide containing sulphur. It is then filtered and washed, and the residuum is treated in the same manner as in the sulphuric acid assay.

II. *Oxidized Substances.*

A. Lead oxides free from earths (*litharge, minium, skimmings* (*Abstrich*), *etc.*).—5 grammes of the sample are fused at not too high a temperature with 12.5 to 15 grammes of potassium carbonate with 30 to 35 per cent. of flour, or black flux, and covering of common salt, with small pieces of coal on the top, in the same manner as in the roasting and reducing assay. If the sample should contain any sulphur, 0.25 to 0.5 gramme of iron wire is added. 20 to 25 minutes are required for fusion in the muffle-furnace after the "*flaming*" in the muffle, and 13 to 15 minutes in the wind-furnace after the flame is under way.

B. Lead oxides with earths.—The charge is the same as in II. A, with the exception that from 25 to 30 per cent. of borax is added, and from 5 to 10 minutes more is required for fusion.

C. Salts of lead oxide, namely :—

1. *Lead carbonate* (*cerussite*), *lead chromate* (*crocoisite*), *lead phosphate* (*pyromorphite*), *mimetene* (*lead arsenate*), *and yellow*

lead ore (wulfenite).—The charge is the same as in II. A, with an addition of from 20 to 30 per cent. of borax, according to the presence of more or less earthy substances. With pyromorphite containing arsenic from 5 to 10 per cent. of iron is added to separate the arsenide of iron. Time for fusion the same as in II. A.

Charges for oxidized ores according to *Percy:* 500 grains of ore, 350 grains of sodium carbonate, 150 grains, or less, of borax, and 50 grains of argol. The charge is fused in an iron pot (p. 82).—*Lead carbonate (cerussite*): 500 grains of ore, 500 grains of sodium carbonate, 100 grains of argol, and 30 grains of borax. The charge is fused for about 20 minutes in a clay crucible in the wind-furnace, and poured out (p. 86).—*Lead phosphate (pyromorphite*): 300 grains of ore, 400 grains of sodium carbonate, 20 grains of powdered charcoal, and 30 grains of borax; or, 350 grains of sodium carbonate, 100 grains of argol, 30 grains of borax, and some metallic iron. Fusing time: 25 to 30 minutes, counting from introducing the charge until it is poured out.[1]

2. *Lead sulphate (anglesite, lead fume, dross, sweepings, tailings, skimmings, etc.*).—The charge is the same as for the assay with sulphuric acid (p. 89). If necessary, a smaller quantity of metallic iron (for instance, only 10 per cent. for fume and skimmings, which carry but little sulphate) is used, and 20 to 30 per cent. of borax is added if earths are present.

3. *Lead silicate (slags*).—10 grammes of slag are ground as fine as possible, and mixed in an assay vessel with 15 grammes of potassium carbonate with 30 to 35 per cent. of flour. Upon this, in case of acid slags are placed 1 to 2 grammes of borax. For basic slags, 2.5 to 5 grammes of borax, or equal parts of glass and borax, are taken, and for slag containing sulphur, and which may not have been previously roasted, 0.25 to 0.5 gramme of iron and a covering of salt with a piece of coal. The charge is melted in a tall crucible (Fig. 49, p. 64), such as is used in the roasting or reducing assays, or in the crucible shown in Fig. 52, p. 67. 1 to 1½ hours are required for smelting acid slags in muffle-furnace, and ¾ to 1 hour for basic slags in the wind-furnace after the flame is free. A thin liquid fusion is absolutely necessary, so

---

[1] Percy, Metallurgy of Lead, pp. 112, 113.

that the lead globules can unite, this being the reason why a longer smelting is required.

III. *Alloys of Lead.*

The wet method must be employed for assaying alloys of lead.[1]

26. WET ASSAYS.

These, if accurate results are to be obtained, are tedious, require much time, and, as a necessary condition, require the absence of certain substances. These processes resemble the manipulations occurring in chemical analysis, and for this reason we shall give here only processes which can be easily carried out.

A. Assay by gravimetric analysis.

1. *Assays in Bleiberg in Carinthia and other places.*[2]—2 grammes of galena are powdered as fine as possible, and heated with nitric acid. When red vapors cease to come off, the mass is evaporated nearly to dryness with a few drops of sulphuric acid. If much lime is present, it is diluted with $\frac{1}{8}$ of a liter of water before the sulphuric acid is added, and the evaporation with the nitric acid is not carried too far in order to keep as much as possible of the lime in solution. It is then allowed to cool off, diluted, filtered, and washed until the acid reaction of the wash-water ceases. The contents of the filter (lead sulphate and insoluble earths, sulphur, etc.) are then rinsed off into a beaker glass, and digested with a concentrated solution of neutral sodium carbonate to convert the lead sulphate into carbonate. It is then filtered and washed until the water does not become clouded by the addition of barium chloride. The residuum is heated with diluted nitric or acetic acid to extract the lead, filtered, and the filter is washed with hot water until acid reaction ceases. The lead is then precipitated from the filtrate with as small a quantity of sulphuric acid as possible, to prevent a precipitation of lime

---

[1] Analysirmethoden für Blei; Fresenius's Ztschr. viii. 118 (Fresenius); Preuss. Ztschr. x. 125 (Hampe).

[2] B. u. h. Ztg. 1871, p. 62.

with it. The lead sulphate is filtered, washed with hot water, and dried. It is then detached from the filter, heated to a red heat and weighed. There should be at the utmost a difference of 0.1 per cent. between assays and counter-assays. Another process is as follows: The ore is dissolved in nitro-muriatic acid (*aqua regia*) and evaporated to dryness. The lead chloride (as also the ferric chloride and copper chloride) is extracted by boiling water and filtered. The filtrate is neutralized with ammonia to a precipitating point. The lead is precipitated with dilute sulphuric acid, and allowed to stand quietly for 6 hours. The lead sulphate is then filtered and washed, the filter dried and burned, and the sulphate heated and weighed.

2. *Mohr's process.*—1 gramme of finely powdered galena is boiled in a glass or porcelain vessel with hydrochloric acid. To this is added a small piece of zinc, and the contents of the vessel are then heated until the fluid becomes clear. The precipitated lead is washed by decantation and dissolved in diluted nitric acid. It is then filtered to free it from adhering insoluble substances (quartz, heavy spar, etc.). The filtrate is very much diluted with water, and the lead is precipitated as lead sulphate with sulphuric acid in the same manner as in 1.

Storer[1] weighs the lead precipitated with zinc, directly or dissolves it, in case it contains insoluble admixtures, in nitric acid, then weighs the residue and obtains the weight of the lead from the difference. Mascazzini[2] fuses the impure lead, which has been dried, with a mixture $1\frac{1}{2}$ to 2 parts of caustic alkali, 5 parts of borax, and 5 parts of starch. Löwe[3] frees the lead sulphate from its admixtures with earths by dissolving it in sodium hyposulphite, precipitates it as sulphide with sulphuretted hydrogen, and converts this into sulphate. Riche[4] determines the lead as superoxide by electrolysis. The galena is dissolved in hydrochloric acid. The clear solution is filtered into diluted sulphuric acid, and the lead determined as sulphate. But this must be ignited gently to

---

[1] Storer, in Fresenius's Ztschr., ix. 514; B. u. h. Ztg. 1870, p. 208; 1873, p. 91.
[2] Mascazzini, in Dingler, ccvii. 46; B. u. h. Ztg. 1878, p. 382.
[3] B. u. H. Ztg. 1874, p. 322.
[4] B. u. h. Ztg. 1878, 382.

prevent the volatilization of traces of lead chloride which may be present, as this would make the result less accurate.

B. Volumetric processes.[1]—These have been frequently proposed without any practical success.

C. Colorimetric processes.—Bischof[2] has given a method for determining small quantities of lead, which is based upon the browning of a solution containing lead, by sulphuretted hydrogen.

D. Electrolytic processes.—Parodi and Mascazzini proceed as follows:[3] The lead is first separated as sulphate and then dissolved by an alkaline solution of sodium tartrate. A measured portion of this solution is acidulated with hydrochloric acid until a white precipitate forms, then neutralized with sodium carbonate, and after 30 c.c. (150 grammes in $\frac{1}{2}$ liter) has been added, the whole is diluted with ammonia to 200 c.c. The platinum electrodes are then placed in the solution and the galvanic current allowed to act. Precipitation being finished, after washing the cone thoroughly without interrupting the current, the current is stopped, and the precipitate upon the cone is twice washed with water and alcohol, and dried at a temperature of from 30° to 40° C. in an atmosphere of illuminating gas. The lead precipitated by electrolysis dissolving with some difficulty in nitric acid, it is recommended to repeat, if necessary, the dissolving, and also to weigh the platinum cone before and after the solution of the lead, in order to be sure that no more adheres to it.

According to Kiliani,[4] the precipitate of superoxide of lead adheres tightly to the anode in large quantities only when the strength of the current is so slight that no evolution of gas takes place along with the separation. If there is an evolution of gas, the precipitate exfoliates in scales from the anode. With such a weak current the precipitation is slow; it is therefore advisable to use a dish as an anode (see Fig. 68) and a cone or disk as a cathode. This arrangement enables the precipitate to be washed with ease and without loss. The precipitate is not a pure superoxide, but a hydrate with

---

[1] Mohr, Lehrbuch der Titrirmethode, 1874 p. 460; Fleischer, Titrirmethode, 1876, pp. 42, 87, 293; Mitchell, Practical Assaying, 1868, p 397.

[2] Fresenius's Ztschr. 1879, p. 43; B. u. h. Ztg. 1879, p. 187.

[3] Gazetta chim. ital. 8. Ztschft. f. anal. Chem. Bd. 18 p. 588.

[4] Bg. u. Httnmsch, Ztg. 1883, p. 401.

a varying content of lead, which, according to Wernicke,[1] contains the more anhydrous hyperoxide the more concentrated the liquid and the longer the electrolytic action has lasted. By heating this precipitate at 200° to 250° C., until the weight is constant, there remains, according to Kiliani, pure superoxide, no decomposition taking place. In drying and cooling the dish should be covered, otherwise small particles of the oxide might be thrown out.

According to Schucht,[2] the complete separation of *all* the lead as superoxide takes place only in the presence of at least 10 per cent. free HNO_3.

Tenney[3] confirms this observation. In the presence of large quantities of lead a platinum dish should always be used as a positive electrode. Too much HNO_3, however, is injurious, as by its decomposition the nitrous acid formed can redissolve only a portion of the lead oxide.

May recommends to rub off the precipitated superoxide from the positive electrode, with a bit of rubber tube on a glass rod, into a beaker containing a little water. The filtered precipitate, after the filter has been burnt separately, is ignited to PbO in a porcelain crucible. If a portion of the superoxide adheres firmly to the positive electrode, this is also converted into PbO by careful heating of the electrode. The determination as superoxide is, however, preferable. It should be remembered that the superoxide must be washed without interrupting the current to avoid loss.[4]

II. Copper.

27. ORES.

Native copper.—Sulphuretted ores: *copper glance*, Cu_2S, with 97.7 Cu; *erubescite (purple copper ore)*, Cu_3FeS_3, with 55.6 Cu; *copper pyrites*, $CuFeS_2$, with 34.6 Cu. Ores with antimony or arsenic: *tetrahedrite (Fahlerz)*, $R_4Q_2S_7$ (in which $R = Cu_2$, Ag_2, Fe, Zn, Hg, etc., $Q = Sb$, As), with 15 to 48 Cu, as much as 30 per cent. silver, and from 0 to 18 per cent. mercury; *bournonite,*

---

[1] Poggendorff, Anal. Bd. 141, p. 109.
[2] Ztschft. f. anal. Chem. Bd. 22, p. 487.
[3] Am. Chem. Journ. vol. 5, p. 415.
[4] Classen, Chem. Analysis by Electrolysis. Trans. 1887, p. 69.

PbCuSbS$_3$, with 13.03 Cu and 42.54 Pb; *enargite*, Cu$_3$AsS$_4$, with 48.6 Cu. Oxidized copper ores: *cuprite (red copper)*, Cu$_2$O, with 88.8 Cu; *malachite*, CuCO$_4$+H$_2$O, with 58 Cu; *azurite*, Cu$_3$C$_2$O$_7$+H$_2$O, with 35.7 Cu; *dioptase* (Kieselmalachit), CuSiO$_3$+2H$_2$O, with 35.7 Cu; *atacamite* Cu$_4$O$_3$Cl$_2$+3H$_2$O, with 59.4 Cu; *phosphates of copper*, with 30 to 56 Cu; *arsenates of copper*, with 25 to 50 Cu; *cupric sulphate* or *blue vitriol*, CuSO$_4$+5H$_2$O, with 25.3 Cu.

28. DRY ASSAYS.[1]

These, besides consuming much time, and being expensive, require great experience and are less accurate than the wet assays, which are now much employed in smelting works for valuing ores. There are two methods, known as the *German* and the *Cornish* assay. Both are based upon the principle that copper has a stronger affinity for sulphur, and less for oxygen, than the foreign metals (iron, zinc, antimony, lead, arsenic, etc.) with which the ores are contaminated, so that, when they are present in an oxidized condition, or have been converted into it by roasting, they are mostly slagged off at comparatively low temperatures on being fused (*reducing fusion*) with reducing agents (black flux free from sulphur, potassium carbonate and flour) and solvent agents (borax, glass), and only a very small part of them passes into the reduced copper (black copper). The latter during the subsequent oxidizing fusion (*refining*) eliminates the foreign metals in an oxidized condition, and is itself transformed into refined copper, while the metallic oxides which have been formed are either slagged off by the solvent agent (borax) or are carried into the cupel by the lead oxide. On account of the high temperature at which copper fuses, it is well to add in the reducing fusion *collecting* and *liquefying* agents (such as antimony and arsenic, but lead is less well adapted as it may cause losses in the refining). The *German* and *Cornish* or *English* methods, which undergo modifications according as the copper is combined with sulphur,

---

[1] [Percy. Metallurgy of Copper, Zinc, and Brass. London, 1861. pp. 454–478.—G.]

antimony, or arsenic, or is oxidized or alloyed, are still practised in smelting works, and give results which suffice for the business of working copper on a large scale.

A. German copper assay.

1. *Ores with sulphur, antimony, or arsenic.*

a. Roasting.—Dead roasting 5 grammes of ore, or enough of it is dead roasted that the refined button of copper, to be turned out on the refining dish, does not weigh much over 0.5 gramme. It is then repeatedly rubbed up and treated with coal and finally with ammonium carbonate for completely removing the sulphur (p. 33) which otherwise would produce copper sulphide and occasion losses during reducing fusion. A well-roasted sample should be earthy (without metallic lustre), have a brownish or black color, should not appear sintered, and should neither fume nor smell.

Pyrites is rubbed up two or three times and treated once with powered charcoal—Tetrahedrite, *Fahlerz* (Hungary), 10 grammes of ore are rubbed up ten to twelve times at a very low roasting temperature, taking them out whenever they commence to fume. The ore is then roasted without charcoal for $\frac{3}{4}$ to 1 hour at a temperature not above red heat, rubbed up, and roasted for 1 hour more at a white heat.

b. Reducing and solvent fusion.—The roasted ore is well mixed in an iron mortar with $\frac{1}{3}$ of the required quantity of black flux (consisting of 2 to $2\frac{1}{2}$ parts of argol and 1 part of saltpetre, or of 3 parts of potassium carbonate and 1 part of flour), a collecting agent (antimony is the best), and, if the ore contains no iron, some iron filings are added to decrease the slagging of copper in refining it on the dish. The mixture is poured by means of the mixing scoop (Fig. 6, p. 30) into a suitable crucible of refractory clay (Fig. 52, p. 65). The remaining $\frac{2}{3}$ of the flux is then added, upon this is placed a mixture of borax and glass, then a covering of common salt and a small piece of charcoal.

Examples of Charges.

	Black flux.	Borax.	Glass.	Antimony (arsenic).	Iron filings.
	Per cent.	Per cent.	Per cent.	Per cent.	Per cent.
Pyrites (copper)	300	50	40	10	
Purple copper ore	300	40	40	10	
Tetrahedrite	300	30	25	6	8
Matt rich in iron	300	40 to 50	40	10	
Matt rich in copper	300	—	50 to 100	5	

Charges suitable in many cases: 5 grammes of ore, 12.5 to 15 grammes of black flux, 0.3 gramme of antimony, 1.25 grammes (1 small assay spoonful) of borax, 1.5 to 2.5 grammes of glass, and 10 to 15 grammes of common salt; or 5 grammes of roasted ore are mixed with the same quantity of black flux and put into the crucible. Upon this are placed 8 to 10 grammes of black flux, 1 gramme of borax, 2.5 grammes of glass, and a covering of common salt six millimeters thick. *Hungarian fahlerz:* Half of the roasted sample—5 grammes are mixed with 7 to 8 grammes of black flux. The mixture is placed upon the same quantity of black flux in the crucible and covered with common salt.

American charge: 10 grammes of ore are roasted and mixed with 20 grammes of black flux, 3 grammes of borax glass, and 10 to 20 grammes of hematite. Upon this are placed 10 grammes of black flux, 3 grammes of wood charcoal, and a covering of common salt. Charge for unroasted ores containing heavy spar and gypsum: 5 grammes of ore, 5 grammes of borax glass, 5 grammes of powdered glass, and ten per cent. of rosin; and, besides for poor ores, 25 per cent. of iron pyrites, if this is not already present in the ore. The whole is covered with common salt and fused to matt, which is then roasted and treated as above.

Time required for fusion: ¾ to 1 hour in a red heat in the muffle-furnace after the "flaming" has ceased, or in a wind-furnace after the flames are free. The charcoal should be piled high in front of the crucibles in the muffle. Indications of a successful assay are: a well-fused regulus (which on account of its brittleness must be carefully freed from slag),[1] without a black coating of brittle matt rich in copper, well-fused slag, whose color may be black or green, but must not be red, and without an admixture of metallic grains.

c. Refining.—This consists in an oxidizing fusion, during which the foreign metallic oxides are sooner oxidized than the copper, and are easily separated, either at once as easily fusible masses, though this is seldom the case, or must be dissolved by fluxes (borax, lead oxide), and then separate with these as slag (refining with borax on the refining-dish), or are absorbed by the cupel (refining with lead).

Lead is pre-eminently oxidizable. The lead oxide yields its oxygen to the foreign metals. It therefore acts as a vigorous oxidizing agent, and produces easily fusible combinations, but it contributes to the slagging off of the copper, and for this reason the refining process with lead is less accurate than that with borax without lead. *Ferrous oxide, antimony,* and *arsenic* protect copper from slagging. *Nickel* and *cobalt* are difficult to separate from copper, and do so only at the expense of considerable copper, which is slagged off (see assay of Nickel). *Tin* and *zinc* give refractory oxides and slags, and, if present in large quantities, the wet method must be employed. Gold and silver remain in the refined copper.

α. *Refining on the dish.* a. *With borax.*—This is the most accurate process (to within 1 to ½ per cent.), and is principally adopted for black copper free from lead, or which contains iron, arsenic, or antimony. According to the size and impurity of the copper button, 1.25 to 2.5 grammes of borax glass are placed, either wrapped up in a cornet, or by means of an iron spoon, in a refining-dish, which stands in the white-hot muffle, and is surrounded by glowing coals. The copper button, weighing from 0.5 to 0.6 gramme, is taken up with the curved tongs (Fig. 49, p. 64) and placed in the dish. The mouth of the muffle is then closed by a plug, or by piling coal in front of it, and the button is fused quickly at as high a temperature as possible. The mouth

---

[1] [As soon as the slag has solidified, or has "set," the crucible is plunged into cold water several times, and then left to cool. This immersion in water cracks the slag, and causes the regulus to be more easily removed.—G.]

of the muffle is then slightly opened to allow of the access of air. On account of the oxidation of the foreign metals (iron, zinc, etc.), the button will at first appear dull, but brightens as soon as the foreign metals (with the exception of antimony and arsenic) have been removed, and fumes from the antimony or arsenic, which may have been originally present, or has been added as a collecting agent. (The first fumes more strongly than the latter, therefore indicating the end of the process more plainly.) The completion of the refining process may be recognized by the button, which *remains bright*, ceasing to fume (should the process be continued the button would become dull from a covering of cuprous oxide); and also, by the subsiding of the slag, that is, if the button is not too large, and does not weigh much over 0.4 to 0.5 gramme. The dish is then taken out with the curved tongs (Fig. 59, p. 72), and carefully cooled off *upon* water until it ceases to glow, and then *in* water, and the button freed from slag. The assay is considered successful if the button, on being flattened, shows itself ductile to a certain extent, and has a flesh-red color exteriorly. (A small residue of antimony and arsenic prevents the button from being entirely ductile, and it causes its fracture to be gray, and equalizes small losses in slagging off.) The slag should not be red, or at least scarcely perceptibly so, except at the place where the button has lain. Red slag indicates, either that the oxidizing process has been carried too far, or an absence of iron which otherwise protects the copper from slagging off during the reducing fusion, as well as during refining.

b. *By itself, without borax and lead (Hungarian speiss assays)*[1] for buttons containing more antimony or lead.—The copper button is placed on the white-hot refining-dish (p. 64) standing in the muffle, and is fused quickly at as high a temperature as possible after the mouth of the muffle has been closed. After the fusion is complete the mouth of the muffle is opened, the register is closed, and the dish turned and lifted. This will cause the slag formed from foreign oxides to remain behind, and the button to roll upon a place free from slag. The refining is continued in this manner until the fuming ceases, and the now refined button

---

[1] B. u. h. Ztg. 1866, No. 28; 1868, No. 12; 1871, p. 255.

assumes a sea-green color. The refining-dish is then pulled slowly towards the mouth of the muffle, and taken out as soon as the button brightens. The dish is then dipped in hot water, when the button can be hammered out without cracking on the edges. It should have a red color. It is now weighed, and the difference in weight between it and the black copper determined. From this the loss of copper, which must be placed to the account of the copper, is calculated by allowing for plumbiferous copper 1 part for every 10 parts of black copper lost, and for antimonial and arsenical copper 2 parts for every 10 parts lost.

If the ore should contain neither lead, antimony, nor arsenic, a little borax is added, and some lead during the fusing process should the copper be very refractory, for instance, if it contains much iron, or cobalt and nickel.

Differences allowed in assays.[1]—In the mining district of Schemnitz in Hungary the assays are made by three different assayers. The differences allowed in the results are as follows:—

Amount of copper in ore.	Differences allowed.
1 to 4 per cent.	0.5 per cent.
4 " 10 "	0.7 "
10 " 20 "	1.0 "
20 " 40 "	2.0 "
40 " 70 "	4.0 "
70 or more "	6.0 "

In the Klausenburg mining district in Transylvania the following differences are allowed:—

Amount of copper in ore.	Differences allowed.
0 to 3.0 per cent.	0.5 per cent.
3.25 " 5.0 "	0.75 "
5.25 " 8.0 "	1.00 "
8.25 " 12.0 "	1.25 "
12.25 " 20.0 "	2.00 "
20.25 " 40.0 "	3.00 "
40.25 " 70.0 "	4.00 "
70.25 or more "	6.00 "

c. With lead and borax (Müsen assay).—The black copper obtained from 5 grammes of ore is fused with 2.5 grammes of granulated lead and a small quantity of borax. The muffle is

[1] Fortschritte im Probirwesen. Balling, Berlin, 1887, p. 91.

slightly opened to give access to the air until the copper brightens in the continually increasing brownish slag. The dish is cooled off in water. A counter-assay with refined copper is made, and the loss of copper occurring thereby is added to the principal assay. 10 parts of lead will slag off about 1 part of copper.—*In Mansfeld*, 2.5 grammes of black copper were formerly refined with 0.4 grammes of lead.

β. *Refining by cupellation.*—This method is the most suitable one for plumbiferous black copper, and especially for quickly obtaining approximate results. But a correction (*counter-assay*) is necessary to determine the amount of copper slagged off by the lead which becomes oxidized during the oxidizing fusion, and then parts with oxygen, not only to the foreign metals, carrying them with it into the cupel, but also slags off a part of the copper.—A quantity of refined copper, equal to the weight of the black copper button—for instance, 1 to 1.25 grammes—is weighed off for the counter-assay. The assay sample and the counter-assay sample are each wrapped up in a cornet. A quantity of granulated lead, amounting to 2 or 2½ times the quantity of copper, is weighed off for each copper sample, and each also wrapped up in a cornet. Two large-sized cupels (Fig. 54, p. 66) are placed alongside each other in the centre of the strongly heated muffle, and are brought to a white heat. Upon each is placed one cornet with lead, and the mouth of the muffle is closed until the lead has begun to "*drive*" (that is, until the dark film of lead has disappeared, and a white strongly fuming surface has made its appearance). The muffle is then opened, the cornet containing the refined copper is placed in one cupel, and that containing the black copper in the other, whereupon the muffle is again closed, and "driving" is quickly renewed at a high temperature. The mouth of the furnace is then slightly opened to admit air for the oxidation of the lead and the foreign metals, the oxides of which are absorbed by the cupels, until the copper buttons brighten. As soon as this has ensued a spoonful of coal-dust is strewed upon the cupels to prevent the slagging off of copper. They are then immediately taken out of the furnace and thrown into water. Both buttons are weighed, and the loss of weight of refined copper is added to the weight of the refined button obtained from the black copper.

γ. *Refining with the blowpipe.*—2.5 grammes of ore are roasted, arsenized, and fused as in the nickel assay (see § 40). The button is weighed, and 0.05 to 0.1 gramme of it is taken and fused with borax glass upon charcoal in the inner flame of the blowpipe. The oxidizing flame is then used until the dull button has become bright; it is then further treated upon charcoal without borax in the reducing flame before the blowpipe until it ceases to fume. The bright copper button is weighed, and the weight calculated to the quantity of ore used (*Rothenbach* smelting works near Müsen).

2. *Oxidized substances without sulphur.*—These are fused to black copper (p. 96) without being roasted, and the black copper is refined (p. 99), during which a percentage of iron will protect the copper from slagging.

	Black flux.	Antimony (arsenic).	Borax.	Glass.	Charcoal dust.
	Per cent.	Per cent.	Per cent.	Per cent.	Per cent.
Poor ores with basic gangue	300	10	30 to 40	20 to 25	
" " acid "	300	10	60	15	
" " basic & acid "	300	12	30	30	
Richer ores	300	5 to 10	30 to 40	30 to 40	
Very rich ores	300	5 to 10	30 to 40	30 to 40	5 to 10
Rich slags	300	.5	30	30	5 to 10
Poor slags	300	25 to 50	30 to 50	15 to 20	

From 1 to 10 per cent. of iron filings may be added for refining in case the copper contains no iron or other easily oxidizable metals.—The process of fusing *very poor copper ores* in larger quantities is as follows: 10 to 15 grammes of unroasted ore are mixed with 15 to 20 per cent. of iron pyrites (free from copper), and 20 per cent. of sulphur, 100 per cent. of borax glass, 100 per cent. of glass, and 20 to 25 per cent. of resin are added, and a covering of common salt. The charge is then fused to matt (matt assay) in a clay crucible. The resulting matt is roasted, etc.— Easily decomposable sulphates (cupric sulphate) are decomposed, before they are subjected to the reducing fusion, by roasting with an addition of charcoal (p. 33); sulphate difficult to decompose (cuprous sulphate with calcium and barium sulphates) by preliminary fusion to matt.

3. *Alloys of copper.*—In case they do not contain too many and difficultly oxidizable components (nickle, tin, etc.), they are refined in the refinishing-dish or cupel; otherwise the wet method of treatment is preferable.

B. Cornish copper assay.[1]—This is an imitation of the English smelting process in the reverberatory furnace, and requires much skill. On account of its inaccurate results (involving a loss of from 20 to 40 per cent. of copper, according to the richness of the ore) the wet method has been substituted for it even in Cornish smelting works for valuing ore in quantity. The following operations are required for ores containing sulphur, antimony, or arsenic: A gentle roasting in a suitable crucible (Fig. 50, p. 64); fusing to crude matt in the same crucible. This is then roasted, and the roasted assay sample is fused to black copper. This is purified by fusing it with oxidizing and solvent agents, and the purified black copper is refined with oxidizing and solvent agents, and finally the slag is fused. All operations are carried out in the Cornish clay crucibles (Fig. 50, p. 64.).

29. WET ASSAYS.[2]

These, on account of their greater accuracy and simpler execution, have, as a general rule, been substituted for dry assays. The choice of one of the numerous wet assays depends chiefly on the foreign admixtures (antimony, arsenic, lead, bismuth, mercury, etc.), and somewhat on the richness of the sample. *Colorimetric* assays are especially adapted for poorer ores, and *volumetric assays*, if many are to be made in quick succession, are preferred to *gravimetric assays*.

A: Gravimetric assays.—The methods by which the copper is determined in the metallic state (Swedish and electrolytic assay) are simpler and more convenient to execute than those by which

---

[1 Percy's Metallurgy of Copper, Zinc, and Brass, pp. 454–478.—G.]

[2 Comparison of Various Methods of Copper Analysis. W. E. C. Eustis, Boston, Mass., Trans. American Institute of Mining Engineers, vol. xi. pp. 120–135.—G.]

the copper is separated and determined in combination (determination of copper as cuprous sulphide, or as subsulphocyanide). The Swedish assay can be executed in less time than the electrolytic assay, but it is done at the expense of accuracy, especially with poorer ores.

1. *Modified Swedish assay.*[1]—The cupriferous substance is brought into solution with sulphuric or hydrochloric acid (nitric acid must not be used, as the precipitated copper is again dissolved in it), and the copper precipitated with *iron* or *zinc*, and determined either as metal or oxide. This plan is not admissible in the presence of metals, which are also precipitated by iron and zinc.

But such metals can be removed during the operation without injurious effect (*lead* or sulphate, *silver* as silver chloride, mercury by igniting the precipitated copper); or they must be removed by a preparatory operation (*arsenic* by roasting the assay sample with charcoal powder), or first by itself, and then with an addition of some iron pyrites at not too high a temperature; *tin* and *antimony* by heating with moderately diluted sulphuric acid, then adding nitric acid, and heating nearly to boiling, and an addition, if necessary, as in the case of metallic sulphides, of a few drops of fuming nitric acid. The solution is then evaporated to dryness until the fuming ceases. The dry mass is dissolved in hot water and filtered, and the solution treated with some hydrochloric or nitric acid, etc.; *or*, the antimony is removed by fusing the assay sample with potassium hydrate or potassium carbonate in a silver crucible, lixiviating the potassium antimoniate, and dissolving the residue; *or*, the assay sample is dissolved, neutralized with soda, and digested with a solution of sodium sulphide to extract the antimony, arsenic, and tin in soluble form. It is now filtered, and the residue washed and dissolved as above. To separate *bismuth* by analytical methods is a very tedious operation. Ores containing *bitumen*, for instance cupriferous schists (Kupferschiefer), must be ignited to remove the bitumen, before they are dissolved. Impure (black) precipitated copper can be further examined according to *Parkes's* and *Fleitmann's* volumetric method to be shortly described.

[1] B. u. h. Ztg. 1869, p. 12.

a. Precipitation with iron.—1 to 5 grammes, according to degree of richness of copper; generally, 2.5 to 5 grammes of the assay sample are decomposed in a suitable flask (Fig. 10, p. 36), which is placed in an oblique position, by heating with sulphuric acid, and adding from time to time some fuming nitric acid, or potassium chlorate, until the separated sulphur, inclosing particles of the ore, is oxidized as much as possible; or, the sample is at once dissolved in *aqua regia* (see also p. 36 for method of decomposing metallic sulphides). It is now evaporated to dryness with some sulphuric acid, or until the sulphuric acid vapors appear in the flask. A few drops of sulphuric acid are added to the dry mass (to dissolve the basic salts), and then water is cautiously added; or, is at once added to the (cooled off) mass, while it still contains free acid. The fluid now *entirely free from nitric acid* is filtered into a glass flask, such as is shown in Fig. 16, p. 39. The residue is washed until the wash-water no longer produces a red stain upon a piece of bright sheet iron. Two pieces of iron wire, 3 to 4 centimeters long, are then added (or, in order to shorten the time required for the assay, the fluid may at once be filtered into a porcelain dish in which the iron wires lie, and copper will then be precipitated during the filtration). The filtrate is sufficiently diluted and gently heated until a pointed iron wire, when dipped into the fluid, shows no reddish stain of copper. The copper is twice decanted with *cold* water into a spacious beaker-glass (to prevent the separation of basic iron salts, which are more easily formed by hot water), and is then decanted three times with boiling water. The flask is now completely filled with cold water; a flat-bottomed porcelain dish, about 80 millimeters wide and 20 millimeters high, is placed bottom upwards on top of it. The flask and dish are then inverted. The mouth of the flask is held in an oblique position, and the water is allowed to run into the dish until it is nearly full. The flask is left standing in the dish until all the copper and the iron wires have fallen into the water in the dish (small particles of carbon separated from the iron will remain floating on the water for some time). The flask is now quietly drawn over the side of the dish, which should be somewhat inclined for the purpose. The iron is freed from copper by rubbing with the fingers, which should be rinsed off in the water. The copper is now decanted twice with boiling water. This is poured off as completely as possible from the copper, which is moistened with absolute alcohol, and dried on the water-bath, until it has assumed a pulverulent condition. It is allowed to cool in the desiccator, and is then brought upon the pan of the balance, or into a tared porcelain crucible with the aid of a fine brush, and quickly weighed. It is now dried for 10 or 15 minutes more, and again weighed until the results agree; or the copper is spread out upon a roasting dish and ignited in the muffle-furnace, and the metal calculated from the amount of cupric oxide formed (100 cupric oxide = 79.88 copper). If the water used in decantation shows a reddish sediment in the beaker-glass, it should be filtered, the filter dried and ignited upon the scorifier, and the percentage of copper resulting from the cupric oxide should be added to the principal yield.

Correction for *iron* that may be contained in the precipitated copper on account of a deposit of basic iron salts: The precipitated copper is ignited upon the scorifier until it becomes black. The cupric and ferric oxides formed are weighed and dissolved in hydrochloric or sulphuric acid. The ferric oxide is precipitated with ammonia. The solution is filtered upon a small filter of paper. The filter is dried and ignited, and the ferric oxide, which may be found, is deducted from the combined weight of the cupric and ferric oxides, and the copper calculated from the quantity of pure cupric oxide found. Instead of decanting the precipitated copper, it may all be filtered, dried, ignited, weighed, and dissolved, as above, for obtaining the percentage of iron.

Pure precipitated copper has a fine copper color. If the solution contains antimony and arsenic, it has first a copper color, which changes to black, by the antimony and arsenic which are precipitated later on. The largest portion of the antimony, after the precipitate has been evaporated to dryness with sulphuric acid and again moistened with water, remains as basic sulphate of antimony, while arsenic passes into solution. The residue from the solution of the ore, etc., is tested for copper by heating it with nitric acid, filtering, and adding ammonia in excess (appearance of a blue color indicates copper.)

Instead of two iron wires, a simple one, bent into the form of a ring with one end projecting vertically, may be used. The ring is dipped into the liquid contained in a beaker-glass in such a manner that the end projects. When the precipitation is complete the copper is rinsed from the ring, decanted, etc., as above.— Or a strip of sheet-iron may be used instead of the wire, but it must be immediately removed from the liquid after the precipitation of the copper is complete to avoid the formation of basic iron salts. In solutions that are too concentrated the copper adheres too strongly to the iron.

b. Precipitation with zinc free from lead and arsenic.[1]—A solution of the assay sample is prepared with sulphuric acid, as described on p. 106, and filtered. A strip of zinc is placed in it, and the solution is then heated until a bright iron wire held into it shows no copper deposit; or until a drop of the solution placed upon a porcelain dish is not browned by sulphuretted hydrogen. The strip of zinc is then taken out, and the precipitated copper is washed off with the wash-bottle. It is filtered, until but a small layer of water covering the copper remains. A few drops of warm hydrochloric acid are then added to dissolve any particles of zinc which may be present. It is now decanted, etc., as in the precipitation with iron (p. 106); or it is filtered as soon as effervescence has ceased, quickly washed with hot water, and dried. The copper is then detached from the filter, ignited on the cover of a porcelain or platinum crucible, or upon a roasting dish in the muffle, and the oxide quickly weighed. The black crust upon the end of the piece of zinc, which has been dipped into fluid, is a spongy layer of zinc colored by a trace of sulphide of copper. *Nickel*, which is not thrown down by iron, is precipitated with zinc, but cobalt is not.

Granulated zinc may be used instead of a strip, but the granules must be completely dissolved by the time the bubbles cease. The copper is then decanted, etc. The cuprous fluid may be filtered into a platinum dish, and some hydrochloric acid added to it. It is then heated, and a few small pieces of zinc added to it,

---

[1] Fresenius's Ztschr. für analyt. Chemie iii. 334 (Mohr und Fresenius). Oest. Ztschr. 1868, No. 48 (von Kripp). Erdmann's J. f. pr. Chemie, cii. 477 (Ullgreen). Darstellung von pulverförmigem Zinke in Dingler, cxxviii. 378.

whereupon the copper will deposit itself, firmly on the platinum, but loosely on the zinc. After precipitation is complete, which fact is to be tested with sulphuretted hydrogen, as above, the copper is rubbed and washed off from the zinc. It is then allowed to settle, is decanted, treated with hot water, to which some hydrochloric acid had been added; then quickly washed with hot water by decantation to prevent the loss of any of the copper by solution. It is finally moistened with some absolute alcohol, dried in a water-bath, or at 110° to 120° C. (230° to 248° Fahr.), and the tared platinum dish, which has been allowed to cool off in the desiccator, is weighed; or the copper, if the utmost accuracy is demanded, is heated in a stream of sulphuretted hydrogen.
—*Cuprous schist* (Kupferschiefer):[1] 5 grammes are heated with 40 to 50 cubic centimeters of hydrochloric acid. When the carbonic acid has been expelled, 6 cubic centimeters of diluted acid, consisting of equal parts of nitric acid, of 1.2 specific gravity, and water, are added (to ores free from bitumen, or which have been ignited, 1 cubic centimeter of nitric acid). This is digested for half an hour, then boiled for a quarter of an hour. The hot liquid, which should contain no free nitric acid, is filtered into a beaker-glass. A small rod of zinc upon a strip of platinum is placed in the filtrate, and the copper is precipitated, which will require from $\frac{1}{2}$ to $\frac{3}{4}$ of an hour. The precipitated copper is then decanted and dissolved in nitric acid, together with that adhering to the platinum, and titrated with a solution of potassium cyanide (see p. 122).

The following gravimetric method for the determination of copper is recommended by Genth as giving very accurate results when proper care has been taken that nothing is lost in dissolving the ore and in the evaporation: The finely-powdered ore is placed in a small beaker, and enough pure H_2SO_4 added to convert all metals present into sulphates and to drive off all the HNO_3, which is used for dissolving. The nearly dry mass, after giving off copious fumes of SO_3, is allowed to cool, and dissolved in a small quantity of water, when all, excepting quartz, etc., is in solution, the trace of Ag which may be present is precipitated by a drop of HCl, when the liquid is clear, the insoluble silicates, $PbSO_4$, AgCl, and the

[1] Fresenius's Ztschr. viii. 9.

greater part of the antimoniate of antimony, are separated by filtration from the Cu, As, etc. The clear liquid is next precipitated by H_2S, which throws down CuS, and traces of Sb_2S_3 and As_2S_3—(the greater part of the As remains in solution as As_2O_5), as but a small amount is reduced to As_2O_3, and precipitated by the H_2S). The CuS is washed, then treated with K_2S to dissolve the Sb_2S_3 and As_2S_3, which may have come down, and, after washing, boiled with strong HCl to dissolve any ZnS which may have been precipitated with the CuS, then diluted with boiling water, and a sufficient quantity of H_2S is added to precipitate any traces of Cu which may have gone into solution. The CuS is washed, dried, carefully roasted, then dissolved with 2 or 3 drops of H_2SO_4, H_2O, and HNO_3. After everything is in solution the latter is evaporated to dryness, carefully heated, and finally ignited to drive off *every trace* of H_2SO_4, and from the resulting CuO the amount of Cu is calculated.

2. *Electrolytic assays.*[1]—The copper is precipitated in a coherent film upon a weighed platinum strip by the galvanic current, from the solution of the nitrate containing free nitric acid, which need not be filtered. From this solution, *zinc, iron, nickel, cobalt,* and *chromium* are *not* precipitated by galvanic action. The following are precipitated in the form of peroxides at the positive electrode: namely, *lead, manganese,* and (partially) *silver. Mercury* is precipitated at the negative electrode in the metallic state before copper; *silver* and *bismuth* at the same time as copper, and *selenium, antimony,* and *arsenic*[2] only a considerable time later; the assay, therefore, as a general rule, requires the absence of the last-named metals and metalloids, which blacken the fine red color of the copper. Antimony remains behind undissolved, when the assay sample is dissolved in nitric acid. A large number of assays can be carried out at the same time according to

---

[1] Fresenius's Ztschr. iii. 334; vii. 253; ix. 102. B. u. h. Ztg. 1869, p. 43, 181; 1872, p. 251; 1875, p. 155; 1877, pp. 5, 32. Preuss. Ztschr. xvii. Lief. 3; xx. Lief. 1. Grothe, polyt. Ztschr. 1877, p. 11. The Electrolytic Determination of Copper, and the Formation and Composition of so-called Allotropic Copper. J. B. Mackintosh, Trans. Am. Inst. Mining Engineers, vol. x. p. 57.

[2] [According to Eustis in sulphate and nitrate solutions, containing 50 per cent. arsenious oxide in solution, no arsenic is precipitated as long as an *insoluble* electrode is used. Trans. Am. Inst. M. E. vol. xi. p. 124.—G.]

this very accurate and simple method, which is adapted for rich as well as poor ores, etc.

It is best to employ for the galvanic current, the *Meidinger-Pinkùs's battery*[1] (with 6 large elements for richer copper ores, and 4 small elements for poor ores with less than 10 per cent. Cu); or *Clamond's thermo-electric battery*[2] (consisting of a number of zinc and antimony elements arranged in the form of a ring, and heated by a gas-flame) may be used. If the current is too strong, the copper does not deposit itself firmly upon the platinum. The copper is not precipitated in as pure and coherent a state, from the solution of the sulphate containing free sulphuric acid, and in case iron should be present a part of this is thrown down with the copper, while another part is reduced to protoxide.

One gramme of the assay sample is dissolved in strong nitric acid and evaporated to dryness in a porcelain dish. If necessary, the dish is heated over a lamp to burn off any separated sulphur (p. 36). The residuum is dissolved in 20 cubic centimeters of nitric acid of 1.2 sp. gr., and filtered into a beaker-glass. The filtrate is diluted to 180 to 200 cubic centimeters, and stirred. The platinum spiral a (Fig. 66), weighing about 16 grammes, is now placed in the beaker-glass. A cone of platinum foil (Fig. 65), weighing about 20 grammes, is suspended over this from a stand in such a manner that, when rich ores are to be tested, the cone hangs, at the utmost, 1 centimeter above the foot ring b of the spiral, but only 0.5 centimeter in case of poorer ores, and that a part of the cone shall project above the liquid. The cone is connected by means of a wire conductor with the negative electrode, and the spiral with the positive. The beaker-glass should be covered with a glass plate, cut into halves, and each half perforated with a hole for the wires to pass through. The strength of the galvanic current used is generally such that 90 to 100

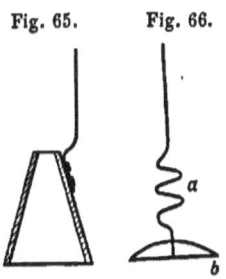

Fig. 65.   Fig. 66.

---

[1] Fresenius's Ztschr. xi. 4 (Mansfeld).
[2] B. u. h. Ztg. 1875, pp. 155, 251, 303.

cubic centimeters of water-gas will be developed in 30 minutes in the voltameter from diluted sulphuric acid (1 : 12), and as much as 180 cubic centimeters from rich cupriferous substances. After an electrolytic action of 12 to 18 hours the liquid is examined in order to ascertain whether the whole of the copper has been precipitated, by adding some water, and stirring.[1] In case it still contains copper, the bright portion of the platinum cone, now partly submerged, will be covered with a red film of copper. If no more copper is separated, the beaker-glass is placed in a spacious porcelain dish, and the acid liquid it contains is displaced by adding water until all acid reaction disappears (or the liquid is removed with a siphon, and water is added from a wash-bottle until gas ceases to develop at the positive electrode). The cone of platinum is then taken out and placed in a beaker-glass with water. It is then rinsed off with hot water, next placed in a beaker-glass filled with absolute alcohol, or washed with it, and finally laid upon blotting paper, and dried in an air-bath at about 94° C. (201.2° Fahr.). (If the operator has some experience, this can be done more quickly upon a piece of sheet-iron heated over a lamp, or by holding the cone in the hot air arising from a large platinum or silver dish heated by the flame.) After it has cooled off the cone is weighed; and, as its weight had been accurately ascertained before the operation, the weight of the copper will be given by the increase in weight of the cone, from which the copper can then be dissolved by hot nitric acid. Dark spots upon the red copper indicate the presence of *arsenic, antimony*, or *selenium*. If only small quantities of the first two are present, they are very slowly precipitated, or not at all, from a strong acid solution, if the current is interrupted, while the fluid possesses still a faint bluish color. The presence of much iron prevents a complete precipitation of the copper, as it (Cu) is dissolved by the ferric sulphate while ferrous oxide is formed, the action of free nitric acid upon which produces blackish-brown circles around the platinum cone. When this is observed it is a sure

[[1] If the solution is heated and maintained at a temperature of from 70° to 80° C. during the reaction, this period may be shortened to 4 or 5 hours. The strength of the current under these circumstances can be reduced to one. Bunsen Cell. (Classen Ber. d. ch. ges. 18. 1796.)—G.]

indication that the process of precipitating the copper has not taken place properly. In this case the assay sample is dissolved in 40 cubic centimeters of nitric acid, and 360 cubic centimeters of water, using a stronger galvanic current, giving 120 cubic centimeters of water-gases in *Volta's* apparatus; or, what is still better, the copper is precipitated from an acid solution by sulphuretted hydrogen, and the copper sulphide dissolved in 30 cubic centimeters of nitric acid of 1.2 specific gravity. It is then digested until the sulphur shows yellow. 200 cubic centimeters of water are added, the fluid is then electrolyzed.

Herpin[1] dissolves 1 gramme or more of the assay sample in nitric acid, evaporates nearly to dryness, dissolves in a small quantity of dilute sulphuric acid, and dilutes the solution to 60 or 70 cubic centimeters. The solution is poured into the platinum dish A (Fig. 67), and the conducting stand B of the dish is connected with the negative electrode, the platinum spiral C with the positive electrode, and the liquid electrolyzed after the funnel D has been placed in position. When the copper has been precipitated, the fluid is poured from the dish. This is rinsed out first with water, and next with alcohol, then dried and weighed, the copper being determined from the increase of weight.—*Hampe's*[2] method of testing *refined copper:* 25 grammes of copper are dissolved in a beaker, at a moderate temperature, in 200 cubic centimeters of water, and 175 to 180 cubic centimeters of nitric acid of 1.2 specific gravity. To this are added 25 grammes of previously diluted sulphuric acid (about 4 cubic centimeters more than is required for transforming the nitrate into sulphate). The liquid is then evaporated to dryness in a porcelain dish on the water-bath. The dry mass is heated upon a sand-bath until the volatilization of the free sulphuric acid is complete. The dish is then covered, and, after the mass has become cool, 20 cubic centimeters of nitric

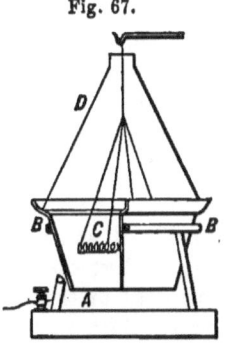

Fig. 67.

[1] B. u. h. Ztg. 1875, p. 394. Dingler, ccxvii. 440.
[2] Preuss. Ztschr. xxi. Lief. 5. Fresenius's Ztschr. xiii. 176.

acid are added to it. Water is now allowed to flow gradually into the dish until the entire volume amounts to 350 cubic centimeters. The silver is removed by the addition of an equivalent quantity of hydrochloric acid. The liquid is electrolyzed in a vessel capable of holding from 400 to 450 cubic centimeters. The strength of the galvanic current used should be such that 130 cubic centimeters of oxyhydrogen gas—it may vary from 90 to 180 cubic centimeters—are developed in 30 minutes in the voltameter from diluted sulphuric acid (1 : 12). The liquid is electrolyzed for about 72 hours, the subsequent manipulations being the same as above described.—The presence of small quantities of *bismuth* in the precipitated copper can be determined by the following process : The copper is dissolved in nitric acid, concentrated hydrochloric acid in large excess is added, and the nitric acid boiled away. The excess of hydrochloric acid is evaporated on the water-bath, a large quantity of boiling water is added, and, after 24 hours, the precipitate, consisting of basic bismuth and copper salts, is filtered off. The filtrate is dissolved in hydrochloric acid, and again precipitated with water. This precipitate is dissolved in nitric acid, and the copper contained therein separated by ammonium carbonate.

Classen's[1] *method for the electrolytic determination of metals* is much used, and is based upon the fact that the separation of the metal is best effected when the oxide is fixed to an acid readily decomposed by the current, so that a secondary reaction cannot take place. Such an acid is oxalic acid, which splits into carbonic acid and hydrogen. Most heavy metals give insoluble precipitates with oxalic acid; their double alkaline salts are, however, readily soluble, and by adding ammonium oxalate in excess the reaction progresses with ease and without the formation of a precipitate. The carbonic acid separated on the positive pole by the decomposition of the ammonium oxalate combines with the ammonium forming ammonium carbonate. In general the process is conducted by converting the neutral chlorides and sulphates of the metals into double oxalates by the addition of a large excess of ammonium oxalate, heating the solution and exposing it to the action of a galvanic current whereby the metals deposit quickly and in a compact form on the negative electrode.

[1] "Quantitat. Analyse auf elektrolytischem, Wege." Aachen, 1885.

The determination of copper, as is well known, can be very easily effected directly from the acid solution. According to Classen's method the solution (concentrated, if necessary, by evaporation)

Fig. 68.

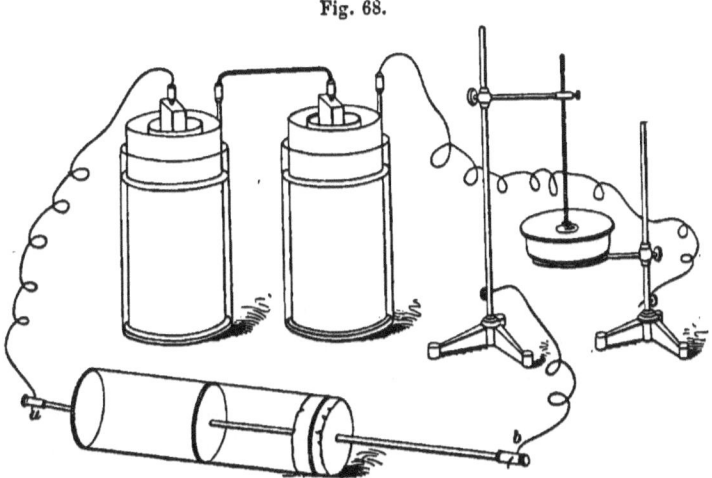

is heated to boiling, and, after adding 3 to 4 grammes of solid ammonium oxalate, subjected to the action of the electric current as soon as everything is dissolved. The copper separates quickly and easily, provided the current is not too weak. With a strength of current corresponding to 300 c.c. of oxyhydrogen gas per hour, 0.15 gramme of metallic copper can be separated in 25 minutes. As the negative electrode, Classen uses a platinum dish, and for the positive electrode a disk 4 to 5 centimeters in diameter of moderately thick platinum sheet (Fig. 69), which is secured by a screw to a medium thick platinum wire. To prevent loss, the platinum dish is covered with a watch crystal perforated in the centre. The entire arrangement of the apparatus is shown in Fig. 68. The glass cylinder shown in front is the resistance arrangement recommended by Classen for the reduction of more considerable strengths of current. It consists of a rod b with

Fig. 69.

a zinc pole, which can be moved to and from the zinc pole *a* until the current reaches the desired strength. The zinc plates (poles) must be amalgamated with mercury and hydrochloric acid, and the contacts at *a*, *b*, kept clean.

As recently precipitated sulphide.[1]—When copper is met with in the course of analysis as *recently* precipitated sulphide, it should be dissolved in potassium cyanide, and the warm solution (70° C.), after adding excess of ammonium carbonate, electrolytically decomposed. The deposit is exceedingly rapid and perfect.

3. *Determination of the copper in the form of cuprous sulphide.*[2]—This demands the absence of metals precipitable from acid solutions with sulphuretted hydrogen, and the metallic sulphides of which are not volatilized in a heated current of hydrogen (*silver, lead, bismuth, cadmium, antimony, tin*). During the dissolving process, *lead* may be separated by sulphuric acid, *antimony* and *tin* by nitric acid; *mercury* and *arsenic sulphide* are volatile. In case a considerable percentage of *nickel* is present, nickel sulphide will be precipitated with sulphuretted hydrogen, which can only be prevented by using a large excess of acid.

One to 5 grammes of the assay sample are decomposed with nitric acid or *aqua regia*. The residue of sulphur is removed by hydrochloric acid and potassium chlorate. The liquid is then diluted with water and the silver precipitated with common salt. The solution is then filtered, heated to about 80° to 100° C. (176° to 212° F.), and saturated with sulphuretted hydrogen, the precipitate filtered, after standing about one hour, and quickly washed with hot water. The filter is dried between blotting paper and next quickly in a hot sand-bath. The precipitate is detached from it and the filter with the addition of some sulphur incinerated upon the cover of a tarred porcelain evaporating dish which contains the copper sulphide. This, after an addition of sulphur, is strongly heated for about half an hour, while a current of hydrogen or illuminating gas is conducted to it through an opening in the cover, or through a perforated mica plate, which has been placed upon the crucible. Cu_2S with 79.85 per cent. of copper

---

[1] Chem. News, 1886, p. 209.

[2] Bestimmung des Kupfers in kupferhaltigen Kiesen, Abbränden und ausgelaugten Abbränden in Fresenius's Ztschr. xvi. 335.

will be formed. (If a current of carbonic acid is used, too much Cu_2S is obtained, in consequence of the less complete decomposition of the cupric sulphide.)

Of *nickel coins*, 0.5 grammes are dissolved in nitric acid and evaporated to dryness with 1 cubic centimeter of sulphuric acid, the residue is dissolved in 200 cubic centimeters of boiling water, and the solution precipitated with sulphuretted hydrogen, etc.

Apparatus[1] *for igniting in a current of hydrogen* (Fig. 70).— *a*, a vessel with water and zinc; *b*, funnel for pouring in sulphuric acid; *c*, discharge-pipe for the gas; *d*, drying tube for the calcium

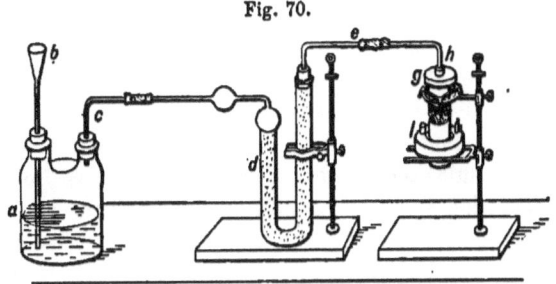

Fig. 70.

chloride; *e*, gas discharge-pipe; *g*, porcelain crucible with perforated cover (*Rose's* crucible); *l*, lamp. Fig. 71—*a*, gas generating

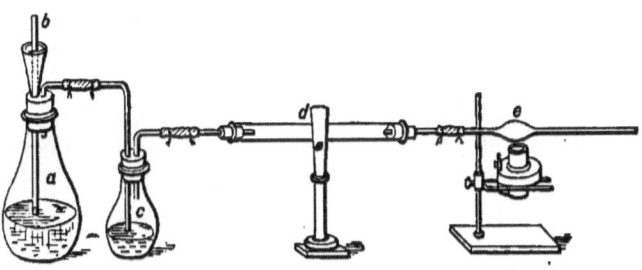

Fig. 71.

flask with funnel tube *b*; *c*, wash vessel with concentrated sulphuric acid; *d*, calcium chloride tube; *e*, bulb-tube for the

[1] Sicherheitsvorrichtung für Wasserstoffentwicklungsapparate in Fresenius's Zeitschr. xvi. 93. Poggendorf's Ann. 1876, Heft 10.

reception of the substance. It is advisable to pass the gas first through a solution of potassium permanganate, and next a solution of sodium hydrate to free the hydrogen from hydrocarbons, etc.

4. *Assay with sulphocyanide.*[1]—This method allows of the presence of nickle, zinc, iron, and arsenic; 0.5 to 1 gramme, or more, of the assay sample is dissolved in nitric acid and evaporated to dryness with sulphuric acid until the free sulphuric acid is *completely* expelled.

The dry mass is now dissolved with a little water and a large quantity of sulphuric acid added to the cold solution. Potassium sulphocyanide is then generally added until white copper sub-sulphocyanide is precipitated (if too much potassium sulphocyanide is added at one time, the black sulphocyanide is formed which is only gradually reduced to sub-sulphocyanide by sulphurous acid). A sufficient quantity of the sulphocyanide salt has been added when the fluid commences to assume a reddish-brown color. This coloration is caused by the presence of iron, but the fluid will, in a short time, become entirely colorless. The precipitate is allowed to settle and is then decanted with cold water until a solution of silver nitrate is not rendered turbid by the wash water. It is then filtered upon a previously weighed filter and dried for 12 hours at 105° to 110° C. (221° to 230° F.), and weighed (100 $Cu_2S_2Cy_2 = 52.2$ Cu). In order to control the correctness of the result, the filter is incinerated by itself, the copper sulphocyanide is heated in a porcelain crucible to decompose the sulphocyanide, and is then ignited with sulphur in a current of hydrogen, and treated as on page 116.

Nickel coins.—1 gramme is dissolved in 10 cubic centimeters of nitric acid of 1.18 specific gravity, and evaporated to dryness with 1 cubic centimeter of concentrated sulphuric acid. The dry mass is dissolved in a little water, and to the solution are added 50 cubic centimeters sulphurous acid solution, and 2 grammes of potassium sulphocyanide. After having stood for 12 hours it is filtered, etc. For the determination of *nickel* in the filtrate see " NICKEL."

[1] Fresenius's Ztschr. xvii. 55.

Copper alloyed with tin (bronze).—1 gramme of the alloy is dissolved in a mixture of 6 cubic centimeters of concentrated nitric acid of 1.5 specific gravity with the addition of 3 cubic centimeters of water. When the action of the acid has ceased the contents of the dish is heated for a time, and next treated with 40 cubic centimeters of boiling water. It is then allowed to settle. The sediment, consisting of stannic oxide, containing 78.7 per cent. of tin (free from copper) is washed and weighed. If other proportions of acid are used the stannic oxide will be cupriferous. If black specks, which will indicate nickel sulphide, show themselves upon solution, some hydrochloric should be added to the nitric acid. *Arsenic,* when copper is precipitated with potassium sulphocyanide, remains in the filtrate. Some sulphurous acid is added, then boiled away, and the arsenic precipitated with sulphuretted hydrogen. *Iron* remains with the nickel, and may be separated by twice dissolving and precipitating with ammonia. *Sulphur* is determined by barium chloride.

B. Volumetric assays.[1]—A large number of precipitating methods have been recommended; according to *Pelouze,* with sodium sulphide; according to *Galetti,* by means of potassium ferrocyanide; according to *Schwarz,* with potassium xanthate; and, according to *Vollhard,* with potassium sulphocyanide. Of reducing methods, the following are recommended, viz: according to *de Haen,* with sodium hyposulphite and potassium iodide; according to *Weil,* with protochloride of tin; according to *Parkes,* with potassium cyanide; according to *Schwarz,* with ferric chloride and potassium permanganate, and others. The method with *potassium cyanide* is largely used in smelting works, for the reason that it is easily executed, and affords a sharp final reaction. *Weil's* method with protochloride of tin is also simple, and both methods are particularly well adapted for the further test of the impure, black, precipitated copper obtained in the Swedish assay (pp. 107, 108).

[1] Mohr, Lehrb. der Titrirmethode, 1874, pp. 214, 262, 319, 473, 665; B. u. h. Ztg. 1871, p. 222 (Pelouze); 1869, p. 19; 1877, p. 207 (Schwarz); 1870, p. 447; 1872, p. 26; Oestr. Zeitschr. 1871, No. xvii.; Oestr. Jahrb. der Bergakademien u. s. w. Bd. xx. p. 133 (Weil); Fresenius's Ztschr. xvii. 53 (Rhodanprobe).

1. *Parkes's assay with potassium cyanide.*[1]—This method is based upon the reduction of an ammoniacal solution of copper by treatment with potassium cyanide. The blue color of the solution disappears and colorless cyanide of copper and ammonium is formed ($4CuN_2O_6 + 8KCy + 2Am_2O = 4\ CuCy + 2AmCy + 8KNO_3 + 2Am,CyO$). Foreign metals which give a solution of peculiar color with ammonia (*nickel, cobalt*) or which form a colorous solution (*zinc, manganese*) exert a disturbing effect, as do also such as produce a precipitate which it is difficult to free from hydrated oxide of copper in washing (*ferric oxide, alumina*).[2] *Lead, silver, tin,* and *antimony* may be removed in advance. The presence of *arsenic*[3] is harmless, provided no ferric oxide is present at the same time, since ferric arsenate is soluble in ammonia. This method is much used in practice, and gives sufficiently accurate results, provided the same conditions are always observed (*i. e.*, uniform dilution, the use of the same quantities of acids and ammonia, etc.).

The following quantities of ore are dissolved:—

10 grammes of ore,	when it contains	0.1	to	1 per cent. Cu.
5 " "	"	"	"	1	to 5	"	"
2.5 " "	"	"	"	5	to 30	"	"
1 to 0.5 " "	"	"	"	30	to 80	"	"

The assay sample is dissolved in nitric acid, and, if it contains lead, is evaporated nearly to dryness with sulphuric acid. It is then diluted with water to the bulk of ¼ liter and (if the above-mentioned injurious foreign metals are present) precipitated from

---

[1] B. u. h. Ztg. 1867, p. 102; 1869, p. 18; 1871, p. 222; 1872, pp. 207, 347, 419. Oestr. Jahrb. xx. 133.

[[2] According to Messrs. Torrey and Eaton of New York, Engineering and Mining Journal, vol. 39, pp. 317, 386, 441. *Three* per cent. of zinc causes an *apparent* increase of ½ per cent. copper. From 5 to 15 per cent. arsenic does no harm, 25 per cent. silver caused an error of $\frac{1}{10}$ per cent. copper, 30 per cent. iron caused a loss of 3.71 per cent. copper. From 5 to 40 per cent. lead had no injurious effect. 20 per cent. bismuth made the same error as silver. —G.]

[[3] It seems doubtful if the presence of arsenic is harmless, as sometimes when it is present the end of the titration is obscured by a yellowish green tint which replaces the blue color of the ammoniacal solution. Trans. Am. Inst. M. E., vol. 9, p. 316.—G.]

the acid solution with sulphuretted hydrogen. The precipitate of copper sulphide is filtered, and washed on the filter with a solution of sulphuretted hydrogen. It is then rinsed off from the filter into the same flask, which was used for the solution, and is heated with 10 cubic centimeters of concentrated nitric acid of 1.41 specific gravity until the sulphur is seen to separate in the form of globules. Ammonia is now added until precipitation takes place, then 20 cubic centimeters of ammonium carbonate solution (1 : 10) are added. The quite clear solution is filtered off from the sulphur into a beaker glass. The sulphur remaining in the flask is treated with hydrochloric acid and potassium chlorate until it is entirely dissolved. It is then evaporated to dryness on the sand-bath, and the residue is treated as above with water, ammonia, and carbonate of ammonium, and filtered into the principal solution. The filter, from which the precipitate of copper sulphide was washed off, is spread out in a beaker glass. Water is poured over it, a few drops of nitric acid are added to it, and it is then boiled for a few minutes. Ammonia and carbonate of ammonia are added as above, and the solution is likewise filtered into the original solution, which is then diluted to the bulk of $\frac{1}{2}$ liter and should have only a faint odor of ammonia. 20 cubic centimeters of the liquid are taken and titrated with a solution of potassium cyanide until the color has so nearly disappeared that only a faint violet tint remains (best seen in a porcelain dish) and which entirely disappears in from 1 to 2 minutes (by titrating at 60° C. (140° F.) the percentage of copper found will be somewhat less, but more accurate).

The potassium cyanide is standardized as follows, and should be used in as fresh a condition as possible, viz: 5 grammes of potassium cyanide are dissolved in $\frac{1}{4}$ liter of water, and 1 gramme of electrolytic copper in nitric acid.[1]

The solution is supersaturated with ammonia and ammonium carbonate, and diluted as above to the bulk of 1 liter. 100 cubic centimeters of the solution will then contain 0.1 gramme of copper from which the potassium cyanide may be standardized.

---

[1] [This copper should invariably be tested for its purity.—G.]

In titrating an ammoniacal solution of copper with potassium cyanide the presence of 4.5 per cent. of zinc does not, according to Peters,[1] exert an injurious effect upon the results of the assay. In a quartz of copper ore containing only copper, zinc, and iron, the presence of 5 per cent. zinc causes a constant error of 0.22 per cent., which increases with the amount of zinc. The presence of arsenic and antimony up to one per cent. causes an error of 0.5 per cent., and if over one per cent. of these metals be present the assay becomes useless.

Steinbeck's modified method of assaying copper.[2]—In the copper works of the Ducktown district in Tennessee, 2 grammes of ore are digested with HCl until the evolution of gas has ceased; 5 c.c. of strong HNO_3 are then added, and after evaporation with H_2SO_4 to dryness, the residue is taken up with water containing H_2SO_4 and filtered into a beaker containing metallic zinc and some water. What remains undissolved is then washed upon the filter with hot water, the zinc is lifted from the copper with a pair of forceps, and the copper filtered off. The latter is then dissolved in a few drops of HNO_3, the solution made ammoniacal, and titrated with potassium cyanide. In the solution freed from copper the iron is titrated with potassium bichromate and the residue remaining undissolved, weighed. These three determinations generally suffice for working assays.

Copper precipitated with iron (p. 106), or with zinc (p. 108). Mansfeld copper schist (Kupferschiefer) may be examined by this method by dissolving 5 grammes of the precipitated copper (pp. 109, 110) in from 8 to 16 cubic centimeters of nitric acid of 1.2 specific gravity. The solution is gently heated, allowed to cool, and is then supersaturated with 10 cubic centimeters of a mixture of 1 volume ammonia and 2 volumes of water. It is titrated with potassium cyanide, 1 cubic centimeter of which represents 0.005 gramme of copper. It is advisable to preserve the solution of potassium cyanide in the dark, in a tightly stoppered flask of green glass.

2. *Fleitmann's method with ferric chloride.*[3]— Copper precipi-

---

[1] Engineering and Mining Journal, 1885, vol. 39.
[2] Bu. u. Httumsch, Ztg. 1886, p. 454.
[3] Ann. d. Chem. u. Pharm. xcviii. 141.

tated by zinc, as described on p. 108, is dissolved in a mixture of ferric chloride and hydrochloric acid. The solution should be made in a glass flask furnished with a rubber valve (Fig. 11, p. 36), and with the addition of a little sodium carbonate to expel the air. Protochloride of iron ($Cu + Fe_2Cl_6 = CuCl_2 + 2FeCl_2$) is formed, which is titrated with potassium permanganate until a pinkish coloration remains permanently.

The standard of the solution of potassium permanganate is fixed as follows: 0.2 to 0.3 gramme of piano wire, containing at an average 0.4 per cent. carbon, is placed in a flask provided with a mark at 200 cubic centimeters and dissolved in diluted sulphuric acid, the air being excluded. It is allowed to cool off, and is then diluted to 200 cubic centimeters with distilled water previously boiled. Solution of potassium permanganate (prepared as below) is then added, drop by drop, to 100 cubic centimeters of the solution of iron until the light rose-red coloration remains permanently. This operation is repeated with the remaining 100 cubic centimeters of iron solution. From these titrations the value of the permanganate may be calculated. The flask, during the operation, should be placed upon white paper, as this will aid in accurately recognizing the tint of the solution which indicates the end of the reaction. The solution of potassium permanganate is made by dissolving 4.5 grammes of potassium permanganate in 1 liter of water, when 1 cubic centimeter of the solution will correspond to 0.008 gramme of iron. It is advisable to have 1 cubic centimeter of the permanganate solution correspond to from 6 to 10 milligrammes of iron, as with this strength one drop of the solution will produce a perceptible tint. 1 equivalent Cu (31.50) = 2 equ. Fe (55.90). Instead of using dissolved iron, the strength of the permanganate solution may be determined with the double sulphate of iron and ammonia, containing 6 eq. H_2O, and representing 14.286 per cent. of metallic iron.

If the original solution of copper contains nitric acid, bismuth, or lead, it is precipitated with ammonia in excess and filtered. The copper is then precipitated with the aid of heat, by means of finely divided zinc, in ammoniacal solution.

3. *Method with sodium sulphide in an ammoniacal solution.*[1]—The solution must be heated to 60° to 80° C.—between these temperatures an oxysulphide is precipitated, while at ordinary temperatures CuS falls, which does not readily subside. If the precipitation is conducted at a higher temperature, another oxysulphide containing more oxygen is precipitated, and copper may remain in solution without coloring the fluid. The solution is placed over a lamp in a flask, and a thermometer inserted in the liquid; when the temperature reaches 75° C. the lamp is withdrawn and sodium sulphide added from a burette until the blue color has completely disappeared. *In an acid solution.*—The solution is diluted with hot water to about 200 c.c. in a stoppered flask, acidified with HCl, and the sodium sulphide run in, replacing the stopper and shaking after each addition. The copper sulphide produced separates rapidly, and leaves a clear solution above it. The sodium sulphide is added until no further precipitate is produced.

Sodium sulphide decomposes slowly, and its strength is determined by means of a solution of copper sulphate containing 39.356 grammes. $CuSO_4 + 5H_2O$ per liter. The determination of the strength of the sodium sulphide must be made as it is to be used; if the copper is to be determined in acid solution, the copper sulphate taken for determination must be acidified. If, on the contrary, the estimation is to be made in an ammoniacal solution, the copper taken is made strongly ammoniacal.

For the better recognition of the final reaction in the determination of copper with sodium sulphide, P. Casamajor[2] recommends the precipitation of the copper from an alkaline solution containing tartaric acid. Dissolve 173 grammes of Rochelle salt (sodium tartrate) in water, add 480 c.c. of caustic soda of specific gravity 1.14, dilute the mixture to form 1 liter of solution, and add a slight excess of the copper solution to it. The deep blue liquid is then heated in a porcelain dish to boiling, and titrated with sodium sulphide.

4. *Method with protochloride of tin.*—Weil has perfected this method by first removing after the solution of the ore in HNO_3, or $HCl + HNO_3$, the greater part of the acid from the solution by evaporating nearly to dryness, diluting to 250 c.c. From 10 to 25 c.c. of the diluted solution are taken out, and after adding an excess of HCl, again evaporated until the vapors no longer give a blue

---

[1] Volumetric Analysis, Hart. New York, 1878, pp. 152–153.
[2] Chem. News, vol. 45, p. 147. 1882.

color to paper saturated with potassium iodide The solution is finally mixed with an equal volume of pure HCl and titrated. The color reaction is extremely sharp and every possible loss from sputtering avoided by evaporation with H_2SO_4.

5. *Fresenius recommends the following method of Haen's.*[1]— Dissolve the compound of copper in H_2SO_4, reduce to a neutral solution. Dilute solution, in a measuring flask, to a definite volume; 100 c.c. should contain from 1 to 2 grms. of Cu. Introduce about 10 c.c of KI solution (1 to 10) into a stoppered bottle, add 10 c.c. of the Cu solution, mix, allow to stand 10 minutes, and then determine the separated iodine, either with sulphurous acid and iodine, or with sodium thiosulphate. The Cu solution must be free from ferric salts and other bodies which decompose KI, also free HNO_3, and free HCl; and the solution must not be allowed to stand too long before titration. With strict attention to these rules the results are accurate.

C. Colorimetric methods.[2]

1. *Heine's assay for poor ores and products (slags, etc.).* Standard solutions, containing, respectively, 0.025, 0.02, 0.015, 0.01, and 0.005 gramme to every 120 cubic centimeters of liquid, are prepared either by dissolving a known weight of copper and diluting this as much as may be necessary to obtain the separate gradations of copper, mentioned above, in equal volumes of the liquid; or by directly weighing off the above-mentioned quantities of electrolytic copper and dissolving each with a few drops of nitric acid in a graduated vessel, adding ammonia in excess and diluting the clear, blue fluid with distilled water to 120 cubic centimeters. The solutions are then introduced into oblong sample-glasses, having exactly the same form and sectional area, about 50 millimeters long, 50 millimeters wide, and 110 millimeters high. They should be closed with ground-glass stoppers, and marked on the outside with figures, representing the strength of the solutions; in this case, 0.025, 0.02, 0.015, 0.01, and 0.005 gramme Cu in 120 cubic centimeters. The solution to be tested is prepared in the following manner: 5 grammes, or more, of ore

---

[1] Quantitative Chemical Analysis, 1881, p. 317.
[2] Dingler, 1857, p. 436. Berggeist, 1867, No. 27. Percy's Metallurgy of Copper, Zinc, and Brass, pp. 487, 488.

are dissolved, from which an ammoniacal copper solution is prepared in the same manner as in the assay with potassium cyanide (p. 120). The volume of the solution is measured to cubic centimeters, and it is then poured into an empty standard-glass. The intensity of the color of the solution is compared with that of the standard solutions, the glasses being held against a sheet of white paper, and it is observed with which of them it corresponds in intensity of color. The percentage of copper is then calculated, due consideration being given to its volume.

Suppose 5 grammes of ore had been used, and 300 cubic centimeters of solution had been obtained, the color of which corresponded to the standard liquid containing 0.02 gramme in 120 cubic centimeters; the copper in 5 grammes would then amount to 1 per cent. $(120:0.02 = 300:x)$.

If the assay solution is darker than the darkest standard solution, it is diluted to a known volume with water, until it corresponds with one of the standards. Should the reverse be the case, the solution under examination is evaporated to a known volume. This method is less accurate where high percentages are involved, as errors in observation will be multiplied. It is best to determine in this way from 1 to 2 per cent. of copper at the utmost.

In precipitating copper solutions containing iron, by means of ammonia, the *ferric hydrate*[1] always retains some copper. This error is equalized by adding, in preparing the standard solutions, a quantity of iron corresponding about to the percentage of iron in the assay sample (assay of slag in Swansea.) Organic substances, in presence of nitric acid, produce with ammonia a greenish tint, which exerts a disturbing effect in comparing the color of the copper solution (therefore, copper-schist should be ignited and filters incinerated before they are brought in contact with nitric acid). Ammoniacal solutions which may become turbid (for instance, if diluted with ordinary water containing lead) are allowed to become clear, and are filtered once more.

2. *Jaquelin-Hubert's method for considerable percentages of copper.*—Only *one* normal or standard solution of known strength

---

[1] B. u. h. Ztg. 1869, p. 302.

is used. This is compared with the solution under examination in a graduated glass tube (calibrated). If richer than the standard, the assay solution is diluted to correspond with the former; and if poorer, the standard solution is correspondingly diluted, until an equal intensity of color has been obtained. From the relative volumes the percentage of copper is then readily calculated. The solution contains 0.5 gramme of copper in 1000 cubic centimeters. As errors of observation may easily occur, a gravimetric or volumetric assay is frequently preferred for larger percentages of copper.

Determination of arsenic in copper.—According to Pattinson,[1] the process is as follows: The copper is dissolved in HNO_3, and sufficient caustic soda added to a little more than neutralize the free acid present, the arsenic is precipitated as copper arsenate. This washed precipitate is dissolved in HNO_3, a slight excess of ammonia added, and the arsenic is determined by magnesia mixture.

Sexton prefers to precipitate the arsenic from the nitric acid solution (by adding sodium acetate and boiling) as basic acetate. The precipitate is dissolved in acid, reprecipitated by H_2S, oxidized to arsenic acid, and determined as the magnesia compound.

Determination of phosphorus in copper.[2]—If phosphorus is also to be determined in the copper, the magnesium salt is redissolved in HCl, the arsenic precipitated by H_2S, and the phosphoric acid in the filtrate again determined as ammonium-magnesium salt, whose weight deducted from that previously found gives as a balance the pure arsenate of magnesium.

Rynoso determines the phosphorus in phosphor copper by adding to the weighed sample at least ten times its weight of tin (*i. e.*, such a quantity as to more than equal the amount of phosphorus supposed to be present and converted by oxidation into phosphoric acid) and boiling with an excess of concentrated HNO_3 for about three hours. The tin in oxidizing absorbs all the phosphoric acid and when decomposition is complete, filter, and heat the phosphate of tin to prevent reduction by any carbonaceous material present. The ash of the filter is always black and must be completely incinerated by moistening it with HNO_3, and igniting. It is also advisable to treat in the

---

[1] Newcastle-on-Tyne Chem. Soc. Feb. 1882. Polytechnisches Notizblatt, 1882, p. 237.
[2] Fortschitte im Probirwesen. Balling. Berlin, 1887, p. 96.

same manner the phosphate of tin. From the weight of the ignited phosphate of tin deduct the weight of the oxide of tin calculated from the weight of metallic tin used in the experiment. The remainder is phosphoric acid formed by oxidation of the phosphorus, from which the content of phosphorus is formed by multiplying by 0.436.

The method is readily performed, but requires the use of perfectly pure tin, as the slightest impurities give rise to serious errors.

A modification of this process by Reisig is found in H. Rose's "Handbuch der Analytischen Chemie," 6 Ausflage, p. 520.

III. SILVER.

30. PRINCIPAL ORES.

Native silver; *amalgam,* Ag and Hg, with 26.5 to 86 Ag; *antimonial silver* Ag and Sb, with 59 to 84 Ag; *silver telluride (hessite),* Ag_2Te, with 62.79 Ag; *silver glance,* Ag_2S, with 87.1 Ag; *brittle silver ore (stephanite),* Ag_4SbS_4, with 68.56 Ag; *ruby silver (pyrargyrite),* $Ag_3(Sb,As)S_3$, with 65.38 to 59.98 Ag; *miargyrite,* $AgSbS_2$, with 35.86 Ag; *polybasite,* $Ag(Cu,Fe,Zn)_9Sb(As)S_6$, with 64 to 75 Ag; *stromeyerite,* CuAgS, with 53 Ag; *horn silver,* AgCl, with 75.26 Ag; *iodyrtte (silver iodide)* AgI, with 46 Ag; *bromyrite (silver bromide)* AgBr, with 57.45 Ag.

31. ASSAYS FOR NON-ALLOYS.[1]

The fire-assay methods are based upon the principle of decomposing the silver ore by means of lead or lead oxide, collecting the silver thus set free with an excess of the lead, the slagging off of foreign substances by suitable fluxes, and the cupellation of the lead button to separate the silver. The collection of silver in a lead button is effected, according to the nature of the foreign admixtures, either in a scorifier (scorification assay) or in a crucible (crucible assay).

---

[1] Blossom, Gold and Silver Assays, in Am. Chemist, Jan. 1871, p. 250; Aaron, Pract. Treatise on Testing and Working Silver Ores, San Francisco, 1877. Percy's Metallurgy of Silver and Gold, vol. 1. John Murray, London, 1880, pp. 224-301.

The crucible assay permits of a large quantity of assay sample being used (which reduces the error from a loss of silver), and for this reason, it may be especially recommended for poor ores and such compounds as are free from antimony and arsenic (chloride, bromide, and iodide of silver and slags), and also for ores or sweepings of a very complex composition. The *scorification assay* is better adapted for ores containing sulphur, antimony, and arsenic, though likewise for other ores, so that this may be called an assay of general applicability; but, nevertheless, for the first-named ores, etc., the crucible assay is simpler and cheaper and the result is more quickly attained. In *America* the *crucible assay* is chiefly used, while the *scorification assay* is preferred in *Germany*, although neither possesses any material advantage over the other.[1]

Wet assays are less commonly used.

I. *Fire Assays.*

A. *Collecting the silver with lead.*

1. *Scorification assay.*—This consists of an oxidizing fusion of the ores with lead, which becomes oxidized and yields up oxygen to the metallic sulphides, arsenides, and antimonides, forms a slag with the oxides thus formed and with the earths which may be present. The slagging off of basic earths is promoted by an addition of borax glass. The following points must be taken into consideration in preparing the charge.

a. *The quantity of lead* to be used will depend on whether the metallic sulphides, arsenides, and antimonides are easily or with difficulty decomposed by lead oxide (p. 76), or whether they are entirely absent. Either granulated lead free from silver, which is measured (p. 29), is used, or argentiferous lead, whose percentage of silver is deducted from the assay by placing the silver button obtained from a corresponding quantity in the balance pan with the weights in weighing.

[1] B. u. h. Ztg. 1867, p. 102; 1874, p. 68; 1877, p. 232.

[The scorification assay is generally preferred for rich ores, while the crucible assay is perhaps better adapted to the poorer ones.—G.]

Lead sulphide is the easiest to decompose, next the *sulphides of iron and zinc*, and then *copper sulphide*. The most difficult are the *sulphurized and arsenized nickel and cobalt ores.*

b. *The quantity of borax* depends on the degree of infusibility of the gangue (silicic acid and aluminous substances require but little, and lime and magnesia, much borax) and of the metallic oxides which are formed (ferric oxide, zinc oxide, stannic oxide, and nickel and cobalt ores require much, oxides of copper, bismuth, etc., but little).

In all cases, but *little* borax should be taken at first, to prevent the entire surface of the charge from being covered, as this would exclude the air. If *more borax* is necessary, it is added before the final heating. Much antimony and zinc oxide cause the cupels to become full of cracks.

c. *The number of samples* to be taken to control the accuracy of results will vary according to the richness of the ores, their want of uniformity, etc., from 2 to 10 or more.

d. If the ore contains less than 1 per cent. of silver, 5 grammes or about $\frac{1}{6}$ A. T.[1] are taken for a charge, if more than 1 per cent., 2.5 grammes (from $\frac{1}{10}$ to $\frac{1}{12}$ A. T.), and of very rich ores, 1 to 0.5 gramme or about $\frac{1}{30}$ to $\frac{1}{60}$ A. T. In assaying very poor ores, the lead buttons obtained by scorification are concentrated by further scorification.

The following table gives some examples of various *charges:*—

---

[1] Assay ton = 29.166 grammes.

SILVER—ASSAYS FOR NON-ALLOYS.

Argentiferous substances.	Test lead: times the quantity of substance.	Borax glass: per cent.	Remarks.
Amalgamation residues	12 to 15	to 15	Two assay samples, each 5 grammes (or say $\frac{1}{8}$ A. T.) are scorified, and the two buttons obtained are cupelled.
Antimonial ores . . .	16	200	Assay sample 1.25 grammes, about $\frac{1}{25}$ A. T.
Antimonial silver . . .	32	300	The same.
German-silver, or China silver	20 to 24	to 40	2.5 grammes ($\frac{1}{10}$ to $\frac{1}{12}$ A. T.).
Arsenical ores . . .	to 16	to 50	Require a high temperature in scorifying.
Galena, pure	6	0 to 15	
" siliceous . . .	} to 11	20 to 30	
" zinciferous . .			
Lead speiss	10 to 20	15 to 25	2.5 grammes ($\frac{1}{10}$ to $\frac{1}{12}$ A. T.) are used for the charge. The buttons obtained from two assays are again fused, with a little borax at first, then with addition of more.
Lead matt	9 to 20	12 to 25	
" nickeliferous . .	11 to 14		
Bronze	20 to 24	20 to 25	2.5 grammes (from $\frac{1}{10}$ to $\frac{1}{12}$ A. T.) are used and repeatedly scorified.
Darrkupfer[1] . . .	18 to 20	10	Charge same as above.
Milling silver ores—			
common . . .	12 to 15	to 15	
basic	8	25 to 50	
acid	8	0 to 20	
siliceous . . .	12 to 14	10 to 15	
Iron, cast-iron . . .	8 to 12	2 to 3, 1 glass	The iron is first oxidized in the muffle by admitting air, or by means of nitric acid.
Fluthafter[2]	12 to 15	15	Several assays, as many as 30, are made, and the lead buttons obtained are scorified into one, and this is cupelled.
Refined copper . . .	18 to 20	1 to 5	Charge 2.5 grammes ($\frac{1}{10}$ to $\frac{1}{12}$ A. T.).
Sweepings, argentiferous and auriferous	8 to 9	0 to 20	See p. 20.
Hearth bottoms . . .	8	10 to 20	
Gun-metal	20 to 24	20 to 25	Charge 2.5 grammes ($\frac{1}{10}$ to $\frac{1}{12}$ A. T.).
Kiehnstöcke[3] . . .	18 to 20	10	The same.
Cobaltiferous ores . .	20	15 to 20	
Cupriferous ores . . .	10 to 20	10 to 15	
Copper matt	12 to 15	10 to 15	

[1 Copper reduced by the liquation or drying process.—G.]
[2 Ore from *placer* deposits (wash ore).—G.]
[3 Residual copper from liquation process.—G.]

Argentiferous substances.	Test lead: times the quantity of substance.	Borax glass: per cent.	Remarks.
Brass	20 to 24	15 to 20	The same.
Nickeliferous ores . .	20	15 to 20	The same.
Furnace deposits . . .	12 to 14	10 to 15	
Raw matt	10 to 12	to 30	Only a small quantity of borax at first.
Slags	12 to 15	10 to 15	The same as fluthafter.
Black copper	18 to 20	10 to 15	Charge 2.5 grammes ($\frac{1}{10}$ to $\frac{1}{12}$ [A. T.]).
Residue from the amalgamation of black copper and extraction of silver	8 to 10	1 to 10	The same.
Zinciferous ores . . .	10 to 16	15 to 25	Fusion at a high temperature.
Argentiferous zinc . .	16	16	Charge 1.25 grammes (about $\frac{1}{25}$ A. T.) of zinc oxide with 16 parts of lead and 16 parts of borax.
Stanniferous ores . .	20 to 30	15 to 25	They are scorified several times
Argentiferous tin . .	16	16	Charge 1.25 grammes (about $\frac{1}{25}$ A. T.) of stannic oxide, with 16 parts of lead and 4 parts borax glass.

The quantity of test lead required according to the foregoing table is measured or weighed off, and divided approximately into two parts. The accurately weighed assay sample is mixed with one-half of the lead in the bottom of the scorifier and the mixture is covered over with the remaining part of the lead, and finally the borax is added. The charged scorifier is placed in a strongly heated muffle, the mouth of which is closed, and a strong draft kept up for the purpose. The lead will soon commence to fuse, and in sinking down absorb silver from the ore. This, on rising to the surface, is roasted off, and is strongly oxidized by the lead oxide formed at the same time. During the oxidation slag is formed from the edges of the scorifier, by the combination of another part of the lead oxide with the metallic oxides and earths that are present, and with the borax. The time required for this first heating (roasting and fusing) is from 25 to 30 minutes when a completely fused ring of slag, without adherence to the edge of the scorifier, will show itself. (Refractory ores, such, for instance, as contain zinc, cobalt, and nickel, or which contain considerable lime, require the strongest heat, and, should they not completely fuse even then, a sufficiently large addition of borax must be made before the final heating.) The second period is that of the "*scorification*."[1] The fire is checked, and the mouth of the muffle is opened, until, by the continued oxidation of the lead and foreign metallic compounds, the entire surface of the lead is covered with slag. This will require from 20 to 30 minutes. The mouth of the muffle is then closed, the heat raised, and a final heating of 10 to 15 minutes is given to render the slag completely fluid. The scorifier is now taken from the furnace, and allowed to cool in the scorifier or poured off. After cooling off, the lead button is carefully freed from slag and *hammered* into the form of a cube, with truncated edges and corners. The time required for the entire operation will be from $\frac{3}{4}$ to $1\frac{1}{4}$ hours, according to the degree of fusibility of the ores. When the ores are very poor, a number of the lead buttons which have been obtained are placed on a scorifier, either with or without borax, and scorified as indicated above. If necessary the *concentration* is repeated until finally *one button* containing the entire percentage of silver is obtained. A second scorification is also advisable, in case the *button be too large*, or when it contains much *antimony, arsenic*, or *copper*. A percentage of *nickel* will exert a disturbing influence in cupellation. A cupel will usually absorb about its own weight of litharge, from which the proper size of the button may be estimated.

Hungary:[2] Two samples each of 2.5 grammes (about $\frac{1}{12}$ A. T.) are each charged with 8 to 16 parts of granulated lead, in such a manner that one-third of it is mixed with the ore and some silver flux (2 parts of melted *Villach* litharge and 1 part of calcined borax), and covered with the remaining two-thirds. *Lower Harz:* 5 grammes (about $\frac{1}{6}$ A. T.) are mixed in the scorifier with 50 grammes of granulated lead and 0.75 to 1 gramme of borax, and covered with 0.5 gramme of borax. The cupels consist of 3 parts wood-ash and 1 part bone-ash.

With chloridized ores: A charge of 5 to 10 grammes ($\frac{1}{6}$ to $\frac{1}{3}$ A. T.) of the ore is scorified with ten times its weight of lead,

---

[1] Percy's Metallurgy of Silver and Gold, vol. i. pp. 242–244.—G.]

[2] B. u. h. Ztg., 1871, p. 254.

and cupelled to determine the percentage of silver. In a second sample the silver chloride is dissolved out by lixiviation with sodium hyposulphite, and the residue is scorified, etc., for the estimation of the unchloridized silver.

2. *Crucible assay.*[1]—In this method of assaying, the ore is fused with lead oxide (litharge, white lead), in order to decompose the metallic sulphides (p. 76), with fluxing agents (potash, borax), for slagging off oxide and earths, and with some carbonaceous substance (charcoal powder, flour, argol, black flux), for reducing the lead which then collects the silver. The quantity of the reducing agent will depend on the reducing power of the ore. It should be so gauged that the lead button produced shall not be too large in order to prevent a notable loss of silver in cupellation. Ores containing a large percentage of antimony, arsenic, and zinc should be previously roasted, to prevent the formation of oxysulphides, etc., which are difficult to decompose, and which carry silver along with them into the slag; 5 grammes (about $\frac{1}{6}$ A. T.) of the finely powdered ore are mixed in a crucible with 40 grammes of a flux consisting of 1.5 parts of litharge, 0.15 part of potassium carbonate, and 0.08 part of flour. This is covered with 25 grammes of litharge, and this in turn with about 4 grammes of borax. The crucible should have smooth sides, a diameter of 45 millimeters at the top, and of 30 millimeters at the bottom, an inside height of 145 millimeters, and outside of 165 millimeters. The charged crucible is then placed in the furnace upon a bed of glowing coke, which should cover the grate to a height of from 100 to 150 millimeters, and is then surrounded up to its rim with wood charcoal. The furnace is left open for the first quarter of an hour. After the coal has been replenished, the cover of the furnace is put on and the fusing is continued for a quarter of an hour longer. The crucible is then taken out, allowed to cool, and the lead button, which should weigh from 20 to 25 grammes, is freed from slag. (Chili.) The same quantity of litharge as used for the assay is fused at the same time, but without ore, with fluxing agents. The lead button is freed from slag and cupelled, to determine the percent-

[[1] Percy's Metallurgy of Silver and Gold, vol. i. pp. 244–248.—G.]

age of silver, which must be deducted from the assay-button. *Or*, 5 grammes (about $\frac{1}{8}$ A. T.) of ore are fused with 50 grammes of litharge, 2 grammes of argol, 12 grammes of sodium carbonate, with a covering of common salt, and the resulting lead button is cupelled. White lead is sufficiently free from silver.

Mexican charge:[1] 20 grammes (about $\frac{2}{3}$ A. T.) of ore, 66 grammes of litharge, the same quantity of sodium carbonate, and 3 grammes of charcoal powder are mixed in a crucible of the above dimensions and covered with 20 grammes of common salt. 40 assays are put in the furnace and fused, first, for a quarter of an hour, during which the furnace is left open. It is then closed, and the assays are fused for half an hour longer. 66 grammes of litharge are reduced, and the amount of silver found is deducted.—*Another charge* is as follows: 16 grammes (246.92 grains, about $\frac{1}{2}$ A. T.) of ore, 48 grammes of litharge, 60 grammes of sodium carbonate, 16 to 20 grammes of powdered charcoal, which is omitted when a large percentage of iron pyrites is present.—*Another charge* is: 2 grammes (about $\frac{1}{15}$ A. T.) of ore, 25 grammes of litharge, 10 grammes of sodium carbonate, and a covering of common salt.—*Spain:*[2] 5 grammes (77.16 grains, about $\frac{1}{8}$ A. T.) of ore are fused in a crucible with 20 grammes of litharge, borax, black flux, or potassium carbonate and flour, with a covering of common salt.—*English charge:* 10 grammes (154.32 grains, about $\frac{1}{3}$ A. T.) of ore, the same quantity of sodium carbonate, 50 grammes (771.61 grains) of litharge, and 1 to 1.5 grammes of argol, with a covering of 10 grammes of common salt and the same quantity of borax.—*Gold and silver sweepings:* 10 grammes of borax and the same quantity of argol are poured into a crucible with smooth sides, 75 millimeters in diameter on the top, and 110 millimeters high; upon this are placed 20 grammes of litharge. The sides of the crucible are moistened by gently breathing upon them, it is then inclined and turned in such manner that litharge adheres to the sides about $\frac{3}{4}$ the way up. 15 grammes of potassium carbonate and 25 grammes of sweepings are then added, and the entire mass is thoroughly mixed together with a broad spatula. It is then covered with 10 grammes of sodium carbonate, and upon this comes

---

[1] B. u. h. Ztg. 1874, p. 86.   [2] B. u. h. Ztg. 1868, p. 26.

a layer of common salt 12 millimeters thick, and finally 5 grammes of litharge are strewed around the sides of the crucible. The furnace is filled with pieces of gas-coke the size of a walnut, the coke, when it is in a glow, is stamped down, and from 6 to 8 crucibles are placed in the fire in such a manner that the edge of the crucible projects but little above the coke. The furnace is then closed, and the heat gradually increased until the charge ceases to swell up. The temperature is then quickly raised for from 15 to 20 minutes, until the charges have become thin fluid and flow bright and uniform. The operation is therefore finished in about half an hour. The crucibles are now allowed to cool, and the buttons weighing about 22 grammes are freed from slag and then cupelled. (In assays of large lots of ore five assays are made.)—*Other charges for sweepings:* 25 grammes (about $\frac{4}{5}$ A. T.) of sweepings, the same quantity of minium, 35 grammes of flux (prepared by mixing 600 parts of potassium carbonate, 200 parts borax, 100 parts glass-galls, 100 parts soda, 30 parts saltpetre, 30 parts powdered charcoal); or, 25 grammes of sweepings, 20 grammes of common salt, the same quantity of sodium carbonate and of potassium carbonate, 25 grammes of litharge, 10 grammes of argol, and the same quantity of powdered glass; or, 25 grammes of sweepings, 20 grammes of litharge, 25 grammes of flux (1 part potassium carbonate and 1 sodium carbonate), and a covering of common salt. The charge is kept in the furnace until quiet fusion, requiring about three-quarters of an hour (Braubach).—*Slags:* 10 grammes (about $\frac{1}{3}$ A. T.) of slag, 150 to 160 grammes of litharge, 2.5 grammes of quartz, and 0.25 gramme of charcoal powder. The charge is fused in a crucible for 20 minutes after the development of gas in the furnace has ceased, and the buttons of two charges are cupelled together (Pribram).—*Freiberg:* 7.5 grammes (about $\frac{1}{4}$ A. T.) of slag are mixed in a crucible with 11 to 15 grammes of potassium carbonate and flour, 19 to 30 grammes of granulated lead are strewed on top, and the charge is fused for 3 hours in the furnace.—*Refuse from stamping mills:* 10 grammes (about $\frac{1}{3}$ A. T.) of substance are mixed in a high crucible (Fig. 49, p. 64) with 60 to 120 grammes of potassium carbonate and flour; upon this is placed 50 to 100 per cent. of borax, next 10 to 15 grammes of granulated lead, and finally a covering of common salt. It is fused in the muffle-furnace for 1½ to 2 hours.

According to Gorz,[1] the results obtained in assaying sweepings and other refuse from the manufactures of silverware frequently show great variations.—This is due largely to the method adapted for the assay, and also to the heterogeneous character of the material, which sometimes contains several per cent. of mechanically mixed carbon. The gold and silver are usually present in the metallic state; only the refuse from smelting works contains chloride of silver.

The following observations are recorded by Gorz:—

1. In samples rich in metal and poor in carbon by the

Scorification assay.	Crucible assay.
8.296 silver	{ 8.270 { 8.000
11.390	{ 10.360 { 11.500
10.555	10.490
10.847	10.400
12.645	12.300
59.775	54.920, an impure chloride of silver
9.753	{ 9.680 { 9.750
9.366	9.190
10.630	9.790
0.028 gold	0.024

2. In samples poor in metal and rich in carbon by the

Scorification assay.	Crucible assay.
0.322 silver	0.315
{ 1.364 silver { 0.021 gold	{ 1.370 silver { 0.021 gold { 1.348 silver { 0.020 gold
{ 0.390 silver { 0.022 gold	{ 0.400 silver { 0.0207 gold { 0.386 silver { 0.023 gold
{ 0.316 silver { 0.075 gold	{ 0.276 silver { 0.079 gold
{ 0.338 silver { 0.336 gold	{ 0.345 silver { 0.340 gold
{ 0.164 silver { 0.063 gold	{ 0.132 silver { 0.068 gold

[1] Bg. u. Httumsch, Ztg. 1886, p. 441.

From the foregoing it will be observed that with rich sweep containing no carbon the highest results are obtained by the scorification assay, and with poor sweep containing carbon the crucible assay gives the best results. Hence, every substance rich in noble metals, containing no carbon and holding the metal chemically fixed, should be tested by the scorification assay, and every substance strongly contaminated with carbon by the crucible assay. Lead assaying crucibles are recommended for the latter on account of their narrowing towards the top; thus loss by sputtering is avoided as much as possible.

3. *Combined lead and silver assay.*—This method of assaying is used for oxidized lead products (litharge, skimmings, dross), and for galena with at least 30 to 40 per cent. of lead, and not over 0.12 per cent. of silver. Such galena, after the assay with potassium carbonate, as described on page 87, is fused to a lead button, which is then cupelled. This method is not satisfactory after the preliminary assay with iron (pp. 81, 82), as the iron sulphide will retain silver in varying quantities in the slag.

Litharge: 20 grammes (about ¾ A. T.) litharge, 15 grammes potassium carbonate and flour, and 5 to 6 per cent. of powdered charcoal, with a covering of common salt, are placed in a crucible and fused in the muffle-furnace. If necessary, several of the buttons are concentrated by scorification (p. 133) and cupelled. *Skimmings* and *dross* are charged in the same manner; but, should they be very impure, the lead buttons must be scorified with 4 to 8 times their weight of granulated lead before they are cupelled.

B. *Cupellation of the argentiferous lead* (*assaying by the cupel or cupellation*).[1]—The lead buttons obtained according to A Nos. 1 to 3 (pp. 129–138) are subjected to oxidizing fusion. During this operation the lead is first oxidized; the lead oxide yields up oxygen partly to the foreign metals, and partly combines with their oxides; and if they are not too refractory (as, for instance, ferric and stannic oxide, etc.), enters with them into the cupel.

If the lead contains much *antimony* and *zinc*, it is apt to cause cracks in the cupels; *copper* colors them green, and the percent-

[1] Percy's Metallurgy of Silver and Gold, vol. i. pp. 279–282.

age of copper may be quantitatively determined within certain narrow limits by the intensity of the color.

The cupels (Fig. 54, p. 66), first carefully wiped out with the fingers, then all extraneous matter blown out, are thoroughly heated (ignited) in the muffle. They are arranged in two rows, six in each row, in the front third of the muffle.[1] The lead buttons are now laid hold of with a pair of tongs (Fig. 59, p. 72) and gently deposited, first in the front and then in the back row of crucibles. The mouth of the muffle is closed, the fire is urged on, and the lead fused as quickly as possible. The lead will at first be covered with a dull, dark film. As soon as this disappears, and the lead shows a lustrous, fuming surface, the mouth of the muffle is opened (with the exception of a low piece of charcoal which is left in it) for the admission of air to oxidize the lead, and the temperature is lowered, by ceasing to stir the fire, to lessen the loss of silver. Cold scorifiers are placed in several rows above each other back of the cupels, and, if the argentiferous lead is very rich, the cooling-iron (Fig. 61, p. 72), which should be frequently cooled off in water, is moved to and fro closely over the cupels. The correct temperature is indicated by the rising lead fumes *whirling* over the assays, and not slowly creeping over them or rising straight up; by the cupels glowing dark brown, by small scales of crystallized litharge (plumose litharge, Federglätte) showing themselves on the inner edge, and by a bright but not too wide border of litharge upon the lead. If the temperature is *too low*, the fume creeps slowly over the cupels; these become too dark, a dark rim of litharge is formed, and the lead ceases to "drive." This is called a "*freezing*" of the assay. Frozen assays, if again brought to "driving" by a higher heat, and generally some addition of lead, cause a considerable loss of silver. By *too high* a temperature, the lead fumes rise up straight, the cupels glow too brightly, neither plumose litharge (Federglätte) nor a rim of litharge shows itself, and the loss of silver increases. At the *correct* temperature small beads of litharge float upon the surface of the lead button, the

---

[1] Hempel's gas-furnace with oxidizing apparatus in Fresenius's Ztschr. xvi. 454; xviii. 404.

heat to which it is subjected causing convection currents, which give the button a motion from below upwards ("driving"), and a convex surface from which the luminous beads and patches of litharge are continually thrown towards the sides, are there absorbed by the cupel. As soon as these patches upon the diminishing lead become larger towards the end, the scorifiers, which had been placed back of the cupels, are removed, the cooling with the cooling-iron is stopped, and the fire is urged on. Towards the end the last of the lead is absorbed, and the silver button presents itself colored with all the tints of the rainbow (*brightening, coruscation*), which gradually disappear, whereupon the button solidifies. (If the temperature has been too low, the surface of the button is dull and yellow, and the unabsorbed litharge forms lumps or scales about it; while otherwise, it is pure silver-white on top and bottom, and very lustrous.) If the bead is large the cupels are allowed to cool off slowly by drawing them to the front of the muffle, to prevent the buttons from "*sprouting*" or "*spitting*." They are then taken out upon a piece of sheet-iron, and the buttons (generally of 99.7 to 99.8 per cent. pure silver) are detached by means of a pair of pliers (p. 74), and brushed off with the button-brush (p. 74). Faultless buttons brightened at a sufficiently high temperature (smaller ones are round, larger ones hemispherical) should have a silvery lustre on the surface, be dull silver-white and crystalline on the bottom, and without rootlets. They are then weighed. The globule of silver obtained from the separately scorified and cupelled granulated lead, if not entirely free from silver, is placed in a scale-pan containing the weights; or, the silver percentage of the lead having been determined once for all, it may be deducted from the results obtained.

Assays of silver should agree very closely, and if properly conducted are of great accuracy. The results of duplicate assays should not differ from each other more than one-half ounce Troy per ton of two thousand pounds. Should a greater difference be found, an additional assay should be made.

Smaller losses of silver occur in scorifying (which for this reason should be continued as long as possible, so as to obtain small lead buttons requiring but a short time for "driving") than in cupelling, by the volatilization of silver, and by the silver oxide passing into the cupel with the lead oxide by which it has been oxidized (loss by cupellation, Kapellenzug). The loss increases with the temperature and the size of the button, and for this reason, with the time required for cupelling, as well as with the porosity of the cupels. The percentage of loss is considerably larger (2 to 4 per cent.) in smaller buttons (poorer ores) than in larger buttons (1 to 1¾ per cent.), but in the first case can generally not be determined by the balance. The smallest loss occurs in gas muffle-furnaces (Fig. 31, p. 50), which have no vent-holes in the muffle. In case the ore contains *tellurium*, the button spots at the moment of solidification, after brightening, and fine globules of the metal are thrown off and lost.

Assay of native silver ores.—The following method is recommended by Lowe.[1] When silver is chemically combined in the ore it is not difficult to determine its assay value, but in ores containing the silver in the metallic state, the assay can yield only approximate results. The following method obviates this difficulty to some extent. The ore is coarsely powdered and sampled down to a half-pound (226.704 grammes), then finely powdered and passed through a 100-mesh sieve. The sifted ore is divided into four parts and four assays made of it, each amount for each assay being taken as nearly as possible from the same relative parts of each quarter. The average of the four assays is taken as the result. The metallic particles or scales of silver which will not pass through the meshes of the sieve are mixed with assay lead, cupelled, and the weight of the resulting button reduced to its proper value added to the result obtained from the assay of the siftings.

Examination of lead ores for traces of silver.—According to Krutzwig,[2] 20 to 25 grammes of galena are fused with a mixture of tartar, soda, and borax in an iron crucible. The reduced lead dissolved out with HNO_3 free from chlorine, dilute the solution, and filter off any lead sulphate formed. Precipitate from the filtrate the iron, lead, and argentic peroxide of lead with excess of caustic soda. Separate the precipitate by decantation, wash thoroughly with an addition of ammonia, evaporate to dryness and take up

---

[1] Engineering and Mining Journal, Sep. 24, 1881, p. 203.
[2] Chemikerztg, 1882, p. 1206 ; Montanztg, 1882, p. 2.

with HNO_3. Remove the lead from this solution with H_2SO_4 and examine the filtrate for silver with HCl. Instead of HCl, caustic soda may be added which in the presence of silver produces an intensely yellow precipitate of argentic peroxide of lead.

II. Wet Assays.

Balling's volumetric assay.[1]—2 to 5 grammes of galena are fused in a porcelain crucible with 3 or 4 times its weight of a mixture of equal parts of saltpetre and soda. This is allowed to cool off; the contents of the crucible are lixiviated with water, heated in a porcelain dish, and filtered. The residue is decomposed with diluted nitric acid, and evaporated to dryness. The dry mass is then taken up with water acidulated with nitric acid, heated and filtered. Ferric sulphate, or iron-alum is added to the cooled-off filtrate, and it is then titrated with a $\frac{1}{10}$ normal solution of ammonium sulpho-cyanide, which is prepared by dissolving 0.7 to 0.75 gramme of the salt in 1 liter of water. The titer is made to correspond with a silver solution of known strength, in such a manner that 1 cubic centimeter of the ammonium sulpho-cyanide solution corresponds exactly to 1 cubic centimeter of silver solution. The latter is obtained by dissolving 1 gramme of chemically pure silver in nitric acid, and diluting it to a bulk of 1 liter. The presence of copper in small quantities is not injurious, and of that of lead is rather favorable, as the white precipitate of lead sulphate, which is formed after the ferric sulphate has been added, makes the recognition of the final reaction sharper. A large percentage of iron gives a brownish colored solution, which does not permit a distinct recognition of the final reaction. Larger quantities of copper must be previously removed, but cobalt and nickel, if the operator has some experience, admit of an easy recognition of the final reaction by a yellowish brown color. The assay requires about three hours, and may be especially recommended when no muffle-furnace is at hand, or if only one sample is to be assayed for which it would not be worth while to heat a furnace. The permissible error allowed

---

[1] Fresenius's Ztschr. xiii. 171; Oestr. Ztschr. 1879, No. 27.

in the smelting works at Pribram is 0.03 per cent. from ores carrying 0.30 to 0.60 per cent. of silver, and the differences obtained by this assay vary within narrower limits than those allowed for dry assays. The assays give equally good results for all degrees of richness, if the galena is pure and contains but little iron.

32. ASSAYS OF ALLOYS.

These are generally executed by the *wet method* only for silver containing copper (as, for instance, coins). The *dry method* is mostly used for other alloys, and they are either assayed by direct cupellation without scorification, with an addition of lead in case the sample does not already contain a sufficient quantity.

A. Dry assays.

1. *Lead bullion.*[1]—10 to 20 grammes, according to the percentage of silver in it, are directly worked off on the cupel; but if the lead is impure (slag lead, zinciferous lead), it must be previously scorified. Poor lead is slagged off on the scorifier in quantities of from 40 to 50 grammes, and the resulting buttons are concentrated into one button by scorifying them once or several times, and the collected button thus obtained is cupelled. (For instance, 10 assays of 50 grammes each of *Pattison's* granulated lead, which is very poor in silver, are concentrated to one button.) The cupels used for a charge of from 10 to 20 grammes of granulated lead, have an outer diameter of 49 millimeters on the top and 39 millimeters on the bottom, a clear width of 37 millimeters, a total height of 23 millimeters, and a depression of 17 millimeters.

2. *Silver amalgam.*—5 grammes of the assay sample are weighed off in a watch-glass, and gradually heated in the cupel for 1½ hours in a moderately heated muffle. After all the mercury is volatilized, 6 or 7 times the quantity of lead is added and the charge cupelled.

3. *Copper poor in silver* (black copper, refined copper).—2.5 grammes are scorified with 18 to 20 times the quantity of lead and cupelled, whereby the cupel will be colored dark green. Pure

---

[1] Abtreiben mit Sauerstoff in B. u. h. Ztg. 1868, p. 351.

or plumbiferous copper may also be immediately cupelled with 16 to 18 times the quantity of lead in one charge.

4. *Cupriferous silver or fine silver*[1] (coins, refined silver, etc.).— The sample is directly cupelled (*mint assay*) with a quantity of lead free from silver (simple weights of lead in the form of small sticks or round or half-round pieces are used, but not granulated lead), corresponding to the percentage of copper. Smaller and finer cupels (*mint cupels*) are used. They consist, like the French cupels, either of powdered bone-ash alone, or of bone-ash and lixiviated wood-ash, which makes them more porous. They are placed in a small muffle-furnace (*mint* or *fine assay furnace*, Fig. 28, p. 54), for the better regulation of the heat; or what is still better, in a gas-furnace (Fig. 31, p. 50). Where the percentage of silver is not known, the approximate percentage[2] is first determined by a preliminary assay (cupelling with 16 times the quantity of lead, or by the touchstone),[3] so that the proper quantity which experience has proven to cause the smallest loss of silver may be used. With silver coins, in which the silver percentage is known, the preliminary assay is unnecessary.

The numbers given in the following table may be taken as a guide:—

Degree of fineness of the alloy. Silver in thousandth parts.	Multiples of lead.
1000 to 950	4
950 " 900	6
900 " 850	8
800 " 750	12
750 " 650	14
600 " 0	16 to 17

[1] Probe über der Lampe in Fresenius's Ztschr. 1879, p. 82.

[2] Coins: The German Reichsmark, German Thaler, Austrian and South German Gulden, 900 thousandth parts Ag; English silver coins, 925; French small silver coins, 835; 5, 2, 1, ½, ¼ franc pieces, 900; German nickel coins, 75 Cu and 25 Ni; German copper coins, 96 Cu, 3 Sn, 1 Zn; American silver coins, 900; American nickel coins, 75 Cu and 25 Ni; American copper coins, 95 Cu, 3 Sn, 2 Ni; French small coins (5 cent.), 95.21 Cu, 3.18 Sn, 0.44 Zn, 0.25 Ni, 0.58 Pb, 0.06 Ag. Swiss coin (5 cent.), Cu 58.920, Zn 23.700; Ni 11.561, Ag 5.146, Pb 0.326, Co 0.286.

[3] Dingler, cxxiii. 366; Ann. de Chemie et Phys. 1875; Bay, Ind. u. Gew. Bl. 1869, p. 130; Kick, techn. Bl. 1873, p. 35; Fresenius's Ztschr. 1878, p. 142.

A sample of the alloy to be assayed, weighing 0.5 gramme, is hammered out and cut up into fine shreds. Generally two assays are made at one time. The samples from bars are taken from the upper and lower sides, and are obtained from pieces weighing about 2.5 grammes, which have been cut out from the lower and upper sides of the bar on opposite ends. The samples are wrapped up in cornets of fine letter-paper, and placed upon a small assay-plate. When lead granules are not at hand, a piece of stick lead is weighed out and placed in two thoroughly glowed-out cupels, standing in the centre of the strongly heated muffle-furnace (Fig. 30, p. 49). The Paris *mint cupels* have been especially recommended for the purpose. Their outer diameter is 26 millimeters on the top, and 22 millimeters on the bottom; clear width, 21 millimetres; total height, 14 millimeters, with a depression of 8 millimeters. The mouth of the muffle is closed with a coal, and as soon as the lead "drives" the cornet containing the sample is placed in the cupel, the mouth of the muffle again closed, and the assay allowed to "drive." The mouth of the muffle is now again opened, with the exception of a low piece of coal or small piece of iron, the cupels are drawn forward towards the mouth of the muffle and the temperature is lowered by partly closing the draught of the furnace, or also by cooling with a small cooling-iron (Fig. 61, p. 72), until a small ring of litharge and some plumose litharge (Federglätte) appear. The assays are now gradually pushed back, and the temperature is raised by opening the draught of the furnace, so that the assay may "brighten" sufficiently hot, during which the ring of litharge will disappear, but the plumose litharge (Federglätte) remain. The cupels are now drawn forward towards the mouth of the muffle, and allowed gradually to cool off to prevent "spitting." When sufficiently cool they are taken from the muffle, and the buttons are removed by means of a pair of pincers and brushed.[1] In successful assays the surface of the button is smooth, with a silvery lustre on the top, and a dull silver-white color on the bottom. If the operation has been conducted at too low a tem-

---

[1] [The button should be strongly squeezed with the pincers to detach any adhering particles of bone-ash or oxide.—G.]

perature, the surface is dull, and has a bluish tint, and the bottom is covered with a yellowish or greenish coating of lead oxide. If the temperature has been too high, the button is dull in some places, very lustrous in others, the surface is sunken, it is liable to spit, exhibits rootlets, adheres more strongly to the cupel, and is porous towards the bottom. The buttons are then weighed, and, in assays of top and bottom samples, either the average percentage or the lowest percentage is given. The loss from absorption by the cupel (Kapellenzug) is added.[1]

Bars with over 980 thousandths of silver show no difference, if the work has been carefully done. To 725 thousandths they show a difference of $\frac{1}{2}$ to 3 thousandths; from 720 to 710 thousandths, again, no difference, or only an infinitely small one (certain chemical combinations seem to be formed at this percentage); but the greatest differences occur at 200 to 400 thousandths fineness. Very considerable differences may occur if the bars or buttons have been badly fused. The silver button contains about 2 thousandths of lead.

[1] [The quantity of silver absorbed by the cupel is very small and need only be estimated when great exactness is required. The white portion of the cupel is broken off and rejected; the stained portion is powdered and mixed with charcoal in the proportion of 4 to 5 parts by weight to every 100 parts of litharge in the cupel, together with from 100 to 250 grains (6.5 to 16.2 grammes) of a mixture of equal parts fluor-spar, borax, and carbonate of soda. If the amount of litharge in the cupel is small, it is best to add from 200 to 300 grains (13.0 to 19.5 grammes) of litharge and about 15 grains (1 gramme) of charcoal to the mixture. Should the silver button thus obtained be large, the cupel used in this case should be examined for silver in a similar manner, and even a third or fourth cupel may be similarly treated.—G.]

Correction Table for the Absorption by the Cupel, determined by the French Commission on Coinage and Medals.

True quantity of silver.	Loss to be added, thousandths.	True quantity of silver.	Loss to be added, thousandths.
1000	1.03	500	4.68
975	1.76	475	4.50
950	2.50	450	4.31
925	3.25	425	4.13
900	4.00	400	3.95
875	4.07	375	3.61
850	4.15	350	3.27
825	4.22	325	2.94
800	4.30	300	2.60
775	4.41	275	2.58
750	4.52	250	2.56
725	4.64	225	2.55
700	4.75	200	2.53
675	4.73	175	2.12
650	4.71	150	1.70
625	4.70	125	1.29
600	4.68	100	0.88
575	4.68	75	0.66
550	4.68	50	0.44
525	4.68	25	0.22

In *Freiberg* somewhat different results have been obtained. With refined silver the loss by absorption by the cupel was found to be 0.0015 to 0.002, and in alloys of medium richness the loss was greater than that stated in the table; for instance, with 750 thousandths and 16 weights of lead, the loss was 5.55 thousandths, but with 11 weights of lead it accorded with the table, 4.52 thousandths. According to *Plattner*, fine silver with 5 times the quantity of lead frequently gives a loss up to 0.009, refined silver with 937 thousandths and 5 times the quantity of lead, 0.0042 to 0.0059; refined silver with 687 to 750 thousandths and 14 times the quantity of lead, 0.0073 to 0.0083.

B. *Wet assays.*[1]

They are used for *refined silver* and *coin alloys* of copper and silver. Compared with the fire assay, they allow of an accurate determination of the degree of richness to within 0.5, and even to 0.1 thousandths. They are more frequently *volumetric* and *gravimetric* assays.

[1] [Percy's Metallurgy of Silver and Gold, vol. i. p. 282.—G.]

1. Volumetric assays.

a. Gay-Lussac's method with sodium chloride.[1]—This method is based upon the precipitation of silver from a nitric acid solution by means of a standard solution of sodium chloride. For this purpose a *normal solution of common salt* is required, 100 cubic centimeters of which will precipitate 1 gramme of chemically pure silver. There is further required a *decinormal solution of common salt*, 10 times weaker than the first, and a *decinormal solution of silver*, consisting of a solution of silver in nitric acid, containing 1 milligramme of silver in 1 cubic centimeter of solution.

Preparation of the assay solution.—The degree of richness of the silver is approximately determined by a preliminary assay, the fine assay (p. 144) being generally chosen for the purpose. 4 to 6 thousandths parts the amount of silver found by this assay are added to the result. It is generally preferred to assume the degree of richness a few thousandths higher than is actually the case, and to base the calculation for the quantity of assay sample required upon this, as, to effect the more rapid settling of the silver chloride, it is preferable to add, during the titration, a few thousandths from the decinormal solution of salt than to be obliged to add from the decinormal solution of silver. The quantity of alloy containing 1 gramme of silver which is to be taken is then calculated (for instance, if the preliminary assay gives a percentage of 897 thousandths, then 1.115 grammes of alloy containing 1.000 gramme of silver should be taken, $1000 : 897 = x : 1000$). The sample in the form of shavings or granules is placed in a numbered flask, together with 6 to 7 cubic centimeters of nitric acid free from chlorine, and dissolved, either on a water or sand bath. The flasks in which the samples are dissolved are from 10 to 15 centimeters high, and 5 to $5\frac{1}{2}$ centimeters wide. If several assays are to be made, it is advisable to dip the flasks, which are arranged upon a stand (Fig. 72),

---

[1] Gay-Lussac, Vollst. Unterricht über das Verfahren, Silber auf nassem Wege zu probiren, Braunschweig, 1833; Mulder, Silberprobirmethode, Leipzig, 1859; Muspratt's Chem., Bd. vi. p. 477; Bolley, Handb. der techn.-chem. Untersuchung, 5 Aufl. pp. 52, 332; Dingler, cxci. 172. Trans. Am. Inst. Mining Engineers, vol. iv. p. 347; vol. x. p. 493.

SILVER—ASSAYS OF ALLOYS.   149

into hot water. (A black residue may be gold or sulphide of silver; should the latter be the case, some concentrated nitric acid is added and the fluid heated, or sulphuric acid used.) The nitrous acid formed is then driven out of the flask by means of a small bellows with curved extremity, and the contents of the flask are treated with the normal solution. But, as the influence of the temperature upon the volume of the normal solution of common salt must be taken into consideration, its titer must always be determined on the same day the assays are to be made, with 1 gramme of pure silver + 1 to 2 cubic centimeters decinormal solution of silver, in order to be

Fig. 72.

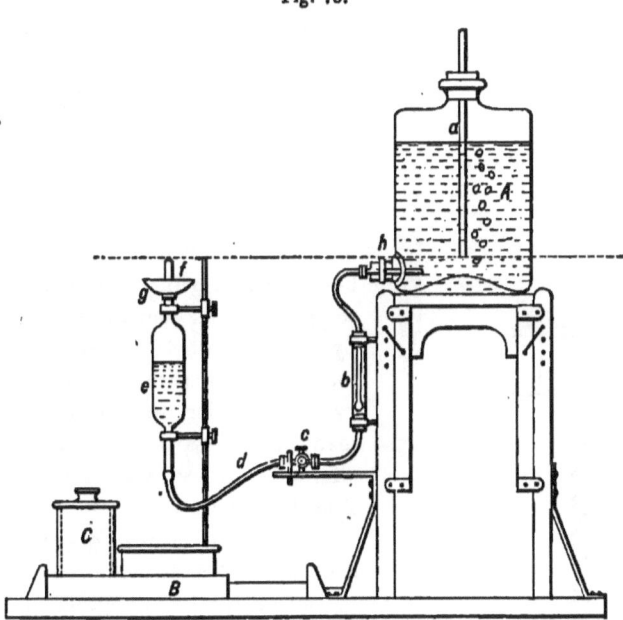

Fig. 73.

able, for the above-mentioned reason, to use decinormal solution of salt for the final titration. The silver solution is then titrated by placing the glass flask in the metal cylinder C (Fig. 73) standing upon the sliding carriage *B* (*Sire's* apparatus[1]). The glass cock *c* (a pinch-cock may be used instead) is then opened, and, accompanied by the admission of air through *a*, the normal solution of sodium chloride flows from the vessel *A* through *h*, the thermometer tube *b*, and the rubber tube *d*, into the burette *e*. It ascends in this, and a small quantity reaches the saucer *g* through the orifice *f*. The cock *c* is now closed (*h* and *e* may be also directly connected by a rubber tube provided with a clip), and the pipette *e*, which is now filled, will contain exactly 100 cubic centimeters of liquid. The index finger of the left hand is now placed upon the mouth *f* of the pipette, the rubber tube *d* is detached from the lower end of the pipette *e*, and the sliding carriage *B*, upon which stands the metal cylinder *C* containing the flask with the solution of silver, is pushed underneath the discharge orifice of the pipette. The index finger is now removed from *f*, and the 100 cubic centimeters of the solution of common salt are allowed to run into the flask, care being taken that the pipette does not rest on the neck of the flask. The sliding carriage is then pushed back, the flask is closed with its ground-glass stopper, and its contents are cleared by shaking, which is best done by inclosing it in a metal cylinder of proper size for the purpose. If many assays are to be made, it is advisable to use *Gay-Lussac's* or *Mulder's* agitator.

Fig. 74.

Gay-Lussac's apparatus (Fig. 74). —*A*, a stand with cylindrical compartments for the reception of the flasks which are provided with well-ground stoppers. The stand is suspended by the handle *e f* to the steel spring *c d*, and is connected

---

[1] B. u. h. Ztg. 1873, p. 189.

below with a spiral spring $a\ b$. The apparatus is shaken by means of the handle $e\ f$.[1]

One cubic centimeter of decinormal solution is now added to the entirely clear fluid, standing over the precipitate of silver chloride, by means of a graduated pipette contained in a flask (Fig. 75), whereby the point of the pipette should be placed against the neck of the flask containing the silver solution. If turbidity is produced, the silver solution is agitated until it is again clear, and 1 cubic centimeter of the decinormal solution of common salt again added, etc., until the last cubic centimeter which is added does not produce any turbidity. This last cubic centimeter is not taken into calculation, and only one-half of the one previously added. (For the reason stated on p. 148, it is more suitable to use decinormal solution of common salt than decinormal solution of silver for the final reaction).

Fig. 75.

Calculation.—Suppose the richness of the alloy was found according to the preliminary assay to be 897 thousandths, 1115 thousandths of the sample containing 1.000 gramme would have to be weighed off. 1000 cubic centimeters of the decinormal solution of salt = 1 gramme of silver. Now suppose 1002.5 cubic centimeters of decinormal solution of common salt had been used, 1000 parts of the alloy would therefore contain 899.1 parts of silver.

In case *mercury*[2] should be present, sodium acetate (0.5 gramme to 5 thousandths of mercury) is added, which will prevent the mercury from being precipitated by the sodium chloride solution; or the mercury is previously volatilized by heating the sample in a small graphite crucible in the muffle. For *bismuth* some tartaric acid is added. In case *tin* is present, the sample is dissolved in sulphuric acid instead of nitric acid. According to *Thorpe*, only 2 parts of silver chloride freshly precipitated, and 0.8 part that has been blackened by exposure to light, are dissolved in 100,000 parts of nitric acid.

[1 A small engine is sometimes used for this purpose. Trans. Am. Inst. Mining Engineers, vol. x. p. 492.—G.]

[2 B. u. h. Ztg. 1870, p. 303.

Preparation of the normal solutions.—A completely saturated solution of common salt is prepared, of which, if the salt used is entirely pure, 170 cubic centimeters contain 54.15 grammes of common salt. These 170 cubic centimeters are diluted to the volume of 10 liters. 100 cubic centimeters of this solution correspond to 0.5415 gramme of common salt, which will completely precipitate 1 gramme of pure silver. The true standard is obtained by pouring 100 cubic centimeters of the solution of common salt into a solution of 1 gramme of chemically pure silver. This is agitated by shaking until it becomes clear, and the number of thousandths of common salt or silver which remain free are exactly determined by the addition of an observed volume of very dilute salt solution of known strength, or of a decinormal solution of silver, and from this the quantity of water or of common salt is calculated which must be added to obtain the correct standard. When this addition has been made, a new test is had with the standard solution and the decinormal solution prepared from it, and this is continued until the solution does not show a perceptible variation from the correct standard. The decinormal solution of common salt is prepared by pouring 100 cubic centimeters of the standard solution of salt into a flask capable of holding 1 liter and filling it with water to the liter mark. For the decinormal solution of silver, 1 gramme of fine silver is dissolved in 5 to 6 grammes of nitric acid, which is then diluted with water to 1 liter.

When a large number and great variety of silver alloys have to be assayed, the following tables, A and B, calculated by Gay-Lussac for his silver assay, will be found to save much time and trouble in determining the fineness of the alloy. If, after the addition of the standard (common) salt solution, no reaction takes place on adding the decimal salt solution or the decimal silver solution, it is an indication that the silver in the alloy assayed amounts to exactly 1000 milligrammes, and it further indicates that the standard solution is correct. This titer is indicated in the columns by the figure 0.[1]

---

[1] Fortschritte im Probirwesen. Balling, Berlin, 1887, p. 101.

SILVER—ASSAYS OF ALLOYS. 153

Table A.—For decimal salt solution.

Weight of alloy assayed, in milligrammes.	Cubic centimeters of decinormal salt solution added.										
	0	1	2	3	4	5	6	7	8	9	10
	correspond to a content of fine silver of										
1000	1000.0	—	—	—	—	—	—	—	—	—	—
1003	997.0	998.0	999.0	1000.0	—	—	—	—	—	—	—
1005	995.0	996.0	997.0	998.0	999.0	1000.0	—	—	—	—	—
1007	993.0	994.0	995.0	996.0	997.0	998.0	999.0	1000.0	—	—	—
1009	991.0	992.0	993.0	994.0	995.0	996.0	997.0	998.0	999.0	1000.0	—
1010	990.1	991.1	992.1	993.1	994.1	995.0	996.0	997.0	998.0	999.0	1000.0
1011	989.1	990.1	991.1	992.1	993.1	994.1	995.1	996.0	997.0	998.0	999.0
1015	985.2	986.2	987.2	988.2	989.2	990.1	991.1	992.1	993.1	994.1	995.1
1020	980.4	981.4	982.4	983.3	984.3	985.3	986.3	987.2	988.2	989.2	990.2
1025	975.6	976.6	977.6	978.5	979.5	980.5	981.5	982.4	983.4	984.4	985.4
1030	970.9	971.8	972.8	973.8	974.8	975.7	976.7	977.7	978.6	979.6	980.6
1035	966.2	967.1	968.1	969.1	970.1	971.0	972.0	972.9	973.9	974.9	975.8
1040	961.5	962.5	963.5	964.4	965.4	966.3	967.3	968.3	969.2	970.2	971.1
1045	956.9	957.9	958.8	959.8	960.8	961.7	962.7	963.6	964.6	965.5	966.5
1050	952.4	953.3	954.3	955.2	956.2	957.1	958.1	959.0	960.0	960.9	961.9
1055	947.9	948.8	949.8	950.7	951.7	952.6	953.5	954.5	955.4	956.4	957.3
1060	943.4	944.3	945.3	946.2	947.2	948.1	949.1	950.0	950.9	951.9	952.8
1065	939.0	939.9	940.8	941.8	942.7	943.7	944.6	945.5	946.5	947.4	948.4
1070	934.6	935.5	936.4	937.4	938.3	939.3	940.2	941.1	942.1	943.0	943.9
1075	930.2	931.2	932.1	933.0	933.9	934.9	935.8	936.7	937.7	938.6	939.5
1080	925.9	926.8	927.8	928.7	929.6	930.6	931.5	932.4	933.3	934.3	935.2
1085	921.7	922.6	923.5	924.4	925.3	926.3	927.2	928.1	929.0	930.0	930.9
1090	917.4	918.3	919.3	920.2	921.1	922.0	922.9	923.8	924.8	925.7	926.6
1095	913.2	914.2	915.1	916.0	917.0	917.8	918.7	919.6	920.5	921.5	922.4
1100	909.1	910.0	910.9	911.8	912.7	913.6	914.5	915.4	916.4	917.3	918.2
1105	905.0	905.9	906.8	907.7	908.6	909.5	910.4	911.3	912.2	913.1	914.0
1110	900.9	901.8	902.7	903.6	904.5	905.4	906.3	907.2	908.1	909.0	909.9
1115	896.9	897.8	898.6	899.5	900.4	901.3	902.2	903.1	904.0	904.9	905.8
1120	892.9	893.7	894.6	895.5	896.4	897.3	898.2	899.1	900.0	900.9	901.8
1125	888.9	889.8	890.7	891.6	892.4	893.3	894.2	895.1	896.0	896.9	897.8
1130	885.0	885.8	886.7	887.6	888.5	889.4	890.3	891.2	892.0	892.9	893.8
1135	881.1	881.9	882.8	883.7	884.6	885.5	886.3	887.2	888.1	889.0	889.9
1140	877.2	878.1	878.9	879.8	880.7	881.6	882.5	883.3	884.2	885.1	886.0
1145	873.4	874.2	875.1	876.0	876.9	877.7	878.6	879.5	880.3	881.2	882.1
1150	869.6	870.4	871.3	872.2	873.0	873.9	874.8	875.7	876.5	877.4	878.3
1155	865.8	866.7	867.5	868.4	869.3	870.1	871.0	871.9	872.7	873.6	874.5
1160	862.1	862.9	863.8	864.7	865.5	866.4	867.2	868.1	869.0	869.8	870.7
1165	858.4	859.2	860.1	860.9	861.8	862.7	863.5	864.4	865.2	866.1	866.9
1170	854.7	855.6	856.4	857.3	858.1	859.0	859.8	860.7	861.5	862.4	863.2
1175	851.1	851.9	852.8	853.6	854.5	855.3	856.2	857.0	857.9	858.7	859.6
1180	847.5	848.3	849.2	850.0	850.8	851.7	852.5	853.4	854.2	855.1	855.9
1185	843.9	844.7	845.6	846.4	847.3	848.1	848.9	849.8	850.6	851.5	852.3
1190	840.3	841.2	842.0	842.9	843.7	844.5	845.4	846.2	847.1	847.9	848.7
1195	836.8	837.7	838.5	839.3	840.2	841.0	841.8	842.7	843.5	844.3	845.2
1200	833.3	834.2	835.0	835.8	836.7	837.5	838.3	839.2	840.0	840.8	841.7
1205	829.9	830.7	831.5	832.4	833.2	834.0	834.8	835.7	836.5	837.3	838.2
1210	826.4	827.3	828.1	828.9	829.7	830.6	831.4	832.2	833.1	833.9	834.7
1215	823.0	823.9	824.7	825.5	826.3	827.2	828.0	828.8	829.6	830.4	831.3
1220	819.7	820.5	821.3	822.1	822.9	823.8	824.6	825.4	826.2	827.0	827.9
1225	816.3	817.1	818.0	818.8	819.6	820.4	821.2	822.0	822.9	823.7	824.5
1230	813.0	813.8	814.6	815.4	816.3	817.1	817.9	818.7	819.5	820.3	821.1
1235	809.7	810.5	811.3	812.1	813.0	813.8	814.6	815.4	816.2	817.0	817.8
1240	806.5	807.3	808.1	808.9	809.7	810.5	811.3	812.1	812.9	813.7	814.5
1245	803.2	804.0	804.8	805.6	806.4	807.2	808.0	808.8	809.6	810.4	811.2
1250	800.0	800.8	801.6	802.4	803.2	804.0	804.8	805.6	806.4	807.2	808.0
1255	796.8	797.6	798.4	799.2	800.0	800.8	801.6	802.4	803.2	804.0	804.8

ASSAYING.

Table B.—For decinormal silver solution.

Weight of alloy assayed, in milligrammes.	Cubic centimeters of decinormal silver solution added.										
	0	1	2	3	4	5	6	7	8	9	10
	correspond to a content of fine silver of										
1000	1000.0	999.0	998.0	997.0	996.0	995.0	994.0	993.0	992.0	991.0	990.0
1003	997.0	996.0	995.0	994.0	993.0	992.0	991.0	990.0	989.0	988.0	987.0
1005	995.0	994.0	993.0	992.0	991.0	990.0	989.0	988.1	987.1	986.1	985.1
1007	993.0	992.0	991.1	990.1	989.1	988.1	987.1	986.1	985.1	984.1	983.1
1009	991.0	990.0	989.0	988.0	987.0	986.0	985.0	984.0	983.0	982.0	981.0
1010	990.1	989.1	988.1	987.1	986.1	985.1	984.2	983.2	982.2	981.2	980.2
1011	989.1	988.1	987.1	986.2	985.2	984.2	983.2	982.2	981.2	980.2	979.2
1015	985.2	984.2	983.2	982.3	981.3	980.3	979.3	978.3	977.3	976.4	975.4
1020	980.4	979.4	978.4	977.4	976.5	975.5	974.5	973.5	972.5	971.6	970.6
1025	975.6	974.6	973.7	972.7	971.7	970.7	969.8	968.8	967.8	966.8	965.8
1030	970.9	969.9	968.9	968.0	967.0	966.0	965.0	964.1	963.1	962.1	961.2
1035	966.2	965.2	964.2	963.3	962.3	961.3	960.4	959.4	958.4	957.5	956.5
1040	961.5	960.6	959.6	958.6	957.7	956.7	955.8	954.8	953.8	952.9	951.9
1045	956.9	956.0	955.0	954.1	953.1	952.1	951.2	950.2	949.3	948.3	947.4
1050	952.4	951.4	950.5	949.5	948.6	947.6	946.7	945.7	944.8	943.8	942.9
1055	947.9	946.9	946.0	945.0	944.1	943.1	942.2	941.2	940.3	939.3	938.4
1060	943.4	942.4	941.5	940.6	939.6	938.7	937.7	936.8	935.8	934.9	934.0
1065	939.0	938.0	937.1	936.1	935.2	934.3	933.3	932.4	931.4	930.5	929.6
1070	934.6	933.6	932.7	931.8	930.8	929.9	929.0	928.0	927.1	926.2	925.2
1075	930.2	929.3	928.4	927.4	926.5	925.6	924.7	923.7	922.8	921.9	920.9
1080	925.9	925.0	924.1	923.1	922.2	921.3	920.4	919.4	918.5	917.6	916.7
1085	921.7	920.7	919.8	918.9	918.0	917.0	916.1	915.2	914.3	913.4	912.4
1090	917.4	916.5	915.6	914.7	913.8	912.8	911.9	911.0	910.1	909.2	908.3
1095	913.2	912.3	911.4	910.5	909.6	908.7	907.8	906.8	905.9	905.0	904.1
1100	909.1	908.2	907.3	906.4	905.4	904.5	903.6	902.7	901.8	900.9	900.0
1105	905.0	904.1	903.2	902.3	901.4	900.4	899.5	898.6	897.7	896.8	895.9
1110	900.9	900.0	899.1	898.2	897.3	896.4	895.5	894.6	893.7	892.8	891.9
1115	896.9	896.0	895.1	894.2	893.3	892.4	891.5	890.6	889.7	888.8	887.9
1120	892.9	892.0	891.1	890.2	889.3	888.4	887.5	886.6	885.7	884.8	885.9
1125	888.9	888.0	887.1	886.2	885.3	884.4	883.6	882.7	881.8	880.9	880.0
1130	885.0	884.1	883.2	882.3	881.4	880.5	879.6	878.8	877.9	877.0	876.1
1135	881.1	880.2	879.3	878.4	877.5	876.7	875.8	874.9	874.0	873.1	872.3
1140	877.2	876.3	875.4	874.6	873.7	872.8	871.9	871.0	870.2	869.3	868.4
1145	873.4	872.5	871.6	870.7	869.9	869.0	868.1	867.2	866.4	865.5	864.6
1150	869.6	868.7	867.8	867.0	866.1	865.2	864.3	863.5	862.6	861.7	860.9
1155	865.8	864.9	864.1	863.2	862.3	861.5	860.6	859.7	858.9	858.0	857.1
1160	862.1	861.2	860.3	859.5	858.6	857.8	856.9	856.0	855.2	854.3	853.4
1165	858.4	857.5	856.6	855.8	854.9	854.1	853.2	852.4	851.5	850.6	849.8
1170	854.7	853.8	853.0	852.1	851.3	850.4	849.6	848.7	847.9	847.0	846.1
1175	851.1	850.2	849.4	848.5	847.7	846.8	846.0	845.1	844.3	843.4	842.5
1180	847.5	846.6	845.8	844.9	844.1	843.2	842.4	841.5	840.7	839.8	839.0
1185	843.9	843.0	842.2	841.3	840.5	839.7	838.8	838.0	837.1	836.3	835.4
1190	840.3	839.5	838.7	837.8	837.0	836.1	835.3	834.5	833.6	832.8	831.9
1195	836.8	836.0	835.1	834.3	833.5	832.6	831.8	831.0	830.1	829.3	828.4
1200	833.3	832.5	831.7	830.8	830.0	829.2	828.3	827.5	826.7	825.8	825.0
1205	829.9	829.0	828.2	827.4	826.6	825.7	824.9	824.1	823.2	822.4	821.6
1210	826.4	825.6	824.8	824.0	823.1	822.3	821.5	820.7	819.8	819.0	818.2
1215	823.0	822.2	821.4	820.6	819.7	818.9	818.1	817.3	816.5	815.6	814.8
1220	819.7	818.8	818.0	817.2	816.4	815.6	814.7	813.9	813.1	812.3	811.5
1225	816.3	815.5	814.7	813.9	813.1	812.2	811.4	810.6	809.8	809.0	808.2
1230	813.0	812.2	811.4	810.6	809.8	809.0	808.1	807.3	806.5	805.7	804.9
1235	809.7	808.9	808.1	807.3	806.5	805.7	804.9	804.0	803.2	802.4	801.6
1240	806.5	805.6	804.8	804.0	803.2	802.4	801.6	800.8	800.0	799.2	798.4
1245	803.2	802.4	801.6	800.8	800.0	799.2	798.4	797.6	796.8	796.0	795.2
1250	800.0	799.2	798.4	797.6	796.8	796.0	795.2	794.4	793.6	792.8	792.0
1255	796.8	796.0	795.2	794.4	793.6	792.8	792.0	791.2	791.4	789.6	788.8

In making Gay-Lussac's assay of silver bullion a great deal of time is necessarily spent in waiting for the suspended chloride to settle and leave the liquid clear in order to observe the action of the next drop of the precipitant. Whittell[1] has reduced this loss of time and insured greater facility in making the assay by dividing the solution (containing the silver) into several, say five, equal parts in separate vessels. They are placed in a row, and 3 c.c. of salt solution added to the first, 4 c.c. to the second, 5 c.c. to the next, and so on. After the precipitate has subsided, $\frac{1}{2}$ c.c. of the same solution is added to each successively. Numbers 1, 2, and 3 will perhaps show traces of silver still in solution, but numbers 4 and 5 none. The amount precipitated from number 3 multiplied by 5 (as it represents only $\frac{1}{5}$ of the original solution of silver) will be the amount of silver contained in the ore or alloy being assayed. A simple means of settling the precipitated chloride almost instantaneously is to agitate the solution with a few drops of chloroform. Its action seems to be entirely mechanical. The agitation displaces the chloroform in minute globules throughout the silver solution, which in settling to the bottom carries with it every particle of the chloride.

b. Volhard's assay with sulpho-cyanide.[2]—The solution of silver, which should be free from nitrous acid, mercury, and palladium, and to which has been added some ferric sulphate, is precipitated in the cold with titrated potassium sulpho-cyanide until a permanent red coloration from iron remains, indicating that all the silver has been precipitated. The assay, which is as accurate as *Gay-Lussac's*, is simpler, and can be executed more quickly, and allows at the same time of a determination of a percentage of gold in the same assay sample.

The standard solution of potassium sulpho-cyanide is prepared by dissolving 10 grammes of chemically pure silver in nitric acid free from chlorine, and diluting it to 1 liter; 50 cubic centimeters of this solution are placed in a beaker-glass and diluted with 3 to 4 times its volume of water; 5 cubic centimeters of a

---

[1] Humid Assay for Silver. Dr. A. P. Whittell, Engineering and Mining Journal, Nov. 26, 1881, p. 356.

[2] Volhard, die Silbertitrirung mit Schwefelcyanammonium, etc., Leipzig, Winter, 1878; Dingler, ccxiv. 399; B. u. h. Ztg. 1875, p. 83; 1876, p. 405 (Lindeman); Fresenius's Ztschr. xiii. 171; 1878, p. 482.

pure solution of ferric sulphate (1 part of the salt in 10 parts of water) are added to it, and the solution of potassium sulpho-cyanide is allowed to flow to it, under constant stirring, from a burette holding 50 cubic centimeters and divided into $\frac{1}{10}$ and filled exactly to the 0 point, until the color of the solution remains permanently red. The assay fluid is prepared by placing 10 grammes of the silver in a long-necked flask, capable of holding from 200 to 250 cubic centimeters. It is then dissolved on the sand-bath in 50 cubic centimeters of nitric acid free from chlorine, of 1.2 specific gravity, and diluted with distilled water. Any gold which may be present is then allowed to settle, and the clear silver solution is poured into a flask capable of holding 1 liter. The residuum is several times digested with a small quantity of nitric acid, and is then decanted with distilled water until the liter flask is nearly full to the liter mark, and the wash-water shows no traces of silver. The flask in which the solution was made is now filled to the rim with water and inverted in a porcelain crucible, to remove the gold contained in it, and the gold, which is weighed to within 0.0002 gramme, is further treated according to the assay method, which will be given further on. The solution of silver in the liter flask is now diluted to 1 liter, 50 cubic centimeters of it are measured out in a beaker-glass, and titrated with the solution of potassium sulpho-cyanide, after addition of ferric sulphate.

Cobalt and nickel produce peculiar tints which can be easily distinguished from those of the reaction of the silver. In case the sample contains more than 80 per cent. of *copper*, the red coloring is not very perceptible, and *Volhard and Fresenius*[1] have given a modification for this emergency; or pure silver may be added to the sample. *Mercury* is removed by previous volatilization, and nitrous acid must be completely removed by boiling, as it decomposes sulpho-cyanic acid, even in the cold, while nitric acid will only do so when heated. A small percentage of chlorine in the solution of potassium sulpho-cyanide does not cause any trouble, but a large amount of it is injurious.

[1] Fresenius's Ztschr. xiii. 175.

c. *Determination of silver and copper in one solution.*—Quessand's[1] method of determining both these metals in one solution is based upon the fact that in the simultaneous presence of copper and silver only the latter is at first precipitated by potassium ferrocyanide; the pure *white* precipitate of ferrocyanide of silver assumes a pale flesh color only when the copper begins to separate out, which cannot begin until all the silver has been precipitated. The precipitation of the copper is finished when the supernatant liquid has acquired a reddish color, which is due to a trace of the cupric ferrocyanide being soluble in potassium ferrocyanide. The assay must be made by a neutral or only slightly acid one per cent. solution of potassium ferrocyanide. For settling the titer a solution is used of one gramme of pure silver and 0.1 gramme of pure copper in 3 c.c. of HNO_3, diluted with water to 400 c.c., and a solution of one gramme of Rochelle salt (sodium tartrate) and 25 c.c. of caustic soda in 500 c.c. of water. The latter solution serves as an additional proof of the precipitation of the copper, the reddish color of the supernatant liquid being changed to bluish-white by its addition.

d. *Assay by iodide of potassium and starch.*[2]—On adding HNO_3, nitrous acid, iodide of potassium, and starch to a solution containing a silver salt, two reactions take place. Iodide of silver is precipitated; and iodide of starch is produced, which colors the liquid blue. So long as the least excess of silver remains undecomposed this coloration disappears at once on shaking; but if by a fresh addition of iodide of potassium the point of saturation has been reached, a single drop of the reagent suffices to give a permanent blue color to the entire solution. The necessary solutions are thus prepared: Ten grammes of commercial iodide of potassium are dissolved in distilled water, so that the volume is 1023.4 c.c. One cubic centimeter of this solution will precipitate 0.01 gramme of silver. The HNO_3, containing nitrous acid, is prepared by adding 1 gramme of pure sulphate of protoxide of iron to 1000 grammes of HNO_3, of sp. gr. 1.200; and a small quantity of the iron salt must be added from time to time. The starch is prepared by treating one part of starch with 100 parts warm water, the solution allowed to settle, and the clear liquid decanted; to this, 20 parts of pure nitrate of potash are added.

---

[1] Chemikerztg., 1884, p. 1655.
[2] Annales des Mines, 1856, vol. 10, p. 83; Chem. News, 1860, vol. 2, p. 17; Pogg. Ann., 1865, 124, p. 347.

In making an assay by this method, 1 c.c. of the HNO_3 is added to 1 c.c. of the solution under examination, then 10 or 12 drops of starch solution, and a few drops of iodide of potassium. The addition of the iodide is carefully continued until a permanent blue coloration is produced, and the number of c.c. added gives a direct measure to the number of centigrammes present. This method yields excellent results in presence of acids and organic bodies, but is not applicable in presence of salts of mercury, protoxide of tin, arsenious acid, etc., which decompose iodide of starch, or substances which color the solution.

2. *Gravimetric analysis.*—As the solution of common salt evaporates too much in a hot climate, the following method is used in the *East Indies:*[1] 1.22 grammes of the alloy are dissolved in nitric acid. The silver is precipitated by hydrochloric acid, and the silver chloride is carefully washed out.[2] The flask is filled with water and inverted in a smooth washing crucible, and then removed. The greatest part of the water is decanted off, and the assay dried, first on the water-bath, and next in an air-bath at 150° to 170° C. (302° to 338° F.), and the silver chloride weighed while still warm.

Determination of selenium in silver.[3]—Silver used for parting alone contains selenium; it is never found in fine silver. Its presence originates from the H_2SO_4 used in parting; this acid is mostly produced from pyrites, which usually contains selenium. In precipitating the silver from the H_2SO_4 solution with copper, the entire content of selenium is also precipitated. If the solution contains no more than .001 per cent. of selenium, the silver becomes brittle, and the surface of the metal is covered with gray stains of selenide of silver which, being distributed throughout the entire mass, cannot be removed by polishing. In melting such silver with copper a vigorous boiling takes place and particles of the alloy are apt to be thrown out and lost. This is caused by the formation of selenious acid, which is formed by the action of the oxygen, present in the copper as protoxide, upon the selenium and escaping as a gas.

---

[1] Dingler, cciii. 97, 203.

[2] [This operation should be very carefully repeated until the precipitate is thoroughly washed. Instead of decanting the supernatant liquid can be siphoned off.—G.]

[3] Fortschritte im Probirwesen. Balling, Berlin, 1887, p. 107.

.The method of determining selenium in silver is, according to Debray, as follows: Dissolve 100 grammes of the silver in HNO_3 of 34° B. (1.3 sp. gr.), decant the silver solution from any gold which may be present, precipitate the silver with HCl, filter and dry filtrate on a water-bath. Then boil with a few drops of HCl in order to convert the selenic acid into selenious acid, and then add sulphurous acid, or, with an excess of acid, sodium sulphite, whereby the selenious acid is reduced. Boil for one hour until the precipitate of selenium, at first red, becomes black, collect upon a weighed filter, dry at a temperature of not over 100° C. (212° F.) and weigh. The filtrate from this precipitate of selenium is evaporated and again treated in the same manner in order that the precipitation of the selenium may be complete.

Selenated silver can also be readily freed from its selenium by remelting with saltpetre.

3. *Electrolytic determination of silver.*—Luckow[1] as early as 1865 gave a method for the quantitative determination of silver by electrolysis. According to his later communications,[2] precipitation of the silver is completely effected by passing the galvanic current through a solution containing at the most 8 to 10 per cent, of free HNO_3. The silver is then precipitated in a very voluminous spongy condition. Some binoxide is simultaneously formed on the positive pole; its formation, however, can be prevented by the addition of glycerin, milk-sugar, or tartaric acid. The precipitation of the silver in this spongy, flaky form is very inconvenient, as it easily drops from the electrode and cannot be readily weighed. Such metallic sponge is more especially obtained from comparatively concentrated solutions. According to H. Fresenius and F. Bergmann,[3] the silver can be precipitated from HNO_3 solutions in a compact and beautiful form, provided the solution be diluted and the galvanic current weak. It is suggested that 200 c.c. of the liquid to be electrolyzed contains not more than 0.03 to 0.04 gramme of metallic silver and not over 3 to 6 grammes of free HNO_3, that the electrodes be not over one centimeter apart, and that the strength of the current equals about 100 to 150 c.c. of oxyhydrogen gas per hour. From neutral or slightly acid solutions there is precipitated, according to

---

[1] Dingler's Journ. Bd. 178, p. 43.
[2] Ztschft. f. Anal. Chem. Bd. 19, p. 15.
[3] Ztschft. f. Anal. Chem. Bd. 19, p. 324.

Kiliani,[1] besides metallic silver on the cathode, silver dioxide on the anode; the amount of the latter varies according to the acidity and the amount of silver in the solution. To prevent the formation of this dioxide Kiliani adds lead nitrate to the solution, the quantity thus added being greater than the amount of silver. Under these circumstances only lead dioxide separates upon the anode, which exerts no influence upon the separation of the silver in the cathode.

Schucht[2] states that from all solutions except nitrates, or such as contain much free HNO_3, nothing but the metallic silver is precipitated, and that the greatest amount of silver dioxide is precipitated from concentrated acid solutions.

According to Classen, an excess of ammonium oxalate is added to the silver solution, the white precipitate thoroughly washed and dissolved in a solution of potassium cyanide. The current should not exceed 80 to 100 c.c of oxyhydrogen gas per hour; one Bunsen cell will therefore be found sufficient. If the silver is to be separated from copper, the latter can be readily determined in the solution after the precipitated silver has been filtered off. To analyze an alloy of copper and silver dissolve about 0.1 gramme in HNO_3, evaporate to dryness, take up with water and add ammonium oxalate until the precipitate appears pure white. Dissolve in potassium cyanide and proceed as above.

C. Hydrostatic assay.—According to *Karmarsch*,[3] the quantity of silver in coins can be determined from the specific gravity L, in thousand parts n, according to the formula—

$$n = \frac{L - 8.833}{0.0016474}$$

This method is not adapted for very fine alloys, nor for such as have been cast and little worked after the casting, as the results obtained are too high.

[1] Bg. u. Httumsch. Ztg. 1883, p. 401.
[2] Ztschft. f. Anal. Chem. Bd. 22, p. 491.
[3] Dingler, ccxxiv. 565.

IV. Gold.

33. GOLD ORES.

Native gold, with 0.1 to 40 per cent. of Ag, occurring in quartzose veins (gold quartz), and in pyrites (iron or copper pyrites, arsenical pyrites), and disseminated in alluvial deposits (auriferous gravel); *sylvanite* (Au, Ag), Te_2, with 24 to 30 Au and 3 to 15 Ag; *nagyagite*, $PbTe_2$ with PbS and $AuTe_2$, with 6 to 9 Au and 50 to 60.5 Pb; *white tellurium* $(Au,Ag,Pb)(Te,Sb)_3$, with 24.8 to 29.6 Au, 2.7 to 14.6 Ag, and 2.5 to 19.5 Pb.

34. NON-ALLOYS.

Sometimes *mechanical wash assays* are made use of for an approximate determination of the metallic gold contained in poor earthy and gravelly ores. *Dry or fire assays (scorification or crucible assays)* are mostly used for a more accurate determination of the percentage of gold in very poor ores; and sometimes the wet assay (*Plattner's assay*) also. The taking of assay samples requires the utmost care on account of the very unequal distribution of the gold in the ores (pp. 19 *et seq.*).

A. *Mechanical assay by washing,* for determining the *approximate* percentage of gold in *earthy and gravelly minerals, poor in gold.* The sample is rubbed as fine as possible and sifted. About 20 grammes of it are washed with water in a vanning shovel (Fig. 5, p. 27), until the pure gold begins to show itself at the upper end. The quantity is either estimated or weighed, or measured by bringing it into a narrow strip about 0.36 millimeter wide (Hungary and Transylvania). Sometimes it is also amalgamated with mercury[1] and ignited in a small crucible (Transylvania, United States).

Montana: 5 kilogrammes of *earthy* gold ore are taken from the heap, powdered, mixed, and sifted. The coarser gold remaining in the sieve is weighed and assayed by itself; 500 grammes of the fine sifted matter are placed in the vanning trough (or

---

[1] B. u. h. Ztg., 1863, p. 271; 1868, p. 127; 1875, p. 311.

shovel) (Fig. 5, p. 27), mixed with some water and 5 grammes of mercury, and slowly washed for two hours (if the water shows an acid reaction, some caustic soda is added), and finally some potassium cyanide is added, and the amalgam completely purified; mercury is removed by glowing the mass gently in a crucible or retort. The residue is cupelled with lead, and the alloy separated by inquartation and parting; 6 to 8 assays are made and the average is taken.—*Australia:* 1 kilogramme of gravelly gold ore is dead-roasted. It is then placed in an iron mortar and mixed with water to a stiff paste. A tablespoonful of mercury is added and thoroughly rubbed together with the paste; and, after a short time, another tablespoonful. The mass is then washed in an enamelled dish, the amalgam collected and distilled off. This method will give from 80 to 90 per cent. of the quantity of gold which would be obtained by a fire assay.

B. Fire or fusion assays.—The object of these assays is to collect the gold in the lead (*smelting with lead by the scorification or crucible assay*), and to separate the gold by *cupelling* the auriferous lead button. In case the gold button should contain any silver, this can be separated by the wet method by means of nitric acid (*inquartation*). Whether the scorification assay or crucible assay is to be chosen depends principally on the foreign admixtures (earth or gravel), and, as a general rule, the same rules hold good here that were given for silver ores (p. 128).

1. *Smelting the gold with lead.*

a. Scorification assay for ores of every kind.—0.5 to 10 grammes, according to the degree of richness of the assay sample, are weighed, and if the material is poor, a sufficient number of assays are made so that the button which is obtained does not weigh less than 0.05 to 0.20 gramme. The same rule in regard to the quantity of granulated lead and borax is observed as in the silver assays, and the assays are executed in the same manner.

b. Crucible assay.—Poor, earthy, and oxidized ores can be assayed by this method without preliminary preparation, but those containing sulphur, antimony, and arsenic must be previously roasted. It is less adapted for ores rich in gold and copper than the scorification assay. It is simpler and more convenient, as it allows of operating with larger quantities, especially when the substances are poor in precious metal, and is more accurate than the scorification assay, as the losses are distributed among larger quantities of assay sample. The assay sample is fused with granulated lead or litharge and reducing and fluxing agents. The same smelting pots or crucibles (Fig. 52, p. 65) are used as in the corresponding silver assays. The assay is fused in the ordinary furnace, or in a gas-furnace.

a. Substances with earths and oxides (gold quartz, slag, gold sweepings).—They are fused in an unroasted condition.

Sweepings, as stated on p. 136. *American gold ores:* 50 grammes of ore, 70 grammes of dry sodium carbonate, 100 to 120 grammes of litharge (or a corresponding quantity of white lead), and 6 to 8 parts of powdered charcoal. The ore, litharge, and charcoal are first mixed together, and then with the fluxing agent; and, in case sulphur should be present, a small piece of iron wire is added. The charge is placed in a smooth French clay crucible and fused for half an hour at an intense heat in the furnace. It is then poured out, after which the crucible can be used several times more.

The results from 100 pounds of gold quartz by the scorification and crucible assay may be given as follows:—

If 100 pounds of gold quartz give by scorification assay:	They give by crucible assay:
parts of pounds.	parts of pounds.
12.5	12.25
1.5	1.6
0.14	0.14
0.09	0.088

Rheinsand: 500 grammes of ore are mixed with 200 grammes of soda, 300 grammes of potassium carbonate, and 50 grammes of borax. Upon this are scattered 20 grammes of granulated lead free from gold; upon this come a thin layer of soda and a covering of common salt.

Assay with nothing but red lead (litharge) as flux.—This method can be used with advantage with very silicious ores, indeed with almost any ores when the ordinary fluxes are not at hand. Charge: 500 grains (32.4 grammes) of ore, 1300 grains (84.4 grammes) of red lead, and 35 grains (2.33 grammes) of charcoal are thoroughly mixed and placed in a crucible. Place in a "cold fire" and raise the temperature very gradually until the charge is thoroughly fluxed and uniform in color. After pouring and detaching slag, the resulting lead button is cupelled in the usual manner. If the crucible is placed in a hot fire at first, the lead will be reduced without fluxing the ore, which will remain intact. The assay is somewhat difficult to make and requires considerable skill to obtain accurate results.

β. Ores, etc., with combinations of sulphur, antimony, or arsenic.—Larger quantities, 0.5 to 1 kilogramme, are roasted so that buttons weighing not less than 0.05 to 0.20 gramme are obtained. The roasting is done in small clay boxes about 200 millimeters long, 70 to 90 millimeters wide, and 40 to 50 millimeters deep. The ore is placed in these boxes and roasted in the muffle, being carefully stirred meanwhile with a stirring rod. Or the ore may be placed upon a plate of sheet iron with upturned edges, which has been previously covered with a coating of clay, reddle, or chalk, and is then roasted over a brazier, or in a furnace until the fumes cease to be evolved (according to Winkler, Tscheffkin, and Merrick, a loss of gold occurs during this operation, which Crookes denies).[1] If copper pyrites, antimony, and arsenic are present, it is best to add charcoal and ammonium carbonate in roasting. The charging and fusing of the roasted sample are done in the same manner as that indicated in the assay for silver (p. 134).

Pyrites poor in gold.—500 grammes of the roasted ore are mixed with the same quantity of granulated lead free from gold, 125 grammes of black flux and the same quantity of glass. The charge is fused for two hours in a Hessian crucible in the furnace. The resulting button is flattened on an anvil and cut up in pieces. The separate pieces are concentrated on a scorifier, and the button thus obtained is cupelled. 500 grammes of ore are roasted and mixed with 125 to 250 grammes of potassa or soda glass, 125 grammes of black flux, or 250 grammes of potassium carbonate,

---

[1] [Such losses are undoubtedly apt to occur in the presence of tellurides, and in some cases in roasting with common salt (sodium chloride). See the Losses in Roasting Gold-Ores and the Volatility of Gold. S. B. Christy, Trans. Am. Inst. M. E. vol. xvii. p. 3.—G.]

and 32 grammes of flour, then covered with 500 grammes of granulated lead free from silver, and a layer of common salt. The entire charge is put in a Hessian crucible and fused for two hours in the furnace, or it is distributed into several smaller crucibles.

Tailings containing a small amount of auriferous pyrites.— Roasting can be dispensed with and the pyrites decomposed by a piece of iron (hoop) placed in the bottom of the crucible.

Ore	.	.	.	.	1000 grains	(64.80 grammes)	
Litharge	.	.	.		500 "	(32.40 "	)
Sodium carbonate		.	.		1200 "	(77.77 "	)
Charcoal	.	.	.		30 "	(1.95 "	)

Instead of using the iron the ore may be roasted and one-third of the sodium carbonate replaced with borax.

Hungarian smelting works: 1 Vienna pound (= 560 grammes) of the auriferous substance is roasted upon a clay plate over glowing coals. The charge consists of 3 pounds (= 1680 grammes) of *Villach* red litharge, 2 pounds (= 1120 grammes) of dry potash, ¼ pound (= 140 grammes) of resin, and 1 loth (= 14.5 grammes) of hard coal. This is mixed and distributed in crucibles in such a manner that on the bottom comes first a spoonful of the mixture, and, upon this, a spoonful of the roasted sample. These are then mixed together; upon this mixture is placed another spoonful of the mixed fluxes, and then a covering of common salt. 115 to 125 crucibles charged in this manner are heated for from 20 to 30 minutes in the furnace, or a small number in the muffle. The resulting buttons are partly cupelled, and those buttons which have not brightened are wrapped up in a cornet of lead foil and cupelled together. The auriferous silver obtained must weigh about 10 mint pounds assay weight, and the gold buttons to be separated from this about 0.1 mint pound. Average difference 0.001 mint pound.

Gold ores containing tellurides.—These ores should be roasted in the most careful manner, as, according to Kustel,[1] as much as 20 per cent. of the gold-content may be lost in this way, due to the volatilization of the tellurium. The difficulties arising from the

[1] Roasting of Gold and Silver Ores, 1880, p. 57.

presence of tellurium in the ores are best overcome by using plenty of lead, the amount, of course, being proportioned to the richness of the ores. In cases of very rich ores as much as from 75 to 100 parts of lead may be used. In such cases unusually large scorifiers must be used or else several of the ordinary size.

Brown[1] recommends the following charge:—

Ore	48 grains	(2.5 grammes)
Granulated lead	960 "	(60.0 ")
Litharge	48 "	(2.5 ")
Borax glass	4 "	(0.25 ")

Sprinkle litharge over the mixed charge. The buttons may need repeated scorifications with plenty of lead.

2. *Cupellation of the auriferous lead.*—The process is the same as for silver with the exception of a hotter "driving" towards the end of the assay, so that no plumose litharge (Federglätte) remains. If the assay sample is poor, the separate lead buttons are either entirely cupelled or only partly. In the latter case they are wrapped in lead foil and cupelled together. The resulting gold button is then weighed, and, in case it contains silver, this is parted by means of nitric acid. We shall only briefly mention the process here, as it will be more thoroughly explained later on in treating of gold and silver alloys (§ 35).

The button is flattened out on the anvil and placed in a flask with a very narrow neck, and then heated with nitric acid of 1.19 specific gravity. *a. When the laminated button breaks up* and brown flakes of gold are separated, this being an indication that a sufficient quantity of silver is present; the heating is interrupted when no more nitrous acid is developed. The gold is allowed to settle, and the liquid is then carefully decanted. It is now washed twice by decantation with boiling distilled water. The flask is then entirely filled with cold water and inverted in a clay crucible, or a small porcelain saucer, and when the gold has dropped into the crucible, the flask is carefully withdrawn over the side. The water is then poured off, the gold dried, the crucible then strongly heated, and finally the adherent gold is removed and weighed. *b. When the flattened button does not break up,*

---

[1] Manual of Assaying, Chicago, 1886, p. 188.

the acid is poured off and the sample decanted with cold water. The flask is filled with cold water and inverted in a porcelain dish and withdrawn over the side. After the water has been poured off, the button is dried and wrapped, with three times the quantity of silver, in a cornet of lead foil or with granulated lead in a cornet, and cupelled. The button, containing now a sufficient quantity of silver, is parted with nitric acid. In the Upper Harz the percentage of gold is not taken into calculation when 10 assay centner (= 50 grammes) contain less than 0.5 part of a pound (= .25 milligramme) of gold. The buttons obtained from gold ores are, as a rule, richer in gold than in silver, and require an addition of 2 to $2\frac{1}{2}$ times the quantity of silver, while those from auriferous silver ores, pyrites, and matt contain generally less than $\frac{1}{8}$ to $\frac{1}{4}$ of gold and require no addition of silver.

C. Wet assay (Plattner's chlorination process[1]).—This is sometimes used for very poor ores. 50 to 200 grammes of earthy or oxidized ore, or *completely* roasted pyrites, are slightly moistened with water and placed in a tubulated glass cylinder, the bottom of the vessel being first covered with pieces of quartz. Here they are treated with chlorine gas for about one hour. The gold chloride formed is lixiviated with hot water, and the solution heated to expel the free chlorine. Solution of ferrous sulphate and some hydrochloric acid is added, which precipitates the gold in a metallic state. It is then filtered and washed, the filtrate is dried and cupelled with 5 to 10 grammes of granulated lead. *Wagner* recommends the decomposition of the ores with bromine[2] instead of chlorine.

35. ALLOYS OF GOLD.

The principal alloys of gold which will be especially considered here are those *with silver*, with *silver and copper*, and with *copper*.

Gold amalgam is distilled in a glass retort, and the residue is carefully scorified with 8 parts of granulated lead (p. 132). *Auriferous lead* and *bismuth* are directly cupelled, but if they contain too small a quantity of gold, they are first slagged off on

---

[1] Plattner-Richter's Löthrohrprobirkunst, 1865, p. 546.
[2] Dingler, ccxix. 544.

a scorifier (p. 64). *Auriferous iron, steel*, etc., are dissolved in nitric acid and evaporated to dryness. The dry mass is scorified with 8 to 10 parts of granulated lead and some borax.

A. Alloys of gold and silver, with or without copper.—The separation of gold from silver (called "*quartation*" on account of the proportion of gold to silver as 1 : 3) is done by means of nitric acid.[1] But the silver is only completely dissolved by boiling the acid three times, and when at least $2\frac{1}{2}$ to 3 parts of silver are present to 1 part of gold. When this proportion exists, the gold will also be obtained in a cohering mass having the same form as that of the alloy used (*a small roll, etc.*). If less silver is present, the gold remains argentiferous, and if more, for instance, 4 to 6 silver to 1 gold, the gold is obtained in brownish flakes or as powder (dust gold), while the silver will be completely dissolved by boiling the assay twice with nitric acid, and there is great liability that mechanical losses will occur. If the silver is to be dissolved by boiling the assay but once with nitric acid, at least 8 parts of silver to 1 part of gold must be present. *A preliminary assay* is therefore required for an approximate determination of the percentage of gold, to enable the assayer to fix the required quantity of silver which must be added, and also for the determination of the percentage of copper, in order to find the quantity of lead required to be added to it in removing it by cupellation.

1. As a *preliminary test for alloys free from copper*, may serve—

a. The color of the alloy.—A deep yellow color requires $2\frac{1}{2}$ to 3 times the quantity; light yellow, twice the quantity; and a white color an equal weight of quartation silver.

For an approximate determination, by color, of the richness of the gold button, sample gold-silver buttons 2 to 3 millimeters in diameter have been prepared with $\frac{10}{10}$, $\frac{9}{10}$, $\frac{8}{10}$, $\frac{7}{10}$, $\frac{6}{10}$, and $\frac{5}{10}$ of gold. They are placed in depression in a box with a cover, and each is surrounded with a black ring and then with a white one. Before the comparison is made, the assay button is breathed

[[1] This proportion varies; some assayers prefer 1 : $2\frac{1}{4}$. According to Pettenkofer, the proportion need not exceed 1 : $1\frac{1}{4}$, provided the subsequent boiling in HNO_3 is sufficiently prolonged.—G.]

on, as otherwise its strong lustre would make the estimation less accurate. *Goldschmidt*[1] has attached similar specimen alloys, in the form of small disks upon porcelain, but it is more difficult to compare the buttons with these than with sample buttons of the same shape. If more than 56 per cent. Ag is present, the gold cannot be recognized. 2 per cent. of Ag imparts a brass color, 50 per cent. a light yellow, and 56 per cent. a white color to the gold.

b. An examination on the touchstone by means of needles, touchstone, and nitric acid requires more experience than the above method, and may also be used for alloys containing copper.

2. *Preliminary assay of cupriferous alloys by cupellation.*

a. With lead alone.—250 milligrammes of the alloy cut up into fine shreds or granulated are weighed off and wrapped up in a cornet. This is placed with 16 to 32 times the quantity of lead (4 to 8 grammes, according to the percentage of copper) in one piece (spherical or hemispherical) in a strongly glowing cupel in the furnace. The cupellation is conducted in the same manner as with the fine assay (p. 144), except that it must "drive" hotter, so that no plumose litharge (Federglätte) remains. The percentage of copper is found from the difference in the weight of the alloy used and the resulting auriferous silver button. An experienced assayer can then estimate the richness of the alloy in gold by the color of the button after breathing on it, and can thus calculate the quantity of quartation silver to be added for the principal assay. The quantity of lead required for removing the copper by cupellation will be indicated from the difference in weight.

The quantity of lead to be taken depends on the *percentage of copper in the alloy*, which must be removed before the quartation. As copper has a greater affinity for gold than for silver, argentiferous gold containing copper requires a larger quantity of lead in cupelling (the maximum is 32 times the quantity) than argentiferous copper (16 to 20 times the quantity).

The following table (Table I.) shows the quantity of lead required for alloys of gold with silver and copper:—

---

[1] Fresenius's Ztschr. xvii. 142. B. u. h. Ztg. 1878, p. 208.

Table I.

If the gold in 1000 parts of the alloy amounts to	Equivalent to gold.	Multiples of lead.
1000	24 carat.	8
980 to 920	23½ to 22	12
920 " 875	22 " 21	16
875 " 750	21 " 18	20
750 " 600	18 " 14	24
600 " 350	14 " 8	28
350 " 0	8 " 0	32

Table II. gives the quantity required if the percentage of gold is very small.

Table II.

If the silver in 1000 parts amounts to	Equivalent to silver.	Multiples of lead.
1000 to 950	15 loth 9 grän.	4
950 " 900	14 " 9 "	6
900 " 850	13 " 9 "	8
850 " 750	12 "	12
750 " 650	11 "	14
650 " 0	10 " and less	16

b. With an addition of lead and silver.—This process is made use of to avoid the estimation of the quantity of gold in the auriferous silver button by the color. 250 milligrammes of the alloy are wrapped up in a cornet, together with 3 times the quantity of silver (750 milligrammes), and 16 to 32 times the quantity (4 to 8 grammes) of lead, and cupelled. The loss of copper is found from the difference in weight between the resulting button and the alloy weighed plus the addition of silver. The auriferous silver button is laminated and placed in a flask with a long and narrow neck which has been previously well cooled off. The matrass should be from 150 to 180 millimeters high, 30 to 50 millimeters wide in the belly, and 6 to 8 millimeters in the neck. The button is boiled in this with pure nitric acid of 1.19 specific gravity until no more red vapors are evolved, and is then washed twice by decantation with hot water. The flask is then entirely filled with water, a small crucible of clay placed over its mouth, and both crucible and flask are inverted, which cause the gold in the form of a small flake or powder to fall into the crucible. The flask is now raised and quickly drawn away over the edge of the crucible. The gold is then thoroughly dried by igniting it in the crucible. Its weight, plus that of the added quartation silver, deducted from the weight of the auriferous silver button originally employed, gives the percentage of silver in the original alloy, according to which, the addition of silver for the assay must be regulated so that the proportion of 1 Au to $2\frac{1}{2}$ or 3 Ag is maintained. For *coins*, the standard of which is known, such preliminary assays are not required.

German,[1] French, and American gold coins contain 900 Au and 100 Cu; Austrian ducats 986, Prussian Friedrichsdor 902, English sovereigns 916, Hanoverian, Brunswick, and Danish pistoles 896 parts of gold. *Pure* gold is prepared by dissolving ducat gold, or gold cupelled with lead and laminated, in cold *aqua regia* (2 parts of hydrochloric and 1 part of nitric acid), by adding the acid gradually, so that, when the solution is complete, there will be no excess of aqua regia. The solution is allowed to stand for several days, for the silver chloride to settle, and is then filtered. It is now diluted, and if necessary, again filtered in a few days. The filtrate is much diluted, and freshly prepared solution of ferrous sulphate added to it until no more gold is precipitated, and then allowed to stand in a warm place. The fluid is then removed by means of a siphon, the gold placed in a porcelain dish and digested with diluted hydrochloric acid. The dried powder is washed, placed in a clean clay crucible, and fused with some borax and saltpetre.

Roberts[2] recommends the following process for the preparation of chemically pure gold.—The purest gold that can be obtained is dissolved in $HCl + HNO_3$. Evaporate to dryness, add potassium chloride and alcohol, to precipitate platinum, take up with pure distilled water and dilute solution until each gallon does not contain over one-half ounce of gold. The solution is allowed to stand for several weeks; some silver chloride will separate out and fall to the bottom of the vessel. Siphon or decant off the supernatant

---

[1] Göldner, Farbe der Zwanzig Mark Stüke in Dingler, cviii. 75.
[2] Fourth Annual Report of Royal Mint, 1874, p. 46.

liquid, and precipitate the gold with carefully washed sulphurous acid. Ferrous sulphate, oxalic acid or formic acid may also be used for this purpose. Wash precipitate repeatedly with hot distilled water, HCl and ammonium hydrate, and finally with pure distilled water. Fuse in clay crucible with potassium bisulphate, and borax. This gold will be (according to Roberts) 999.96 fine. Gold precipitate from an acid solution containing copper by oxalic acid is apt to be contaminated with cupric oxalate. According to Purgotti,[1] however, if the solution is heated with potash, a soluble double oxalate of copper and potash is formed, and the gold left pure.

I. *Roll assay for argentiferous gold.*—This requires the following manipulations:—

a. Preliminary assay as described on p. 168 for determining the percentage of gold and copper in order to fix the quantity of quartation silver and lead to be added.

b. Weighing the assay sample.—Two samples of the alloy, granulated or laminated and cut into fine shreds, each 250 milligrammes, are accurately weighed out upon an assay balance which must be sensitive to 0.1 milligramme. If bars are to be assayed, a sample of 250 milligrammes each is cut from the upper and lower side from opposite ends. The samples are wrapped in cornets.

c. Charging.—The quantity of silver required is weighed off, cut up into fine shreds, and added to the sample. The quantity must be calculated or found from tables according to the results of the preliminary assay. The lead is next weighed off in one piece according to Table I. p. 170.

d. Cupelling.—The lead is placed in thoroughly ignited fine cupels standing alongside of each other in the centre, or more towards the back of a strongly heated mint furnace (Fig. 30, p. 49). The mouth of the muffle is closed until the lead "drives," when it is opened, and the cornets containing the alloy are placed in the cupel. The mouth of the muffle is again closed, and the assay is allowed to "drive," and the operation further conducted in the same manner as in the fine assay (p. 144), with the exception of

---

[1] Zeitschr. Anal. Chem., vol. 9, p. 127.

a stronger heat towards the end. Of the loss of gold in cupelling we will speak later on.

If fine gold with 990 thousandths "drives" too hot, or too cold, the resulting gold button will be one-thousandth too heavy. This is very likely caused by some lead which remains with the gold, and which cannot be completely removed by nitric acid. For this reason a sample of pure gold is generally cupelled with the same quantity of lead at the same time as the principal assay, and the gain in weight of the fine gold is then deducted from the gold percentage of the principal assay. If the button has been brightened too hot, it is apt to crack in laminating.

c. Flattening (laminating) the button.—The button is removed by means of a pair of pliers, brushed off, and the edges carefully pinched with the pliers to remove any adhering particles of bone ash of cupel.[1] It is then laminated with a hammer on a polished steel anvil, having a diameter of 6 to 8 centimeters. The head of the hammer has on one end a round or square face, smoothly polished, and about 4 centimeters (1.57 inches) in diameter, and on the other end a rounded-off edge. Or, the button after it has been somewhat flattened on the anvil is passed between rollers, being frequently annealed meanwhile on a cupel, or dish, in the muffle. The small oval leaf, into which it is laminated, is about 25 millimeters long, 12 millimeters wide, and 0.5 millimeter thick. The frequent annealing of the button and hammering of the edges are required to prevent the leaf from cracking on the edges. If necessary, the leaves are numbered by means of a punch and hammer. After having been annealed once more, the leaves are rolled with a pair of pliers and dry-fingers, into the form of a spiral, or over a glass-rod, into a small roll.

This "flattening" operation of the button a (Fig. 76) should be very carefully done and requires some skill. The first blow is struck on the centre, then on the edge, and a third blow on the opposite edge, which elongates the metal. After annealing, the flattened button is rolled into a thin leaf c, and then into a cornet or spiral d. After boiling in HNO_3 the bulk is considerably reduced

---

[1] [Immersion in warm dilute hydrochloric acid will also answer.—G.]

as *e*. Care should be observed in rolling up the cornets that the outer surface should be that which, before flattening, was in contact with the cupels.

Fig. 76.

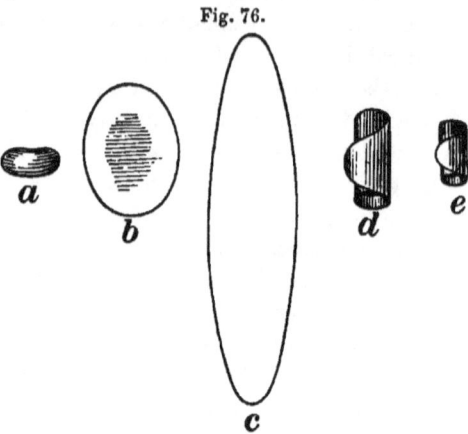

f. Boiling in nitric acid.—One or more of the numbered rolls are placed in a long-necked glass flask, about 150 to 180 millimeters long, with a body width of 40 to 50 millimeters, and a neck width of 15 to 20 millimeters. Here they are heated with a quantity of nitric acid (about 10 grammes) sufficient to fill the body of the flask half-full. The acid should be of 1.2 specific gravity, as, if it is stronger, its action might be too violent and tear the leaf. It should be free from nitrous acid, sulphuric acid, and chlorine, and, if necessary, is freed from chlorine by adding some solution of silver nitrate. The leaves are heated until the fumes of nitrous acid have disappeared. Bumping during the ebullition is prevented by throwing a small splinter of coal, or, what is still better, a completely carbonized pepper-corn, into the flask.[1]

The flask is lifted from the fire by means of a wooden clamp, and the solution of silver carefully poured into a porcelain dish. Nitric acid of 1.3 specific gravity, previously heated to boiling, is now poured upon the leaves, and they are again boiled for 10 minutes. The pouring off of the solution is repeated, and the leaves are again boiled for 10 minutes, with fresh strong nitric

---

[1] Polyt. Ctrbl. 1857, p. 314. B. u. h. Ztg. 1861, p. 407.
[Small pellets or peas of burnt clay are better.—G.]

acid, previously boiled. This third boiling is sometimes omitted, if the percentage of gold is below 750 thousandths. After the third boiling, according to *Kandelhard's* experiments, such a small residue of silver remains in the gold, that it is equalized by the loss caused by cupellation, and the result will be correct.

(The heating may be done by placing a single flask upon a support, with three legs, and provided with a handle, over a lamp, or upon glowing coals. If several flasks are used at one time, they may be placed in depressions in the periphery of a sheet-metal disk, which are filled with sand, and furnished with clamps for holding the necks of the flasks; or, they are placed upon a movable support, with a gas-pipe and burners throwing out lateral flames; or upon *Levol's* gas-heating apparatus. *Matthey* and *Johnson's*[1] new platinum apparatus permits of many rolls (10 to 100) being boiled at one time in small thimble-like crucibles, which are immersed in the acid. It is highly recommended on account of its great convenience and cleanliness, and saving of acid. *Tookey*[2] uses a platinum tube for heating.)

By dissolving silver in nitric acid, nitrous acid is formed which does not attack the gold as long as silver is present. But it will do so if it is developed after the silver has been removed. Such development may be caused by the action of the splinter of wood-charcoal placed in the flask to prevent bumping during ebullition, in case it should contain any woody substance. For this reason, it is best to avoid the use of charcoal for this purpose.

g. Washing (rinsing off) the rolls.—After the last boiling with nitric acid, the acid is poured off, and hot distilled water is allowed to flow slowly into the flask from a copper kettle, the spout of which is introduced into the neck of the flask, or, what is still better, from a glass pitcher. While the water is running into the flask the latter should be constantly turned, until it is about ⅔ full of water. The water is then poured out, and this operation is twice repeated, until the last traces of silver nitrate are removed from the roll and the sides of the flask. The flask is now filled entirely full, a small crucible glazed inside, or a porcelain cup or

---

[1] B. u. h. Ztg. 1870, p. 325.
[See Percy, Metallurgy of Silver and Gold, vol. i. p. 263.—G.]
[2] Dingler, cxcvii. 93. B. u. h. Ztg. 1870, p. 283.

saucer, is placed over its mouth, and both cup and matrass in this position are slowly inverted, which causes the roll to gradually slide down into the cup or crucible. The flask is then drawn away over the edge of the crucible, after which the water, now no longer showing a silver reaction with hydrochloric acid, is poured from the crucible.

h. Drying and annealing of the rolls.—They are thoroughly dried in crucibles, which are covered and placed on the shelf in front of the muffle, or into the round holes of a metal plate, into which they fit. The feet of this plate stand upon a sheet-metal plate, heated by glowing coals from below. The rolls, which are of a brownish color and porous, are now heated in the same crucibles in the muffle-furnace at a white heat, whereby they must assume the lustre and color of gold, after which they are taken from the crucibles.

i. Weighing of the rolls.—After the rolls have become cool, they must be quickly weighed, as they easily absorb gases. A roll, and another from a check assay are laid on the opposite pans of a balance. If they agree (in assays from the upper and lower side of a bar, differences may occur), they are both weighed, and the percentage of gold is thus determined. The average is taken of the upper and lower assay. (It is customary to give the lowest percentage of fine assays of silver, p. 146.)

A loss of gold occurs in cupelling, partly on account of the volatilization of gold with other metals and partly by absorption by the cupel (Kapellenzug). According to *Kandelhard*, this loss is equalized by the retention of a residue of silver in the roll, but this, according to Rössler,[1] is not always the case, as differences may occur between the found and actual percentage of gold, from a cooler or hotter "brightening" and in alloys of different proportions. According to *Kandelhard's* method, the residue of silver in the roll is 1 thousandth when $2\frac{1}{2}$ parts of silver to 1 part gold have been used in the quartation, and the alloy has been boiled three times; 1.5 to 2.5 thousandths if boiled not quite so thoroughly, and up to 5 thousandths if boiled but once. According to *Rössler*, the loss of gold by cupelling increases with

---

[1] Dingler, ccvi. 185. B. u. h. Ztg. 1873, p. 26.

the quantity of lead used (when ¼ gramme of gold is cupelled with 1 to 2 grammes of lead, the loss is only fractions of a thousandth; when 4 to 8 grammes of lead are used, it is over 2 thousandths, even if a residue of silver of nearly 1 thousandth is taken into consideration). The loss is also greater the smaller the gold button is (therefore also when a smaller quantity of the sample is weighed) and the smaller the quantity of silver (according to *Rössler*, 1 to 3 thousandths of pure gold are lost if 4 times the quantity of lead is used in the quartating cupellation; if the button contains more than $2\frac{1}{2}$ times the quantity of silver, the residue of silver commences to preponderate, and with a large quantity of silver it seems to be almost in excess). According to this, assays with a small percentage of gold cupelled with much lead would, under otherwise equal conditions, come out somewhat worse than those with a high percentage, and, if losses of gold and residue of silver equalize each other, the loss would preponderate in all smaller assays.

An English commission[1] having caused an examination to be made of samples of different coins with an accurately determined percentage of gold, the errors in the assay were found to amount to from $\frac{1}{10000}$ to $\frac{2}{10000}$ per cent.

Platinum renders the surface of the auriferous silver button, after it has been cupelled, crystalline, porous, and rough, and, if much of it is present, gray. It is removed by cupelling the button resulting from quartation, after it has been weighed, with 8 times the quantity of silver and lead, and treating the resulting button, after it has been laminated, with nitric acid until the weight of the roll remains constant and platinum is no longer dissolved with the silver.[2] *Rhodium* and *iridium* produce black stains upon the auriferous silver buttons after the brightening, and, if a large percentage of iridium is present, the rolls break and black iridium powder will be found between the gold. The gold is then dissolved in aqua regia and precipitated with ferrous sulphate.

---

[1] Report of the British Association, 1875, p. 127.

[2] Winkler, Löslichkeit von Platinsilber in Salpetersäure in Fresenius's Ztschr. 1874, p. 369. B. u. h. Ztg. 1845, p. 145.

D'Hennin[1] claims to separate the iridium by fusing 12.5 grammes of gold containing it with 3 grammes of sodium arsenate, 18 grammes of black flux, and 20 grammes of a flux consisting of a mixture of borax, argol, litharge, and charcoal, into a speiss containing iron and arsenic, while the gold and silver are collected in the lead.

Palladium passes with the silver into solution, if the gold containing it is alloyed with 3 times the quantity of silver.

In cupelling pure gold at a temperature above its melting-point and removing it from the muffle whilst still in a fluid state, it shows, according to Van Riemsdijk,[2] the phenomenon of brightening (Blickens) very plainly. The glowing gold cools at a red heat without changing its condition, and in the course of 20 to 40 seconds the button suddenly emits vivid flashes of green light which abate and disappear as it cools and sets.

To obtain good results in cupelling gold the following conditions must be observed:—

1. The cupelling must be effected at least at the melting-point of silver.

2. The gold when taken from the muffle must be in a liquid state.

3. It must show a smooth and even surface and in removing the cupel from the muffle it must not be shaken.

4. The button must cool uniformly.

The phenomenon of brightening being due to the superheating of the gold, it is desirable that it contain a small amount of copper which promotes superheating. A content of more or less silver is detrimental to the appearance of brightening on account of the absorption of oxygen by the silver, which it again yields up in cooling, thus disturbing the internal quietness of the button and rendering the gold less ductile. In this case, however, the brightening can easily be produced by the addition of a known quantity of copper. An alloy consisting of 250 milligrammes of gold, 625 of silver, and 25 of copper brightens very perceptibly in cupelling with from 3 to $3\frac{1}{2}$ grammes of lead. According to the same author, gold containing osmium, ruthenium, rhodium, and iridium does not brighten even if

---

[1] Dingler, cxxxvii. 443.

[2] Archiv. Néerlandaises, tom. xv. p. 185. Bg. u. Hüttenm. Ztg. 1880, p. 247.

cupelled at a very high heat; thus the appearance of brightening in this case is a useful indication of the purity of the gold.

Van Riemsdijk[1] further states,

1. That the smallest quantity of ruthenium or iridosmium present in the original alloy will remain in the gold after parting from silver and cupelling with lead.

2. The presence of about 0.001 per cent. of iridium is not detrimental to the accuracy of the assay, such a small quantity being removed with the silver in parting with HNO_3. A content of over 0.001 per cent. iridium, however, prevents the appearance of brightening and makes the buttons too heavy.

3. No perceptible influence upon the brightening is caused by the presence of 0.002 per cent. of rhodium; a larger amount, however, prevents its appearance. Such buttons adhere tightly to the cupel, congeal immediately on removing from the muffle, have a rose color, and, in proportion to the amount of rhodium present, are red brown or blackish.

4. A very small content of osmium does not affect the brightening, but if the amount in the original alloy is over 0.0025 per cent. it is not entirely removed by cupelling. In cupelling such gold, considerable loss is experienced by the formation of volatile osmic acid, which carries off so much gold mechanically that even an approximately accurate determination of the gold cannot be made.

Bock[2] draws the conclusion from his experiments that the failure to superheat and the brittleness of silver or auriferous silver can only be attributed to remaining traces of lead and bismuth. He confirms the statement that the presence of copper in cupelling promotes superheating and also the malleability of silver or auriferous silver. In assaying he always adds 25 grammes of electrically deposited copper to 500 milligrammes of gold, and by cupelling this quantity with 3 grammes of lead obtains very ductile gold buttons which can be rolled out to smooth-edged ribbons. This method originated with Augendrée, it is used in the mints at Brussels and Utrecht.[3] In assaying gold coins or bullion only 0.25 gramme is used and cupelled. If the button when taken from the muffle does not brighten, it is

---

[1] Archiv. Néerlandaises, tom. xv. Bg. u. Hüttenm. Ztg. 1880, p. 247.
[2] Bg. u. Hüttenm. Ztg. 1880, p. 409.
[3] Dingler, Bd. cxix., p. 112.

taken as an indication of the presence of the platinum group metals or of a residue of lead in the gold. It seems to be a fact that most gold coins as well as commercial gold contain an appreciable quantity of the platinum group metals—probably iridosmium. Such buttons present an uneven appearance and the liquid metal bubbles before becoming solid, due, in the presence of osmium, to osmic acid, or, in the presence of iridium and ruthenium, to oxygen.

Samples from different parts of the same piece or ingot sometimes act in a quite different manner after cupelling, the phenomenon of brightening occurring in one and not in the other. This is due to the fact of some of the platinum group metals not being uniformly distributed or alloyed throughout the mass of gold. A cupelled button of pure gold which has been superheated and shown brightening can be hammered and rolled without cracking the edges. If the gold is not superheated in cupelling, *i. e.*, has not brightened, traces of lead may remain in it which affect its malleability in the same way as some of the platinum group metals.

II. *Pulverulent assay (Strubprobe) of auriferous silver.*—*a.* A sample, about 5 grammes, is taken on opposite ends from the upper and lower sides of a bar. Duplicates of the sample of 0.5 gramme are weighed out, and cupelled in the mint furnace with 8 times the quantity of silver (p. 168) and the quantity of lead which is found to be necessary according to Table II. p. 170, after a preliminary assay has been made. The resulting button is laminated by hammering or passing it between rollers. It is then placed in a flask, the neck of which should not be wider than 6 to 8 millimeters. Here it is boiled with nitric acid, of 1.2 specific gravity if the percentage of gold is small, and of 1.3 specific gravity if the percentage is larger, for instance 100 thousandths, until the vapors of nitrous acid have disappeared. The bumping is prevented by throwing a carbonized grain of pepper, etc. (p. 174) into the flask. The pulverulent gold is allowed to settle in the flask, which has been placed in a revolving stand. The acid is then poured off into a porcelain saucer or cup, and the gold is rinsed three times with hot distilled water. The flask is now filled with water and inverted in a small unglazed porcelain crucible. In this position it is placed upon the revolving stand until the gold has descended into the crucible. When this is the case, the matrass[1] is carefully drawn away over the side of the crucible. The water is now poured out of the crucible by allowing it to run down on a small rod, and that which remains behind is soaked up with filter-paper. The gold is then dried and strongly ignited in the crucible so that the particles of gold form a coherent mass. It is then weighed.

If the button in brightening is less white, and does not "spit," or if, as is the case where platinum is present, it is crystalline, grayish, and has flat edges, the gold is again cupelled with 8 times the quantity of silver and 3 times that of lead and boiled with acid; these operations must be repeated until the weight of the pulverulent gold remains constant.

The gold may also be determined in connection with Volhard's assay of silver with potassium sulpho-cyanide (p. 155). Jüptner[2] fuses alloys of gold and silver rich in gold, with 5 to 8 times the quantity of zinc, and dissolves the alloys in nitric acid, whereby the gold remains behind.

b. Separation of auriferous silver grains from samples of ores.—This is done in the manner indicated on pp. 166 *et seq.*

B. Alloys of gold with copper.—The metals are separated by cupelling the alloy with 32 times the quantity of lead, and adding 3 times the quantity of silver, otherwise the process is the same as given on p. 170.

Quartation of gold with cadmium.[3] *Balling's method.*—Cadmium can advantageously be used in quartation instead of silver, as it is very fusible, readily unites with the gold, and renders it possible to effect the quartation in a short time in a porcelain crucible over a lamp. To obtain coherent gold, $2\frac{1}{2}$ times its weight of cadmium should be added. The gold-cadmium button is brittle and cannot be flattened. After quartation the gold retains its original form, but shows upon its surface numerous cracks from which the cadmium has been dissolved.

---

[1] The narrow neck of the flasks used in the assay of gold prevents the air from entering while the flask is being removed, which otherwise might stir up the gold dust.

[2] Fresenius's Ztschr. 1879, p. 104; B. u. h. Ztg. 1879, p. 187.

[3] Oestrr. Ztschft. f. Bg. u. Httnwsn. 1879, p. 597; u. 1880, p. 182. Ztschft. f. Anal. Chem., Bd. 19, p. 201. Dingler, ccxxxvi. 323.

The gold or gold alloy is melted with cadmium under a flux of potassium cyanide. After the fusion is complete, the crucible is allowed to cool, and the potassium cyanide dissolved out with water. The button is then boiled *once* with HNO_3 of 1.2 specific gravity, and *twice* with acid of 1.3 specific gravity, thoroughly washed, dried, annealed, and weighed.

This method of quartation offers the following advantages:—

1. It is very rapid and avoids the trouble of heating up a muffle-furnace.

2. There is no loss of gold by volatilization or cupellation; loss by the latter means increases with the content of copper in the gold.

3. The addition of $2\frac{1}{2}$ times as much cadmium as gold suffices to obtain the latter in a compact form after parting.

4. It can be used equally as well with gold-silver alloys as with gold-copper alloys, and the separate cupelling of the gold-copper alloy is omitted.

5. The thoroughly boiled and washed gold always shows after annealing the pure gold color.

6. This method is especially well adapted for the immediate assay of finished alloys.

7. The content of silver in the parting solution can be determined volumetrically after it has been carefully poured off from the gold.

In performing this assay Kraus[1] recommends to weigh off two separate quantities of alloy of 0.25 gramme each. Melt a piece of potassium cyanide in a small porcelain crucible, and place the alloy and cadmium in it. The fusion takes place quickly; from 20 to 30 fusions may be made in an hour by using several crucibles and dissolving out the potassium cyanide as soon as the fusion is complete. The buttons should be boiled in HNO_3 of 1.3 specific gravity for not less than half an hour; or longer, according to the amount of gold present. Boil a second time 10 minutes in acid of same strength and a third time in water 5 minutes.

As it can be quickly executed, Volhard's method is the most convenient for the determination of the silver in the solutions poured off from the quartation. Of course, the wash-water as well as the nitric acid solutions must be so titrated after cooling.

Juptner's[2] *volumetric assay of gold.*—If the gold alloys to be assayed contain tin, antimony, or some platinum metals, all of

---

[1] Dingler, ccxxxvi. 323.
[2] Oestrr. Ztschft. f. Bg. u. Httnwsn. 1880, p. 182.

which are insoluble in HNO_3, such insoluble residues are treated with $HCl + HNO_3$. The platinum metals are precipitated with ammonium chloride, evaporated to dryness, taken up with water and filtered off. Mix the filtrate with a measured excess of the ferrous sulphate, reducing the chloride to metallic gold, then titrate the non-oxidized portion of the ferrous oxide with permanganate.

The reaction is expressed by the following equation—

$$Au_2O_3 + 6FeO = 2Au + 3Fe_2O_3$$

More correctly $2AuCl_3 + 6FeSO_4 = 2Au + Fe_2Cl_6 + 2[Fe_2(SO_4)_3]$.

Hence, 2 gold correspond to 6 iron, or 1 gold to 3 iron; designating the portion of iron oxidized in titrating with permanganate by m, and the content of gold sought by x, we have the proportion—

$$3\ Fe : Au = 168 : 196.7 = m : x$$
$$x = \frac{196.7}{168},\ m = 1.172\ m.$$

Preparation of the ferrous ammonium sulphate solution.—Dissolve 6 grammes of Mohr's salt in one liter of water acidulated with H_2SO_4. These 6 grammes contain $6 \times 0.14285 = 0.857$ gramme of iron, and 1 c.c. of this solution precipitates exactly one milligramme of gold.

Preparation of the permanganate solution.—This solution should be so prepared that 1 c.c. will exactly correspond to 1 c.c. of Mohr's salt solution. It is then only necessary to deduct the number of c.c. of permanganate used in titrating from the amount of ferrous ammonium sulphate solution added to the gold solution, in order to determine the amount of gold, in milligrammes, in the solution of the alloy.

Separation of gold from platinum by electrolysis.[1]—By bringing gold in connection with the positive pole of a battery in a bath of chloride of gold, it dissolves and firmly deposits itself on the negative electrode without the bath undergoing any change. The gold to be so treated should be in the form of a sheet and connected to the positive pole by means of a copper wire. An accurately weighed sheet of the purest gold should be connected with the negative pole. Both sheets are placed in the solution of chloride of gold and subjected to the action of the electric current. The gold disappears from the positive pole and is deposited on the negative

[1] Fortschritte im Probirwesen. Balling, Berlin, 1887, p. 116.

pole, while the platinum metals are liberated and fall to the bottom as a gray-black powder. When all the gold has been deposited, the sheet of gold at the negative pole is washed with water and alcohol, dried, and weighed. The increase in weight equals the amount of pure gold in the alloy. This method is used on a large scale for the production of fine gold in the North German refining works at Hamburg.[1]

Toughening brittle gold.—This is sometimes necessary when the metal has been rendered brittle by the presence of minute quantities of other metals or impurities.

Roberts[2] removes such impurities by converting them into volatile chlorides by a stream of chlorine gas.

According to Wagner,[3] bromine may be used instead of chlorine for this purpose. Brittle gold may also be toughened by throwing a small quantity of corrosive sublimate (mercuric chloride) on the surface of the molten metal.

It can, according to Warington,[4] be toughened by the addition of about 10 per cent. of black oxide of copper. If the gold is but slightly brittle, it may be toughened by simply pouring it in a thin stream through atmospheric air into a crucible lined with borax, or by the addition of a small amount of chloride of copper.[5]

Booth[6] uses a flux of soda-ash and borax glass. When a quiet fusion has taken place a small quantity of nitre is added. The moment that the visible oxidizing action begins to slacken, the flux is skimmed off. The gold in the pot will then be found toughened.

V. PLATINUM.

36. ORES.

Native platinum is almost always combined with other metals of the platinum group (Rh, Ir, Pd, Os), with precious (Au) and base

---

[1] Bg. u. h. Ztg., 1880, p. 411.

[2] First and Second Annual Reports of Deputy-Master of Mint, 1870–72, pp. 93 and 34, respectively. See Percy's Metallurgy, Silver and Gold, pp. 405, 407, and 437.

[3] Bull. Chem. Soc., Paris, t. xxv. 1876, p. 138.

[4] Journ. Chem. Soc., xiii. 1860, p. 31.

[5] Encyclopædia Britannica, vol. x. p. 750.

[6] Engineering and Mining Journ., July 12, 1884, p. 22.

metals (Fe, Cu), and admixed with iridosmium, earthy and metallic minerals.

37. ASSAY OF PLATINIFEROUS ORES.

A. Fire assays.[1]—These extend to the determination of—

1. *Percentage of sand.*—2 grammes of ore are mixed with 10 grammes of granulated silver and placed in a clay crucible glazed with fused borax. It is covered with 10 grammes of borax glass, with a small piece of wood charcoal, and fused. The resulting button is weighed, and its difference in weight, as compared with that of the ore, giving due consideration to the added silver, indicates the percentage of sand.

2. *Percentage of gold.*—10 grammes of the ore are boiled with mercury for several hours. It is then washed out with hot mercury, and the gold amalgam distilled in a small retort.

3. *Percentage of platinum.*—The platinum is combined with lead by fusing 50 grammes of ore with 75 grammes of granulated lead, 50 grammes of galena, 10 to 15 grammes of borax, and adding to the fused mass 50 grammes of litharge; or 20 grammes of ore are fused with 15 grammes of borax, 30 grammes of soda, 1 gramme of powdered charcoal, and 50 grammes of litharge. The fused mass is allowed to cool off, and the lead button separated from the matt (sulphur compounds of copper, iron, and lead) above it, and the iridosmium below it. The lead button, if too large, is scorified with some borax (p. 64), and cupelled at as high a temperature as possible.[2] The platinum remaining is purified by a further fusion with 6 to 7 per cent. of lead in a lime crucible heated with illuminating gas and oxygen.[3]

B. Wet assay.—5 to 10 grammes of ore are treated with hydrochloric acid, and the residue is washed out and digested with *aqua regia* for from 8 to 12 hours. The platiniferous solution is filtered from the residue (sand, iridosmium) and evaporated almost to dryness. Absolute alcohol and solution of sal-ammoniac are then

---

[1] Mispratt's Chem.; v. 1151.

[2] [The surface of the button after this operation is dull and the fracture crystalline, even when very small quantities of platinum are present.—G.]

[3] [The amount of lead retained in the *unpurified* button varies from $\frac{1}{8}$ to $\frac{1}{4}$ of its weight.—G.]

added until a precipitate ceases to form. This is again filtered and washed, and the yellow ammonio-platinic chloride is dried. This is highly heated, and the resulting spongy platinum weighed. If the ore contains gold, it is precipitated from the filtrate of ammonio-platinic chloride with ferrous sulphate. The precipitated gold is then digested with hydrochloric acid, filtered, washed, and dried, and fused with the addition of some borax glass.

38. ALLOYS OF PLATINUM.

These may be:—

1. *Gold with platinum.*—The alloy is cupelled with three times the quantity of silver, and sufficient lead to remove any copper which may be present (8 to 30 times the quantity of lead if from 200 to 500 thousandths or more of copper are present.) The laminated button is treated with nitric acid, as in the gold assay (p. 170), whereby the platinum will be dissolved with the silver, the gold remaining behind. The silver is precipitated from the solution by means of common salt, and the platinum is separated from the filtrate as ammonio-platinic chloride, and further treated as on p. 185. If the percentage of gold is large, the alloy is dissolved in *aqua regia,* and the separation conducted in the manner indicated for platinum ores (p. 185).

2. *Silver with platinum.*—0.5 gramme of the alloy is cupelled with a sufficient quantity of lead to remove the copper which may be present, and with such a quantity of silver (to be determined by a preliminary assay) as to make the ratio 1 part platinum to 2 parts of silver. The alloy is laminated and boiled twice (each time from 10 to 12 minutes) with concentrated sulphuric acid of 1.85 specific gravity, whereby platinum will remain behind in the form of a small roll, and, if a larger quantity of silver is present, as a powder. It is then washed with hot water, dried, ignited, and weighed.

3. *Silver and gold with platinum.*—200 milligrammes are cupelled with sufficient silver, for instance, 100 milligrammes, to make the ratio 1 part gold to 3 parts silver, and with lead to remove the *base metals.* The button is laminated, being frequently heated during the operation. It is then made into a roll and boiled with concentrated sulphuric acid. The residue is washed,

ignited, and weighed; the difference in weight represents the *silver* originally present plus that added to it. The residue, containing gold, platinum, and iridosmium, is cupelled with lead, and with a quantity of silver at least 12 times that of the platinum (the effect of less silver would be to leave a residue of platinum, and if more is taken the residue will be pulverulent instead of in the form of a roll). The roll is first boiled with nitric acid of 1.16 specific gravity, and then with acid of 1.26 specific gravity, after which the residuum (gold and iridosmium) is washed and ignited. The dissolved platinum is determined from the difference. The residue is digested with *aqua regia*, and the gold precipitated with ferrous sulphate, while iridosmium remains behind. About 3 hours are required for two assays.

Electrolytic assay.[1]—Compounds of platinum are decomposed with the greatest ease by the galvanic current, with the deposition of the metal on the negative electrode. A current of two Bunsen cells produces decomposition so rapidly that the platinum separates as platinum black, and cannot be determined. If one cell is used it separates in so dense a form that it cannot be separated from hammered platinum. It is possible, in this way, to gradually deposit considerable quantities of platinum on the negative electrode without changing its appearance. For the determination of platinum in its salts, the solution may be slightly acidified with HCl or H_2SO_4, or treated with ammonium or potassium oxalate, gently warmed and electrolyzed. The platinum separates in a comparatively short time; for example, a solution of platinum chloride diluted to 200 c.c., containing 0.6 gramme platinum, deposited 0.5 gramme in five hours.

Iridium is not reduced from its solutions by a single Bunsen cell; this fact may be used for the separation of platinum from iridium.

VI. NICKEL.

39. ORES.

Copper nickel, NiAs, with 44 Ni; *antimonial nickel* (*breithauptite*), NiSb, with 31.4 Ni; *rammelsbergite*, $NiAs_2$, with 28.2 Ni;

---

[1] Classen's Chem. Analysis by Electrolysis. [Trans.] New York. 1887. p. 71.

nickel sulphide (millerite), NiS, with 64.5 Ni; *antimonial nickel ore (ullmannite)*, NiSbS, with 27.6 Ni; *nickel glance (gersdorffite)*, NiAsS, with 35.1 Ni; *nickel silicates* as *revdanskite* and *garnierite*, with 10 to 20 Ni; *nickel arseniate*, $Ni_3As_2O_8 + 8H_2O$, with 29.5 Ni; *nickeliferous iron, copper*, and *magnetic pyrites*.

40. FIRE ASSAY (*Plattner's Assay*).

This is based upon the formation of constant combinations of Ni_2As, and Co_2As, with respectively 60.7 Ni and 61.1 Co, and their subsequent treatment, in a manner to be indicated with borax, after other foreign admixtures have been removed. The process requires certain modifications if *copper, lead, bismuth*, and *antimony* are present.

A. Compounds free from copper.

1. A sufficient quantity of assay sample is weighed out so that the resulting buttons of Ni_2As and Co_2As will weigh from 0.4 to 0.6 gramme. Thus, about 5 grammes of poor ores will be required, 1.5 to 2.5 grammes of medium, and 0.5 to 0.6 gramme of rich ores.

2. Compounds containing *metallic sulphides* must be *completely roasted* with charcoal and ammonium carbonate; otherwise, the buttons cannot be properly slagged with borax. Substances free from sulphur need not to be roasted.

Five grammes of ore containing sulphates which cannot be decomposed by roasting (gypsum, barytes, etc.), with 10 to 15 grammes of borax, 5 to 10 grammes of glass, and 0.5 gramme of resin are placed in a suitable crucible (Fig. 52, p. 65), covered with a layer of common salt, and fused to (brittle) matt (p. 99). This is dead-roasted. 0.5 to 1.5 grammes of arsenic are added if the nickel was not combined with sulphur or arsenic.

3. *Arsenizing.*—The roasted assay sample is intimately rubbed together with 1 to 1½ times the quantity of metallic arsenic in an iron mortar, placed in a suitable covered crucible (Fig. 52, p. 65), and heated in the muffle, kept at orange-red heat (for 10 to 15 minutes) until the arsenical flame and vapors have ceased to appear. When this is the case, the metallic oxides contained in the roasted sample will have been reduced by a part of the arsenic,

and converted by another part into metallic arsenides of variable composition (of iron, nickel, cobalt, etc.), which are either only sintered together (especially if the charge is rich in cobalt), or fused.

Compounds free from sulphur and rich in arsenic, containing more arsenic than is necessary for the formation of Co_2As and Ni_2As, do not require roasting and arsenizing. Alloys (argentan, nickel coins, nickeliferous black copper, etc.) must be laminated, and several times arsenized with an equal quantity of arsenic; and also substances rich in cobalt (for instance, mixtures of nickel and cobalt oxides), which may have been precipitated in the wet way.

4. *Reducing and solvent fusion* for the purpose of collecting the metallic arsenides into a button (arsenical iron, arsenical nickel, arsenical cobalt), and of slagging off earths and foreign oxides, zinc being entirely, and antimony partly volatilized in the operation. The mass is placed in a sound crucible, and 10 to 12.5 grammes of potassium carbonate and flour are added to it. Upon this is placed one small spoonful of borax and two of powdered glass, and a covering of common salt with a small piece of coal. The crucible is then placed in the muffle-furnace, wood-charcoal is piled high around it, the mouth of the muffle is closed, and the charge, after the "flaming" has ceased, is fused for one-half to three-quarters of an hour at an orange-red heat. The assay is then taken out, allowed to cool off, and the brittle button very carefully freed from slag.

Modifications which may occur.

a. *Addition of iron filings*: 0.5 to 0.75 gramme of iron filings must be added during arsenizing, if the ores, etc., are rich in cobalt, and refractory; 0.05 to 0.20 gramme, if they are entirely free from iron, or contain but little of it, in order to prevent the slagging off of cobalt too soon, by the borax. If *lead* is present, 0.5 to 0.75 gramme of iron in the form of a thick wire is added, or the mass is fused in an iron crucible, which will allow a better regulation of the consumption of iron. The lead attaches itself upon the button of arsenides, and its weight is found by weighing lead and button together, cutting the former off and reweighing. If *bismuth* is present, it will attach itself upon the brittle button of arsenides, in the form of brittle metal, which cannot be detached from it. In this case, it is necessary to add 0.5 to 0.6 gramme of granulated lead to the charge, when a ductile alloy of both metals will separate on the button of arsenides, which can be easily disconnected from it. The approximate percentage of bismuth can be calculated after deducting the added granulated lead, minus 4 per cent. loss.

b. Arsenizing and fusion in one operation.—The roasted assay sample is rubbed together with arsenic in the same manner as previously stated, and the mixture is wrapped up in a cylinder of soda paper. The cylinder is formed over a wooden stick of 16 millimeters diameter, by closing the lapping edges with lac. It is pressed firmly into a crucible (Fig. 52, p. 65) and covered with 15 grammes of black flux, 1 small spoonful of borax, 1 small spoonful of glass, 15 grammes of common salt, and a small piece of charcoal. Accurate results are obtained by this process.—Or, the roasted sample is rubbed together with an equal quantity of arsenic and 15 per cent. of arsenical iron (Fe_2As), and fused with the above fluxes.

5. *Slagging off of the arsenical iron.*—Wood charcoal is placed all about the inside of the muffle, and one or two refining dishes are placed in the centre of it. The muffle is then closed, and the dishes are brought to a white heat by a strong fire. 1.5 to 2 grammes of borax glass are then placed in the dishes by means of an iron spoon (or wrapped up in a cornet), the muffle is closed and the borax fused. The button of arsenides is now placed in the dish, the mouth of the muffle is again closed, and the button fused *as quickly as possible* at a *very high* temperature (if the temperature is too low and the fusing takes too much time, obalt also will be slagged off). The mouth of the muffle is now opened, placing a piece of glowing charcoal in front, to allow the entrance of air, whereby the arsenide of iron is oxidixed to basic iron arseniate. This covers the button with a crust or scale (the scaling of the button) which is continuously dissolved by the borax until the surface of the dull button appears bright, when this operation is finished. The dish is lifted out by means of the tongs, and the lower part of it is first dipped into water until its contents have ceased to glow, when the entire dish is submerged.

The following are indications of a successful assay: The button is bright, the slag black or green, with a bluish tint, which is a sure indication of all the iron having been removed.

Modifications.—If the button is very rich in iron it is repeatedly treated with fresh borax, as this is saturated and becomes stiff and the button no longer "drives."

Separation of *copper-red scales* of iron arseniate from strongly saturated borax. *More cobalt* will slag off (the slag has a strong blue tint) if the temperature is too low or the slagging off is continued too long, or when no iron, or but little of it, was present in the button.

6. *Dearsenizing.*—An excess of arsenic is volatilized by heating the button in a small covered crucible (Fig. 49, p. 64) in charcoal powder, in the muffle heated to bright redness for one-fourth to one-half hour, in order that constant combinations of Ni_2As and Co_2As shall be formed. The resulting button is weighed, and the operation is repeated until its weight remains constant.

7. *Slagging off the cobalt arsenide.*—The process is the same as in slagging off the iron arsenide (p. 190), but at a higher temperature, the quiet button remaining *bright* during the slagging off of the cobalt. The process is interrupted as soon as a film of apple-green basic nickel arseniate forms on the surface of the button. The dish is taken and cooled off in the same manner as in the slagging off of iron arsenide (p. 190). If the assay has been properly done, the bright, white button will show on its surface small green patches of nickel arseniate, the slag is blue with a violet tint (from the blue of the cobalt and the brown of the nickel), and a green stain will be perceptible on the place where the button has rested. The button consisting of Ni_2As is weighed, and the percentage of nickel calculated therefrom (p. 188), the Co_2As being determined from the difference of $Co_2As + Ni_2As_2$.

B. Cupriferous compounds.

A percentage of copper[1] remains behind with Ni_2As as a constant combination of Cu_2As, and can be determined according to *Plattner's* method:

---

[1] B. u. h. Ztg. 1868, p. 24; 1868, p. 94 (Kleinschmidt); 1878, p. 88 (Schweder).

1. *If the percentage of copper is small*, and does not exceed that of nickel, by the addition to the weighed button ($Ni_2As + Cu_3As$) of 6 to 8 times the quantity of gold accurately weighed (to prevent a slagging off of the copper in the subsequent operation). The arsenide button with gold addition wrapped in a cornet is placed in salt of phosphorus which has been fused in a suitable shallow dish. This salt exerts a more vigorous effect than borax in slagging off with yellowish-brown color the nickel arseniate which will be formed. If necessary, the oxidizing process is continued by renewing the saturated salt of phosphorus until the button appears bright, a proof that the nickel is slagged off, the complete volatilization of the arsenic being indicated later on by the button ceasing to fume. The remaining alloy of Au and Cu is weighed, and the weight of the copper, which is obtained by deducting that of the added gold, is calculated to Cu_3As, with 71.7 per cent. Cu, and deducted from the total weight of the $Ni_2As + Cu_3As$, from which the percentage of nickel is calculated.

This assay becomes less accurate with an increase in the percentage of copper, as, during the slagging off of the last portions of the nickel arseniate, the copper also commences to slag off. For this reason—

2. The *wet method* is partially made use of when the *percentage of copper is large*. The processes are as follows:—

a. The button consisting of Ni_2As and Cu_3As is dissolved in nitric acid, and evaporated to dryness with sulphuric acid. The residue is digested with aqueous sulphurous acid until no odor of the latter remains. The copper and arsenic (also antimony) are now precipitated from acid solution by sulphuretted hydrogen, and the arsenic sulphide (also antimony sulphide) is extracted with a warm solution of sodium sulphide. The residue remaining in the filter is washed and dried in the roasting dish in front of the muffle, and is then ignited. The copper sulphide is rubbed up and strongly heated, ammonium carbonate being added towards the end. The copper oxide which has been produced is weighed, and calculated to Cu_3As. This is deducted from $Ni_2As + Cu_3As$ to determine the Ni_2As (*Patera*).

b. By another method, the ore, etc., is dissolved, and the copper precipitated by the galvanic current (pp. 110 *et seq.*), and the remaining solution with potassium hydrate. The precipitate, containing iron, nickel, and cobalt, is washed, dried, ignited, arsenized, and the further process conducted as given in the dry method, p. 188. When *much iron* is present, it is better, on account of the labor of washing the iron precipitate, to prepare an assay according to Plattner, for $Ni_2As + Cu_3As$, to separate the copper from a second fresh charge by electrolysis, to calculate the copper to Cu_3As, and deduct this from $Ni_2As + Cu_3As$, which will give the Ni_2As (Schweder).

Nickeliferous pyrrhotine, with 0.82 per cent. Cu and 1.72 per cent Ni and Co : 2 grammes are dissolved as above, p. 111, and the copper is precipitated by electrolysis from a solution of 40 cubic centimeters of nitric acid, and 360 cubic centimeters of water (p. 110); or 5 grammes of ore are roasted and charged with arsenic in a soda paper cylinder (p. 190), and fused with fluxing agents in the crucible (p. 191); the iron is slagged off twice and the button then dearsenized. If cobalt is absent, $Ni_2As + Cu_3As$ will remain behind, and the nickel is then calculated as previously stated (p. 191). Cobalt and nickel may also be determined by electrolysis, and calculated to $(Ni,Co)_2As$ and the Cu_3As determined from the difference.

c. Compounds soluble with difficulty, as, for instance, slags.—These are roasted, arsenized, and fused according to the process given on p. 188. If they are poor in nickel, several buttons (say five) are wrapped in a cornet and treated with borax, as before described. The iron is slagged off, the excess of arsenic removed, and the button ($Ni_2As + Cu_3As$) weighed. It is then dissolved in 20 cubic centimeters of nitric acid, 200 cubic centimeters of water are added, and the copper is precipitated by electrolysis until it commences to be colored black by the arsenic (p. 112). The Cu_3As is calculated from the precipitated copper, and deducted from the $Ni_2As + Cu_3As$, etc., or, what is still better, in order to avoid constantly watching the precipitation of copper, lest arsenic be precipitated with it, the button of $Ni_2As + Cu_3As$ is dissolved in nitric acid, in a covered beaker-glass, then evaporated to dryness, and the copper and arsenic are precipitated with sulphuretted hydrogen. The filtrate is heated in order to drive off the sulphuretted hydrogen, ammonium sulphate and ammonia are added, and the nickel determined by electrolysis (see later on). This is calculated to Ni_2As, which is deducted from $NiAs + Cu_3As$ and Cu_3As is found from the difference.

C. Compounds containing antimony.

If a large amount of antimony is present, it becomes necessary to remove it from the dissolved ore by sulphuretted hydrogen. It is then filtered, and the filtrate boiled in order to expel the sulphuretted hydrogen. The filtrate is now oxidized with potassium chlorate, the iron, nickel, and cobalt are precipitated with potassium hydrate, and the precipitate is filtered, dried, ignited, and arsenized (p. 188).

41. WET ASSAY.

The *gravimetric* analysis, being more accurate, and, especially the electrolytic assay, more simple than the *volumetric analysis*, it is more frequently used for the wet assay of nickel.

A. Gravimetric assay.

1. *Electrolytic assay.*[1]—This is based upon the precipitation of nickel (and at the same time of cobalt, if present) from ammoniacal solution (copper from acid solution) by the galvanic current. When *copper* and *lead* are present, 1 gramme of ore is dissolved in 20 cubic centimeters of nitric acid, and evaporated with a few drops of sulphuric acid in order to form lead sulphate, and the copper, antimony, and arsenic in acid solution are precipitated with sulphuretted hydrogen. (This method of precipitation is to be preferred, as, if the copper is previously precipitated from acid solution by the galvanic current, antimony and arsenic remain in the filtrate.)

The filtrate is evaporated in a porcelain dish, first over the lamp, then on the water-bath, with the addition of a few drops of nitric acid; and, in case dust should float in the fluid, some hydrochlorate acid is also added, when the liberated chlorine will destroy the organic substances which otherwise, being converted into sugar by sulphuric acid, would, in the subsequent precipita-

---

[1] Fresenius's Ztschr. 1872, p. 1; B. u. h. Ztg. 1877, p. 5 (Schweder).

tion of the iron, hold a part of it in solution as ferrous oxide, which would then be precipitated with the nickel. The free sulphuric acid is expelled by heating the sand-bath. The mass is next dissolved in water and supersaturated with ammonia, in order to prepare an ammoniacal solution of copper.

If but little iron is present, ammonia may be used directly; but should there be a larger quantity, the precipitated ferric hydrate remains nickeliferous. If this is the case, the residue *completely* freed from *free* sulphuric acid by evaporation, is dissolved in 100 cubic centimeters of hot water, and after the solution has become entirely cold, 200 to 300 cubic centimeters of cold water are added, according as more or less iron is present. Next a solution of ammonium sesquicarbonate in 12 parts of water is added drop by drop from a pipette, under constant stirring, until the liquid appears dark brown, but without being turbid (if this is the case, a few drops of sulphuric acid must be added), and without developing carbonic acid. The vessel is then covered with a watch-crystal, and the liquid *slowly* heated to the boiling point, whereby the greatest part of the iron will be separated as basic sulphate of a leather-yellow color. The vessel is taken from the fire, the watch-crystal rinsed off, and the liquid placed in the water-bath and allowed to settle. It is then filtered, and the precipitate, which is free from cobalt and nickel, washed with hot water. If the assay sample is rich in iron, the filtrate will still contain Fe. In this case it is allowed to become quite cold, and is then again precipitated with ammonium carbonate, as above, etc., but the filtrate will always contain notable traces of iron. A *few* drops of ammonium acetate are then added, and the filtrate is evaporated so far that the solution will be contained in the beaker-glass, in which it is intended to make the electrolytic precipitation. The fluid is then filtered into this beaker from the small quantity of iron sediment, supersaturated with ammonia, etc.

Instead of supersaturating the solution with ammonia, it is better to add 20 cubic centimeters of ammonia, and the same quantity of ammonium sulphate, and to electrolyze it in the apparatus used for the galvanic assay of copper (p. 110). The binding screws of the electrode should be well coated with shellac to protect them against corrosion by the ammoniacal vapors.

In the presence of much iron, and of ammonium chloride, chlorine, which attacks the platinum, will be developed on the positive electrode. For this reason a sulphuric acid solution is to be preferred, and ammonium sulphate exerts a favorable effect upon the process.

As the ammoniacal nickel solutions offer greater *resistance* to the galvanic current than the acid copper solutions, the nickel will be principally deposited on the inner sides and lower end of the platinum cone, where small bubbles of hydrogen may also make their appearance, causing the nickel to be deposited in fine, non-coherent laminæ, which spring off in the subsequent rinsing and drying, thus occasioning losses. This must be avoided by suspending the platinum cone in such a manner that its lower edges shall be about 1.5 centimeters above the bottom of the glass; further, by increasing the electrical conductivity of the solution by an addition of ammonium sulphate, and by using a current sufficiently strong to give at least 100 cubic centimeters of oxyhydrogen gas in the voltameter in half an hour. A black coating of nickel sesquioxide, which may be formed on the positive electrode, will generally disappear by strongly supersaturating the fluid with ammonia. Should this not produce the desired effect, the platinum cone is first taken out, placed in water, the spiral is then removed, and a few drops of hydrochloric acid are allowed to drip down on it, which will cause the coating to dissolve with evolution of chlorine. Some ammonia is then added, and the apparatus again put in place and allowed to operate for a few hours longer. Should ferric hydrate be separated, it is collected, in case it is to be determined by volumetric assay, and added to that first obtained. After an electrolyzation of about 18 hours, the platinum cone is taken out, placed in a beaker-glass containing water, and rinsed off with hot water. It is then taken out and placed upon several thicknesses of filter paper. From here it is brought into a beaker-glass containing alcohol, alcohol being likewise allowed to run down on the wire. The cone is next placed upon blotting paper and then completely dried upon a sheet-iron plate heated over a lamp. After it has become cool, it is weighed and the nickel (or nickel and cobalt) is ascertained from the increase in weight.

When *zinc* is present, some of which is precipitated along with nickel (hydrochloric and nitric acid, and their salts, prevent this precipitation, but sulphuric acid does not), the hydrochloric or nitric acid solution is treated, after driving off the sulphuretted hydrogen, with sodium carbonate until it exhibits only a slight acid reaction. More sulphuretted hydrogen is then introduced, in order to precipitate zinc sulphide, until no further increase of the precipitate is perceptible; a drop of a diluted solution of sodium acetate is then added. Sulphuretted hydrogen is now again introduced for some time, and the liquid then allowed to stand quietly for twelve hours. It is then filtered and washed with sulphuretted hydrogen water. The zinc sulphide is dissolved in hydrochloric acid, and precipitated, after the sulphuretted hydrogen has been expelled, by sodium carbonate; or the zinc sulphide, with the addition of some sulphur, is ignited in a current of hydrogen gas (p. 117). The nickeliferous filtrate, from the precipitate with sulphuretted hydrogen, is heated, some ammonia added, and it is then electrolyzed, after the iron has been separated in the manner indicated on p. 195.

The platinum cone coated with cobalt and nickel is placed in a beaker-glass, and covered with a mixture of 1 part of nitric acid, of 1.2 specific gravity, and 3 parts of water. This is heated until solution of the metallic deposit is complete. The cone is then taken out and rinsed off with hot water. The solution is evaporated to a small volume, and some potassium hydrate added until precipitation just begins. The precipitate is dissolved by the addition of some acetic acid, a concentrated solution of potassium nitrate is added, and the fluid allowed to stand for 24 hours. The yellow precipitate of $CO(NO_2)_3 + 3KNO_2$ is filtered off, and washed first with potassium acetate and then with alcohol. It is next dissolved in sulphuric acid, and digested until the odor of nitrous acid has been entirely dispelled. It is then neutralized with ammonia, 20 cubic centimeters of ammonium sulphate and the same quantity of ammonia are added, and the solution again subjected to electrolyzation. The cobalt is weighed, and the nickel determined from the difference.

Determination of nickel in pyrites and matt.[1]—2 to 5 grammes of the substance are dissolved in hydrochloric acid, to which some nitric acid has been added, and the solution precipitated with sulphuretted hydrogen (p. 194). The filtrate is boiled in order to expel the sulphuretted hydrogen, and nitric acid or potassium chlorate is added to oxidize ferrous to ferric oxide. Ammonia is then added until some precipitate forms, but without precipitation being complete. Some acetic acid is now added until a deep-red solution is formed. This is boiled, and a solution of concentrated sodium phosphate is added *in excess*. The precipitate of iron is filtered off and washed with hot water containing some acetic acid. The filtrate is boiled, and potassium hydrate added until an odor of ammonia is clearly perceptible. The apple-green nickel phosphate is washed out, dissolved in diluted sulphuric acid, and made strongly alkaline by addition of ammonia. The solution is then subjected to electrolyzation. When more than 3 per cent. of nickel is present, the precipitate of iron is again dissolved, reprecipitated, etc. A trace of iron, but not enough to influence the precipitation of nickel by the battery, may remain in the solution if too small a quantity of sodium phosphate is present, or if the solution was made alkaline before the addition of acetic acid.

Assay of New Caledonia ores.—Allen[2] recommends the following process for the examination of these ores (which are a silicate of magnesium containing nickel), or, in general, for ores containing neither sulphur nor arsenic: Fuse 2 grammes of ore in a platinum crucible with potassium bisulphate and nitre, lixiviate with water, boil the residue with some HCl and filter the whole. Neutralize carefully with ammonia, add excess of ammonium acetate and boil, thus precipitating the iron, alumina, and chromic oxide. Dissolve the precipitate in HCl and again precipitate in same manner. Boil the combined filtrates, add sufficient ammonia to nearly neutralize the acetic acid, then pass H_2S through the solution whilst hot. Wash the precipitate of nickel sulphide (perhaps also some cobalt sulphide) with water containing H_2S and ammonium acetate, then dissolve this precipitate in HNO_3 and add some H_2SO_4. Add

---

[1] B. u. h. Ztg. 1878, p. 41.
[2] Bullet. de la Société d'Encouragement, t. vi. p. 36 (1879).

excess of ammonia, filter off the slight precipitate which may form and subject the ammoniacal solution, in a platinum dish, to electrolysis, care being observed that the solution constantly shows an ammoniacal reaction. In all the electrolytic precipitations of nickel (cobalt) the solution must always contain an excess of free ammonia. The presence of ammonium sulphate promotes the separation of the metals, but that of ammonium chloride or nitrate hinders or prevents it. Too much ammonia, however, is not good, as it opposes considerable resistance to the electric current. Braun recommends the use of potassium sulpho-carbonate for indicating the end of the precipitation; its reaction with nickel is very sensitive. The precipitation may be regarded as finished when a sample of the nickel solution acquires only a very slight rose color. Classen's method for the determination of nickel by electrolysis is exactly the same as for zinc.

Electrolytic determination of copper, nickel, and cobalt in speiss. —According to Ohl,[1] the process is as follows: Digest one gramme of the finely pulverized speiss in a covered beaker (of 200 c.c. capacity) with HNO_3 or $HNO_3 + HCl$ and evaporate to dryness. Add enough concentrated HCl to cover the dry residue to a height of 5 millimeters, and after solution has taken place dilute with water up to half the height of beaker. Heat and pass H_2S through solution until it is cold, then reheat and repeat this operation. If only arsenious sulphide has been precipitated, again the warm solution until the odor of H_2S can no longer be detected. If the precipitate is not pure yellow, but is discolored (probably cupreous), filter until the solution has no further odor of H_2S and wash with water containing H_2S, using, however, only pure cold water if arsenic alone is precipitated. After adding potassium chlorate to oxidize any iron present, evaporate the filtrate containing nickel and cobalt to dryness, take up with water and HCl and mix with pure soda solution until alkaline reaction. The precipitate is then dissolved in acetic acid, and the solution largely diluted and boiled. Filter and wash with hot water until a drop of wash water produces no turbidity with ammonium sulphide. If zinc is present, acetic acid must now be added and the zinc precipitated with H_2S. The zinc sulphide filtered off, filtrate evaporated to dryness, taken up with water and H_2SO_4, saturated with ammonia and the solution subjected to the electric current. When the solution of nickel and cobalt becomes discolored it is

[1] Ztschft. f. Anal. Chem. Bd. 18, p. 523.

tested with ammonium sulphide to see if the precipitation is complete. It is better after dissolving the metals from the electrode to separate the cobalt by potassium nitrite and after dissolving the yellow salt formed and saturating with ammonia to determine the cobalt by electrolysis. In the simultaneous presence of both arsenic and copper they are separated in the following manner:[1] Pass H_2S slowly through the original solution (previously filtered and cooled if necessary), whereby the copper sulphide separates out in flakes and the supernatant liquid is clear when held to the light. It is claimed that in this manner the copper can be sharply separated from the arsenic. After filtering and washing, the precipitate is dissolved in HNO_3, evaporated to dryness, taken up with 20 c.c. of HNO_3, and after diluting to 200 c.c. with water the solution subjected to action of electric current. The filter paper may be allowed to remain in the solution so treated without doing harm. In the presence of only a small amount of arsenic the copper is first precipitated. The solution thus freed from copper is added to the filtrate from the copper sulphide, the arsenic completely precipitated with H_2S, the arsenious sulphide filtered off, dissolved in *aqua regia*, saturated with ammonia and precipitated as ammonium-magnesium arsenate. If in the electrolytic precipitation of the nickel and cobalt some iron appears it is filtered off, added to that already precipitated out, the whole dissolved in HCl, and the iron determined by titration with proto-chloride of tin. If antimony is also present, it is separated as antimonic oxide by twice evaporating the precipitated metallic sulphide with HNO_3.

2. *Other assays.*[2]

a. The ore, etc., is dissolved in *aqua regia* and evaporated with the addition of sulphuric acid. Aqueous sulphurous acid is added to this in order to reduce the arsenic acid. The sulphurous acid is then removed by boiling, and the copper, arsenic, etc., are precipitated with sulphuretted hydrogen. The liquid is then filtered, the sulphuretted hydrogen removed by evaporation, and the ferrous oxide oxidized by potassium chlorate. Sodium carbonate, boiling hot, is then added until a permanent precipitate is formed. A drop of hydrochloric acid is now added, until the precipitate

---

[1] Fortschritte im Probirwesen. Balling, Berlin, 1887, p. 151.

[2] Fresenius's Verfahren für Erze, Leche und Speisen in Fresenius's Ztschr. xii. 70.

just disappears, and then a large quantity of sodium acetate added to the hot solution. This is boiled for 10 to 15 minutes, and the precipitate of iron and alumina quickly filtered off, washed, and again dissolved in hydrochloric acid, etc., in order to separate such traces of nickel as it may contain. Chlorine water is added to the nickeliferous filtrate, followed by potassium hydrate (some zinc also will be precipitated), and, if necessary, more chlorine water. This is boiled for a quarter of an hour, or until the precipitate of nickel and cobalt sesquioxides has assumed a black color. It is then filtered, the precipitate dried, and the oxides reduced in a current of hydrogen gas (p. 117). For *the separation of nickel and cobalt,* the weighed metals are dissolved in hydrochloric acid. The solution is slightly supersaturated with sodium carbonate. A concentrated solution of potassium nitrite is added, and then some acetic acid until a weak acid reaction is established. The liquid is then allowed to stand for 24 hours, when the yellow cobalt precipitate is filtered off and washed with a concentrated solution of potassium chloride, or potassium sulphate. The precipitate is now dissolved in hydrochloric acid, the solution precipitated with sodium hydrate and chlorine water, etc., and the cobalt sesquioxide reduced in a current of hydrogen gas. The solution carrying the nickel is precipitated with potassium hydrate and chlorine water, etc.

In order to *separate the iron,* the oxidized iron is partly precipitated with sodium carbonate (p. 200). Some acetic acid is added, and the liquid is heated to 30° to 40° C. (86° to 104° F.) in order to dissolve the precipitated iron. It is then boiled, the precipitated iron is again dissolved two or three times, and the precipitation repeated to separate the nickel from it. Finally, by adding sodium sulphide the ferric oxide is converted into iron sulphide, and this is treated with diluted hydrochloric acid, whereby a trace of nickel sulphide may still be found in the residue.

b. Nickeliferous solution from the assay with sulpho-cyanide for determining copper in nickel coins.—In order to decompose the sulpho-cyanide, the solution is evaporated with the addition of 10 cubic centimeters of nitric acid, which will cause the fluid to become first red, and then colorless. The nickel is precipitated by pouring the nickeliferous liquid into a boiling solution of 100 cubic centimetres of sodium hydrate, 10 per cent. strength, which is boiling in a platinum evaporating dish. It is allowed to boil, then diluted with water, and again heated to the boiling point. It is allowed to settle, and then decant through a filter. The precipitate is again boiled three times, each time with 200 cubic centimeters of water, and then filtered. The precipitate is dried, heated to redness, and rubbed fine, again washed with boiling water, dried, ignited, and the nickel determined from the nickel protoxide[1] (now free from alkali) with 78.38 per cent. nickel.

Inaccurate results will follow if less, or less concentrated, sodium hydrate is used, or if the washing is continued too long, as under these circumstances some nickel is dissolved as hydrated oxide.

B. Volumetric assay with sodium sulphide.[2]—The copper which may be present is removed by means of sulphuretted hydrogen, and its percentage determined by the potassium cyanide (p. 120). The filtrate from the precipitation with sulphuretted hydrogen is evaporated with nitric acid, and the iron precipitated with ammonia. The precipitate is dissolved twice or three times, and again precipitated with ammonia (the methods with ammonium sesquicarbonate or with sodium acetate (p. 201) give a more accurate separation); a standard solution of sodium sulphide, 50 cubic centimeters of which precipitate 0.25 gramme of nickel, is then added to the vigorously boiling solution, until all the nickel (with cobalt) has been separated. The addition of sodium sulphide is made in quantities of not more than ¼ cubic centimeter at a time, the solution being maintained at the boiling point, until a filtered drop becomes brown when brought in contact with lead solution upon a porcelain plate.—The solution of lead is prepared by dissolving equal quantities of lead acetate and potassium tartrate in caustic potassa. The normal solution is prepared by dissolving 0.25 gramme of pure nickel (or cobalt), or an equivalent quantity of pure oxide or salt, in 5 cubic centimeters of concentrated nitric acid. This is diluted with water,

---

[1] Fresenius's Ztschr., 1878, p. 58.
[2] Journal für prakt. Chemie, lxxxviii. 486 (Küntzel); xcii. 450 (Winkler); Fresenius's Ztschr. vi. 66 (Braun).

supersaturated with ammonia, tested with a saturated solution of sodium sulphide, which is thereupon suitably diluted.

Separation of cobalt.—The metallic sulphides of nickel and cobalt, which have been precipitated with sodium sulphide, are filtered off, and the precipitate is washed with sulphuretted hydrogen water. It is then dissolved in *aqua regia* and evaporated with hydrochloric acid to expel the nitric acid. It is now strongly diluted with water in a suitable flask, and nearly neutralized. Elutriated barium carbonate is added, chlorine gas introduced, and the precipitated black cobalt sesquioxide dissolved in hydrochloric acid. The barium is precipitated with sulphuric acid, ammonia in excess is added, and the cobalt titrated with solution of sodium sulphide. The barium is then precipitated from the nickeliferous filtrate, ammonia is added, and the nickel similarly titrated with solution of sodium sulphide. When *zinc* and *manganese* are present, the ores are fused with potassium cyanide, arsenious acid, soda, and black flux, to a button containing all the nickel and cobalt, as well as a part of the copper and iron, while zinc and manganese are slagged off. The button is dissolved in *aqua regia*, the copper and arsenic are precipitated, hot, with sulphuretted hydrogen, and the nickel and cobalt determined as above.

Volumetric assay of nickel and cobalt according to Donath.[1]—This determination is based upon the following behavior of the two metallic oxides. In mixing a solution of nickel and cobalt with an excess of caustic soda or potash and iodine, and boiling, the cobalt is completely converted into oxide, while the nickel protoxide remains unchanged. This nickel, however, can be more highly oxidized with bromine. The process is as follows: divide the solution of the two metals into equal parts, mix one part with caustic potash and bromine, and boil, whereby both metals are precipitated as sesquioxides. The other part is treated with caustic potash and iodine, the cobaltous oxide alone is precipitated. The two precipitates are thoroughly washed, transferred into separate flasks, and boiled with HCl. The chlorine developed is taken up by a solution of potassium iodide, the well-known apparatus used by Mohr for the volumetric determination of manganese (Fig. 80) being used for the purpose. The reaction is expressed as follows:—

$$R_2O_3 + 6HCl = 2RCl_2 + 3H_2O + 2Cl.$$

[1] Oesterr. Ztschft. f. Bg. u. Httnwsn, 1879, p. 550.

Hence, 1 atom of iodine corresponds to 1 atom of cobalt or nickel. The iodine liberated by boiling both precipitates is determined with $\frac{1}{10}$ normal solution of sodium hyposulphite. The difference between the two determinations corresponds to the content of nickel, its weight being computed by multiplying by 0.0059. With one portion, however, only the precipitated cobalt is determined, and the number of c.c. of the hyposulphite used is multiplied by 0.0059 (since the atomic weight of both metals is the same).

C. Colorimetric assay.[1]—A colorimetric method for determining nickel has been proposed by Winkler, but it is of little practical value.

VII. COBALT.

42. ORES.

Smaltine, $CoAs_2$, with 28.19 Co; *cobalt glace (cobaltine),* CoAsS, with 35.5 Co; *cobalt pyrites (linnaeite)* (NiS.CoS.FeS) ($Ni_2S_3.Co_2-S_3.Fe_2S_3$), with 14.6 Ni and 11 to 40.7 Co; *glaucodot* (FeCo) AsS, with 24.77 Co; *earthy cobalt,* $(CoO.CuO) 2MnO_2 + 4H_2O$; *cobalt bloom,* $Co_3As_2O_8 + 8H_2O$, with 37.5 CoO = 29.5 Co.

43. ASSAYS OF COBALT.

The object of the assays is—

1. The *determination of cobalt* by the dry (p. 188) or the wet method; in the latter case by gravimetric (p. 200) and volumetric analysis (p. 202), of which the details have already been given in the foregoing chapter on Nickel.

2. The determination *of the blue coloring powder* (density), and *the beauty of the colors* (smalt colors) which are formed in fusing ores and products containing cobaltous oxide with different quantities of potassium silicates (*smalt assay*).

44. SMALT ASSAY.

Cobaltous oxide, either contained as such in the ores (earthy cobalt, cobalt bloom), or produced by roasting sulphurized and

---

[1] Journ. f. prakt. Chem., xcvii. 414.

arsenized ores, gives a blue color to fused potassium silicate (smalt glass, very likely $CoO.3SiO_2 + K_2O.3SiO_2$), the color, under otherwise equal conditions, being the more intense, the richer the ore in cobaltous oxide. The presence of foreign metallic oxides, which also dissolve in potassium silicate, exerts an injurious effect upon the beauty of the blue tint.

Nickel protoxide, the most injurious foreign substance, produces an objectionable reddish or violet tint; *ferrous oxide*, if present in small quantities, gives a greenish shade, but *ferric oxide, bismuth oxide, lead oxide*, and *manganous oxide* affect the cobalt color to a very small extent only. *Manganic oxide* gives a violet tint; *cupric oxide*, a green; *cuprous oxide*, a red tint; while the coloring power of *manganous oxide* is counteracted by that of *ferrous oxide*, if both are present.

In roasting arsenized and sulphurized cobalt ores, the various metals become oxidized in succession, cobalt the soonest of them all, so that it becomes necessary to conduct the roasting of the ores in such a manner that cobaltous oxide only is produced, which, on being fused with silicic acid and potash, gives the beautifully colored smalt. The foreign metals should not become oxidized, but remain combined with arsenic or sulphur, forming a cobalt speiss.

As, in roasting the above-named arsenides and sulphides, cobalt is first oxidized, and iron and bismuth earlier than copper and nickel, *ores containing copper* and *nickel* must not be roasted too strongly, and least of all dead-roasted; this may, however, be done with *entirely pure* ores of cobalt, or such as contain only *iron*, as, in the latter case, ferric oxide is formed, which only slightly affects the cobalt color. If impure ores are *roasted too little*, a beautiful smalt is formed, but much cobalt is lost in the speiss.

The object of the smalt assay is either, *a*, the determination of the coloring power of a sample (*assay for determining the intensity of the color*), or, *b*, to ascertain the degree of roasting to which the ore must be subjected in order to obtain a color of pure quality (*assay to determine the color tone*), or, *c*, to determine how much of an already known sample must be taken to produce a certain shade of color.

Several lots of ore, each from 1 to 5 grammes according to the richness of the ore, are weighed off. The separate lots are each roasted for a different length of time (for instance, the first lot ¼ of an hour, and the subsequent lots each from 10 to 15 minutes longer), while one of the samples is left unroasted. Each sample is divided by weighing into two equal parts, of which one is tested with fluxes to determine the *quality* of color, and the other its *intensity*.

A. Assay to determine the quality of color.—Each sample is mixed with three times the quantity, by weight, of quartz free from iron and manganese, and with a quantity of pure potash corresponding to half the combined weight of quartz and ore. The mixture is placed in flat dishes of white fire-clay (smalt-dishes). It is then fused in the muffle (Fig. 27, p. 47), which should be heated as strongly as possible, until a completely homogeneous glass has been formed (this will require 4 hours or longer). A sample is then taken from the charge by means of a pair of tongs and cooled off in water. The mass is dried and pounded in a bright steel mortar, in order to obtain angular fragments (powdered smalt is apt to appear dirty). The fragments are sifted upon white paper, and, without taking intensity into consideration, a judgment is formed at which degree of roasting the most beautiful color has been obtained.

B. Assay to determine the intensity.—The assay sample is fused in the above manner with different quantities of quartz (for instance, 1 to 10 times the quantity), and half the quantity of potash, to a homogeneous cobalt glass. Generally 2.5 grammes of the assay sample are taken when it has been mixed with once or twice the quantity of quartz, and 1.25 grammes if with more quartz, to prevent the crucibles from becoming too full. Too much quartz makes it difficult to fuse the mixture, and too much potash gives smeary colors. A sample is taken from the smelted mass, cooled off and dried. It is then pounded and sifted, or washed in spitz-glasses (Fig. 4, p. 26), and its color compared with that of a standard, as to color and grain. This is done by spreading some of the standard color with a knife evenly upon a board, and placing a quantity, as large as a pea, of the assay sample upon it and pressing the latter into the former, when an experienced eye can tell in a well lighted room (not exposed to the direct rays of the sun) whether the assay sample corresponds with the standard in color-tone and grain, or not. If it should correspond in all respects, a confirmatory test is made by pressing some of the standard into the assay sample, when the same conditions must exist, and by examining the grain in both cases with a magnifying glass.

As moist smalt appears darker than dry, the standard and assay sample must stand for from 6 to 8 hours alongside of each other in a somewhat damp place, before the examination is made. The intensity of the color also increases with the coarseness of the grain, and an experienced eye and skilful hand are required to give the assay sample taken from the fused mass, the same grain as the standard with which it is to be compared. If the color of the assay sample is lighter than the standard with which it is compared, the assay must be repeated, with a larger quantity of ore, and *vice versa*. If the color is not exactly the same, recourse is frequently had to mixing the product with lighter or darker varieties of smalt. An ore is the more valuable, the more quartz it requires for the production of a certain intensity of color.

VIII. ZINC.

45. ORES.

Smithsonite (zinc carbonate), $ZnCO_3$, with 52Zn; *calamine (zinc silicate)*, $Zn_2SiO_4 + 3H_2O$, with 53.7 Zn; *willemite*, Zn_2SiO_4, with 58.1 Zn; *zinc bloom*, $Zn_3CO_5 + 2H_2O$, with 57.1 Zn; *zinc blende*, ZnS, with 67.01 Zn; *zinkite*, ZnO, with 80.24 Zn; *franklinite*, $(Zn, Fe)(Fe_2Mn_2)O_4$, with 21 Zn.

46. DRY ASSAYS.

These are inaccurate, but give approximately the quality of the metal which may be expected from an ore (*distillation assay*), or the approximate percentage of zinc. For many purposes they are sufficiently accurate (*indirect assay*).

A. *Assay by distillation.*—A mixture of 400 to 500 grammes of the comminuted assay sample with 80 to 100 per cent. of powdered charcoal, and, in case calamine is to be assayed, with 80 to 100 grammes of potash or calcined soda, is placed in a retort of refractory clay and heated in a furnace with a strong draught (Fig. 32, p. 51). The neck of the retort should project about 10 centimeters. In it is luted a tube of glass or porcelain about 30 centimeters long, which is kept cool by moistened rags wrapped around it. The burning gases and the fumes escape from the end of the tube, while the zinc, mixed with oxide, which is distilled over, is mostly deposited in the neck of the retort and but little of it in the tube. After the "flaming" ceases, the tube should be frequently poked with an iron wire. The retort, after having been exposed to a white heat for several hours and the "flaming" having entirely ceased, is taken from the furnace and allowed to cool off. The metallic zinc is scraped from the neck of the retort, is placed in a crucible and fused with black flux and a covering of common salt. It is then poured out into an ingot mould and weighed. The retort is carefully broken into pieces, and all those having adhering to them either zinc or zinc oxide are collected together and placed in a porcelain dish, and the metal dissolved off by digestion with nitric acid. The solution is filtered and evaporated to dryness. The residue is heated to redness, the zinc calculated from the resulting zinc oxide, with 80.24 per cent. of zinc and added to the metallic zinc.

B. Indirect assay.—5 grammes of *zinc blende* are placed in a *covered* crucible (Fig. 49, p. 64) and heated under the muffle in order to expel the volatile substances. After it has been cooled off, it is weighed and the ignition repeated until a constant weight is obtained. The accurately weighed refractory residue is mixed with 3 grammes of iron filings free from rust, and 2.5 grammes of blast-furnace slag. The mixture is brought into a charcoal-lined iron crucible (Fig. 53, p. 65), covered with 2.5 grammes of blast-furnace slag, and the remaining space filled up with powdered charcoal. A loosely fitting cover is luted on. The charge is then heated at a white heat in the blast-furnace for ¾ to 1 hour, whereby iron sulphide is formed, while zinc volatilizes and the earthy substances fuse together with the iron slag. After the charge has become cool, the button consisting of brittle matt and slag is weighed, and its weight deducted from that of the refractory residue of the zinc blende *plus* 3 grammes iron filings, *plus* 5 grammes of iron slag. The difference will give the percentage of zinc. The fewer the metallic impurities in the blende, which would be separated by the iron and volatilized (such as antimony, lead, bismuth, etc.), the more accurate will this method be.

47. WET ASSAYS.[1]

Gravimetric and *volumetric* methods are used. Although the latter assay is somewhat less accurate than the former, it is nevertheless available in many cases for practical purposes.

A. *Gravimetric assays.*

1. *Determination of zinc as zinc sulphide.*—1 gramme of the finely powdered assay sample, previously dried at 100° C. (212° F.), is placed in a long-necked flask and dissolved in nitric acid. Every trace of nitrous acid is removed by boiling, and the fluid is then strongly evaporated, 30 cubic centimeters of nitric acid and about 200 cubic centimeters of water are added, and the solution is then precipitated with sulphuretted hydrogen without previous filtration. The whole is now filtered off (metallic sulphides, quartz, etc.) and washed. A flask is now placed beneath the funnel, and its contents are treated with hot nitric acid not too concentrated. The filter is then perforated, the undissolved residue rinsed off into the flask, and the filter washed out. The fluid is strongly reduced by boiling, and then water and 30 cubic centimeters of nitric acid are added. It is again precipitated with sulphuretted hydrogen, filtered, and the filtrate which carries some zinc is added to the principal zinc solution. The entire filtrate is then boiled nearly to dryness in the long-necked flask in order to remove the sulphuretted hydrogen, some potassium chlorate being added for the higher oxidation of any ferrous oxide. The mass is then supersaturated with pure ammonia, the iron precipitate filtered off, washed and dissolved in hot, medium strong nitric acid. It is now again precipitated with ammonia to extract any residue of zinc, and filtered through the same filter, these manipulations being repeated once or twice more. The

---

[1] B. u. h. Ztg. 1876, pp. 148, 173 (Laur).

entire filtrate is now acidulated with acetic acid, diluted to the volume of at least 2 liters, sulphuretted hydrogen is introduced (in the absence of nickel and cobalt), and the liquid is then allowed to stand for 24 hours. The clear liquid is then poured off upon a filter, and finally the zinc sulphide. The flask is rinsed out with sulphuretted hydrogen water, the precipitate is also washed with it with the addition of some ammonium acetate. The filtrate from the zinc sulphide is supersaturated with ammonia and allowed to stand for at least 24 hours in a covered glass in order to see whether any zinc sulphide is still deposited. The filter is dried, and the zinc sulphide adhering to it detached by rubbing, the filter being entirely closed during the operation. The zinc sulphide, together with the ash of the filter and some distilled sulphur, is heated in one of *Rose's* crucibles (Fig. 70, p. 117) until it commences to frit, and then in a current of dry hydrogen until two weighings correspond ($ZnS = 67.01$ Zn). The determination of zinc as zinc sulphide is *very* accurate.

2. *Determination of zinc as zinc oxide.*—1 gramme of ore is dissolved in *aqua regia*, and ammonia in excess and ammonium carbonate are added, whereby zinc (and copper) passes into solution. The resulting precipitate (iron, lead, etc.) is again dissolved and precipitated, and ammonia added to extract any remaining traces of zinc. The zinc (and copper) is then precipitated from the filtrate with sodium sulphide, and filtered. The zinc sulphide is separated from the copper sulphide upon the filter by treating with diluted hydrochloric acid, and the copper sulphide remaining behind is washed. The zinc is precipitated in the boiling filtrate by means of sodium carbonate, washed, dried, ignited, and weighed as zinc oxide containing 80.24 per cent. of zinc.

The zinc may also be precipitated from the neutralized filtrate from the copper sulphide by means of sodium sulphide, and determined as zinc sulphide (p. 209). This assay is less accurate than the foregoing one.

3. *Electrolytic assay*, according to Beilstein and Jawein.[1]—0.5

---

[1] Liebig's Jahresber. 1865, p. 686 (Luckow); Fresenius's Ztschr. xv. 303 (Wrightson); xvi. 469 (Parodi und Mascazzini). B. u. h. Ztg. 1878, p. 26 (Riche). Ber. d. deutsch. chem. Ges. 1879, No. 5, p. 446 (Beilstein und Jawein).

to 1 gramme of ore is dissolved in nitric or sulphuric acid, and caustic soda added until a precipitate is formed. Solution of potassium cyanide is gradually added until the solution becomes clear, and then the platinum electrodes (p. 111) are immersed in the solution. The current of 4 *Bunsen* elements (the cylinder of zinc 15.5 centimeters high, the carbon dipping into nitric acid, with which 0.1 gramme of zinc will be precipitated per hour) is then passed through the liquid. The beaker-glass containing the solution should be placed in a dish of cold water, to prevent it, in case but a small quantity is being operated on, from becoming strongly heated by the action of the electric current. When precipitation is supposed to be complete, the electrodes are lifted out of the fluid, the zinc is first washed off with water, next with alcohol, and finally with ether, and then dried in the desiccator. After being weighed, the zinc is dissolved from the platinum with hydrochloric or nitric acid, and the electrodes are again placed in the fluid in order to test it for any residual zinc. Black stains, which may be perceived upon the electrodes, after the zinc has been removed, originate from finely divided platinum.

When *copper* is present (as for instance with brass), the sample is dissolved in nitric acid and evaporated to dryness. The residue is taken up with water, and the copper precipitated by electrolysis from the solution previously acidulated with nitric acid, when *lead* will be deposited on the platinum spiral as peroxide. The zinc is then precipitated as above described.

According to Parodi and Mascazzini[1] when zinc is precipitated from its sulphate solution, the quantity of zinc in this solution should not be over 0.10 to 0.25 gramme. The solution should be mixed with 4 c.c. of ammonium acetate, 2 c.c. of citric acid, and diluted with water to 200 c.c. The electrodes are then placed in position and the whole covered to prevent loss by sputtering. A current of 250 to 300 c.c. of oxyhydrogen gas per hour is used. The separation is complete when a sample of the solution no longer reacts with potassium ferrocyanide. After siphoning off the solution, wash precipitate, interrupt current, and again wash with absolute alcohol, and dry at about 50° C.

[1] Gazz. Chim. ital. 8. Ztschft. f. Anal. Chem. Bd. 18, p. 587.

According to Rheinhardt and Ihle,[1] the solution of sulphate or chloride of zinc, which should be as neutral as possible, is mixed with an excess of potassium oxalate, the precipitate of zinc oxalate is dissolved and the solution electrolized. The oxalate of zinc is decomposed to metallic zinc and carbonic acid, and the potassium oxalate, to potassium and oxalic acid. The potassium acts upon the water with an evolution of hydrogen on the negative pole, at the same time the free alkali formed is converted into potassium carbonate by the carbonic acid at the positive pole. The reaction is nearly complete when the evolution of gas at the positive pole abates. To avoid the peculiar stains produced by electrically deposited zinc on bright platinum it is well to previously deposit a slight coating of copper on the negative electrode, which also facilitates the recognition of the end of the precipitation. The coating of copper must, however, be perfectly bright; otherwise the zinc will not adhere properly. The precipitated zinc is bluish white and adheres firmly to the electrode. The precipitate is first washed with hot water, then in cold water free from air, then in alcohol, and finally in pure ether and dried in a desiccator. The precipitated zinc is redissolved in dilute cold HNO_3, the layer of copper is but slightly affected and reappears with a bright surface. As some of the copper is dissolved with the zinc it is useless to weigh the electrode freed from the precipitate. The use of ammonium oxalate instead of potassium oxalate is not recommended, as it renders the zinc precipitate somewhat pulverulent on the lower edge of the electrode. In the presence of sufficient potassium oxalate the presence of free oxalic acid or almost any other acid is not injurious, although precipitation takes place more readily from a neutral solution. The presence of HNO_3 in any form should be avoided, as it is apt to cause the formation of ammoniacal salts which prevent the compact precipitation of the zinc. As the potassium carbonate formed offers a strong resistance to the current, it is recommended to add pure neutral potassium sulphate to increase the conductivity of the solution. Moore[2] states that from a H_2SO_4 or HCl solution of zinc the zinc is readily precipitated in the metallic state with a current of 5 c.c. of oxyhydrogen gas per minute. The solution must contain 15 per cent. of metaphosphoric acid and heated to 70° C. with the addition of an excess of ammonium carbonate.

[1] Journ. of Pract. Chem. vol. 24, p. 193.
[2] Chem. News, 1886, p. 219. Chemikerztg, 1886, p. 119.

By adding sodium phosphate to a potassium cyanide solution and dissolving the phosphate of zinc formed, then heating to 80° C. with an excess of ammonium carbonate and using a current of 160 c.c. of oxyhydrogen gas per hour, the zinc is precipitated more rapidly and quite as completely. It is claimed, in this connection, that the zinc deposits better upon a silver-plated electrode.

According to Millot,[1] the electrolytic determination of zinc in ores is easily effected by the following method:—

Dissolve 2.5 grammes of the ore in 50 c.c. of HCl, add potassium chlorate to oxidize the iron and evaporate to dryness to separate the silica. Take up with water and precipitate the lime and lead with 100 c.c. of ammonia and 5 c.c. of ammonium carbonate, filter, dilute to 1 liter and mix this solution with a solution of 1 gramme of KCY in 100 c.c. of water. Place electrode in this solution and pass current through. The electrodes should not be more than a few millimeters apart. A current from two Bunsen cells will finish the precipitation in nine or ten hours. The precipitated zinc adheres tightly and can be readily washed.

According to Classen, a H_2SO_4 solution containing about 0.1 gramme of zinc is concentrated to about 50 c.c., the free acid neutralized with caustic potash, excess of ammonium oxalate added, the solution heated and 3 or 4 grammes of solid ammonium oxalate dissolved in it, and the hot solution subjected to the current of two Bunsen cells. The zinc separates very easily as a bluish or gray white coating on the negative electrode. The precipitation is complete when no more gas is developed on the positive electrode, and a sample of the solution shows no zinc reaction with ammonium sulphide. The precipitated zinc dissolves with difficulty in acids; there generally remains a dark coating on the platinum electrode, which can only be removed by igniting and repeated treatment with acids. The use of HNO_3 for dissolving the ore is not advisable, as its presence, or even that of a nitrate, hinders the separation of the zinc in a compact form.

In the electrolytic determination of zinc from a slightly acid solution in the presence of a strong mineral acid, Luckow[2] recommends first placing in the platinum dish, used as an electrode, about 0.5 to 0.7 gramme of mercury. The platinum dish and its contents are then weighed, after which the solution to be decom-

---

[1] Soc. Chim. Paris, 1882, p. 339. Chemikerztg, 1882, p. 410.
[2] Chemikerztg, 1885, p. 338.

posed (containing from 0.10 to 0.15 gramme zinc) is placed in it. A platinum spiral is used as the positive pole and the dish as the negative. A current of 120 to 150 c.c. of oxyhydrogen gas per hour from 6 to 8 Meidinger cells is used. When the zinc separates it forms an amalgam which coats the inside of the dish. When the precipitation is complete the amalgam is washed with water and alcohol, dried, and the dish again weighed. Platinum, iron, nickel, cobalt, and manganese do not form an amalgam with mercury in this manner; consequently this method can be used in the separation of zinc from these metals. This method is also very useful in the electrolytic determination of silver.

B. Volumetric assays.—Of the volumetric assays which have been recommended—

1. *Schaffner's assay with sodium sulphide*[1] is used more than any other. This is based upon the precipitation of zinc from ammoniacal solution by means of sodium sulphide. The termination of precipitation is indicated by the blackening of ferric hydrate as below described. This assay, if certain precautionary measures are adopted, allows of a determination of the zinc to within 0.5 per cent. The presence of metals soluble in ammonia (*copper* and *manganese; nickel* and *cobalt* occur but seldom) requires modifications of the method.

0.5 gramme of *oxidized ores* (smithsonite, calamine), with over 35 per cent. of zinc, and more if the ore is poorer, are dissolved in heated hydrochloric acid, with an addition of a few drops of nitric acid, to oxidize the iron; the solution is then supersaturated with ammonia: or, raw, or roasted zinc blende is dissolved in *aqua regia*, evaporated to dryness, and the residue dissolved in 5 cubic centimeters of hydrochloric acid and some water. *Copper* (also lead, antimony, etc.) being frequently present, the solution is precipitated with sulphuretted hydrogen, filtered, and the gas driven off by boiling. 10 cubic centimeters of *aqua regia* are

---

[1] B. u. h. Ztg. 1856, p. 231, 306; 1857, p. 60 (Schaffner); 1876, p. 148, 174 (Laur); p. 225 (Thum); p. 304 (Tobler). Journ. f. prakf. Chem. lxxxviii. 486 (Küntzel). Fresenius's Ztschr. 1870, p. 465 (Deus); 1871, p. 209 (Schott). Mohr, Titrirmethode, 1874, p. 466 (F. Mohr). Dingler, cxlviii. 115 (C. Mohr). Preuss. Ztschr. Bd. 25 (Hampe). Berggeist, 1874, No. 3 (Allenberger Pr.). Berichte der deutsch. chem. Ges. 1879, No. 3, p. 270 (Aarland). Percy's Metallurgy. Copper, Zinc, and Brass, p. 598.

added (or some chlorine water, or a few drops of bromine may be added to the acid solution, or potassium permanganate to faint reddish coloration to the ammoniacal solution, and allowing it to stand for one hour) for the higher oxidation of the *iron* and *manganese* (if iron alone is present, an addition of nitric acid or potassium chlorate suffices), which, when ammonia in excess is added, are precipitated as hydrated oxides, while the zinc remains in solution. The hydrated oxides are again dissolved in hydrochloric acid and precipitated with ammonia in excess, in order to extract any residue of zinc. Both filtrates are then united and diluted (to 500 cubic centimeters if 5 grammes, and to 175 to 225 cubic centimeters if 0.5 gramme of ore have been used). 50 cubic centimeters of the fluid are then placed in a beaker-glass, 1 to 2 drops of a solution of ferric chloride of the concentration given below are dropped into 1 cubic centimeter of ammonia contained in a porcelain saucer. The precipitated hydrated ferric oxide is carefully rinsed in the beaker-glass, where it settles on the bottom. Titrated solution of sodium sulphide (1 cubic centimeter = 0.008 to 0.009 gramme of zinc) is then added, the contents of the beaker-glass being given a spiral motion in the meanwhile, so that the flakes remain on the bottom. The addition of the sodium sulphide is continued until the flakes of ferric oxide become discolored, and finally assume a brown color, which indicates that all the zinc has been precipitated.[1]

The following have also been recommended as *indicators* for recognizing the final reaction; though none of them have taken the place of the hydrated ferric oxide. Porcelain saturated with ferric chloride (*Barreswill*), or paper (*Streng*) which is weighted with platinum wire and laid upon the bottom of the beaker-glass; drop samples (*Tupfproben*) with nickel chloride (*Küntzel*), cobaltous chloride (*Deus*), alkaline solution of lead tartrate (*F. Mohr*); blotting-paper saturated with sugar of lead, and then treated with ammonium carbonate (*Fresenius*); nitro-prussiate of sodium (*C. Mohr*); sized paper coated with white lead, so-called

---

[1] It is advisable to have a stand with three burettes respectively for solution of zinc, sodium sulphide, and ferric chloride.

"polka" paper (*Schott*), over which filtered drops of the fluid are allowed to run.

The following points must be observed in order to make the assay successful.

a. *The quantity of the hydrated ferric oxide*, which is added, must not vary too much, and must possess a uniform coherence. This may be produced by dissolving 3 grammes of piano wire in *aqua regia*, and diluting this to a bulk of 100 cubic centimeters. The same number of drops of this solution are allowed to fall each time, for instance from a burette, into 1 cubic centimeter of concentrated ammonia (p. 215). The ring-like clot of hydrated ferric oxide which will be formed in about one minute may then be rinsed into the fluid which is to be titrated.

b. *The same shade of coloration of the hydrated ferric oxide* used in the operation should always be taken as closely as possible to indicate the end of the reaction, as the quantities of sodium sulphide consumed in producing various shades vary considerably.

c. *The quantity of fluid*, into which the added excess of sodium sulphide is divided in titrating, exerts a material influence in respect to its action upon the hydrated ferric oxide, as, when the quantity of fluid becomes smaller, a smaller excess of sodium sulphide suffices for blackening the hydrated ferric oxide.

In consideration of this circumstance, the volume of liquid is measured, according to *Tobler*, in the smelting works of the *Vieille Montagne*, at the termination of the titration, and for each 100 cubic centimeters of it, the volume of sodium sulphide consumed is decreased 0.7 and 0.5 cubic centimeters, its standard being 0.008 to 0.009 gramme of zinc to 1 cubic centimeter. According to *Thum*, this correction can be avoided, by always bringing all the liquids before titration to an equal volume, and by using so much zinc for standardizing as corresponds to the average percentage, and the quantity of zinc ore used. This zinc is dissolved and the solution diluted to a volume equal to that of the solution of the ore, and the titer of this is taken. Under these conditions there can be no material difference in the quantities of sodium sulphide consumed in fixing the standard of the solution and of those consumed in the analysis, and therefore the quantities of liquid must be nearly equal after the titration, and in consequence of this no errors can occur on account of differences of volumes.

d. It does not make any difference, as far as the accuracy of the assay is concerned, whether the hydrated ferric oxide is introduced at the commencement, or towards the end of the titration, if the same order is always observed. The heating of the liquid exerts also but little influence.

e. Admixtures having a disturbing effect must be removed. *Copper, silver, cadmium, cobalt, nickel chromium, manganese, arsenic,* and *antimony* are soluble in ammonia, and also *lead* in small quantities, which last, after the precipitation with ammonia, may be present either as carbonate, sulphate, and basic chloride, or as oxy-salt. Of these, the last two are the more soluble in ammonia. *Copper, lead, manganese,* and *iron* occur most frequently.

Copper, together with cadmium, silver, arsenic, and antimony, is removed by sulphuretted hydrogen. If a small quantity only is present, it can be determined by *Heine's* colorimetric assay, and titrated with sodium sulphide, notwithstanding the blue coloring of the ammoniacal zinc solution, and the total consumption of sodium sulphide, reduced by the volume corresponding to the percentage of copper found. *Manganese* is the least injurious of all, and can be easily removed as above described, p. 215. It becomes sulphurized later than the hydrated ferric oxide, and the percentage of zinc can be accurately determined by taking the commencement of the blackening of the hydrated ferric oxide as indicator of the final reaction, and not continuing until it has turned entirely black. But it is best to first remove the manganese by precipitation in case a considerable percentage of it is present. This fact may be recognized by a concentrated acid solution, notably losing its original dark color, upon dilution. *Lead* is removed by sulphuretted hydrogen with the copper, or by evaporating the substance to dryness with sulphuric acid, after the preliminary treatment with acids, then taking up with diluted sulphuric acid, and removing the lead sulphate by filtration, etc. When more than 5 per cent. of iron is present, the ammoniacal precipitate must always be redissolved at least once. Organic substances, which are sometimes found in foreign zinc ores, are destroyed by heating to redness with free access of air, as otherwise they might easily reduce the ferric oxide to ferrous oxide, which is soluble to some extent in ammonia, and decomposes sodium sulphide.

f. The *zinc* used for fixing the standard solution must be sufficiently *pure*.

Good commercial zinc is purified by smelting 50 grammes in a porcelain crucible over a lamp; keeping it in this state for a quarter of an hour, poling it with a wooden stick. The layer of oxide is then taken off, and, after the metal has stood for a few minutes, one-half of it is poured into another crucible, where the same operation is repeated. Half of the metal is again poured off, these manipulations being repeated several times. The zinc is then dropped upon a cold zinc plate (the surface of which must be free from oxide) so that each drop forms a small thin disk that may be readily broken to fragments. It should be kept under a bell-glass with calcium chloride. The *American Passaic* zinc is very pure.

g. There should be a *uniform light* in the room during the titration, so as to enable the operator to observe correctly the coloring of the iron. This is best arranged by covering the windows with tracing linen or tissue paper (Altenberg).

2. *Assay with potassium ferrocyanide.*[1]—The zinc is precipitated from an acid solution by potassium ferrocyanide, uranic salt being used as an indicator. When all the zinc is precipitated, the uranium salt produces a brownish stain, when a test is made by taking a drop upon a white plate. 5 grammes of ore are dissolved in aqua regia, and evaporated. Hydrochloric acid in excess is added to the residue, the copper, etc., are precipitated with sulphuretted hydrogen and filtrated. The filtrate is boiled, the ferrous oxide is oxidized by potassium chlorate, and ammonia is added. The ferric oxide and alumina are then filtered, and again dissolved and precipitated. The filtrate is neutralized with hydrochloric acid and an additional 10 to 15 cubic centimeters of hydrochloric acid of 1.12 specific gravity are added. The solution is then titrated with the solution of potas-

---

[1] Dingler, cxcv. 260 (Galletti); Fresenius's Ztschr. 1875, pp. 189, 343 (Lyte and Galletti); 1874, 379 (Fahlberg); Dingler, cxc. 229 (Renard); cxc. 395 (Reindl).

sium ferrocyanide (1 cubic centimeter)=0.01 gramme of zinc until the appearance of the first brownish stain by the drop test with solution of uranic salt.

In determining zinc with potassium ferrocyanide the presence of manganese exerts, according to Galetti, an injurious influence, as it is precipitated with the zinc. It is, therefore, recommended by Mason,[1] after the separation of the metals precipitated by H_2S, to acidulate the solution with acetic acid and treat it with H_2S. Dissolve the washed precipitate of sulphide of zinc in HCl, and use *this solution alone* for titration. According to Giudice,[2] in the presence of tartaric acid the zinc alone is precipitated from an ammoniacal solution by potassium ferrocyanide, whilst the iron, lead, and alumina remain in solution, thus rendering the separation of the zinc from these metals unnecessary. Copper and manganese, however, must not be present. A ferric salt is used as an indicator; when it is brought in contact with a drop of the solution acidulated with acetic acid, the ferrocyanide of zinc is not affected by the acetic acid.

Dissolve 0.5 to 1.0 gramme of the ore in HNO_3 or HNO_3+HCl, dilute to three times the volume of solution, and add potassium ammonium tartrate. Add excess of ammonia and dilute to 300 or 400 c.c. Although usually the solution is not entirely clear, filtering is not necessary. The titer is best set for a solution of pure zinc, or of a pure zinc salt of known metallic content. Galletti's titrating solution may also be used.

Determination of zinc by decomposing the sulphide of zinc with chloride of silver, and determining the zinc from the equivalent content of chlorine, according to Mann.[3]—Dissolve 1 gramme of the assay material in HNO_3, pass H_2S through solution, filter, boil the filtrate and precipitate the ferric oxide and alumina with ammonia, redissolving, however, the larger part of this precipitate in HNO_3, and again precipitating. Add to the two filtrates, which contain zinc and some manganese, acetic acid, and again pass H_2S through. Boil off excess of H_2S, filter while hot, and place the filter with the sulphide of zinc in a small beaker, add 30 to 50 c.c. of hot water, stir with glass rod, and add an excess of thoroughly pure moist chloride of silver, which should be kept on hand protected from the

---

[1] Am. Chem. Jour. vol. 4, p. 53. Ztschft. f. anal. Chem. Bd. 22, p. 246.

[2] Giorn. Farm. Chem. 31, p. 337. Chemikerztg. 1882, p. 1034.

[3] Oesterr. Ztschft. f. Bg. u. Httnwsn. 1879, p. 426. Ztschft. f. anal. Chem. Bd. 18, p. 162.

light. Boil until the sulphide of zinc is entirely decomposed, and the supernatant liquid perfectly clear, then add a few drops of dilute (1 : 6) H_2SO_4, filter and wash cold. The filtrate contains all the zinc as chloride. Now add an excess of nitrate of silver, shake well, and mix with a few c.c. of iron-ammonium-alum, or ferric sulphate, and determine excess of silver by Volhard's method, by which the amount of silver *not precipitated* by the chlorine is determined. From the amount of silver *precipitated* by the chlorine the content of zinc is calculated.

Preparation of the titrating solutions.—Dissolve 33.2 grammes of pure silver in HNO_3, evaporate to dryness, take up with water, and dilute to 1 liter. One c.c. of this solution contains 0.0332 gramme of silver, and corresponds to 0.01 gramme of zinc. The standard solution of sulpho-cyanide is prepared so that 1 c.c. exactly equals 1 c.c. of the silver solution.

Determination of zinc from its combinations with sulphur by decomposition with nitrate of silver. Balling's method.[1]—The pure sulphide of zinc, obtained by separation from the other metals by the wet method, is filtered off, thoroughly washed from the filter into a beaker, and an excess of nitrate of silver poured over it—the amount of silver present must be in excess. The whole is boiled for about a half hour. Since 216 of silver correspond to 65.2 of zinc, at least ($\frac{216}{65.2} =$) 3.312 centigrammes of silver must be added for every centigramme of zinc expected. The decomposition of the sulphide of zinc and nitrate of silver is plainly perceptible at ordinary temperatures, and at a boiling heat it is effected in a half hour. Filter off the sulphide of silver, wash the filter, and place it and its contents in a flask containing boiling HNO_3 of 1.2 sp. gr. Boil until no more brown vapors escape, dilute with water, and, after cooling, titrate with ammonium sulpho-cyanide. Use the normal silver solution prepared for Volhard's silver assay, and the corresponding solution of ammonium sulpho-cyanide (see page 155) as the titrating solutions. Since the zinc is determined by the quantity of silver equivalent to it, each c.c. of the cyanide salt used in titrating has to be multiplied by 0.30185 to get the content in the assay material. This method is well adapted for the determination of lead and copper in their sulphur combinations, and still better for the

---

[1] Oesterr. Ztschft. f. Bg. u. Httnwsn. 1881, p. 35. Ztschft. f. anal. Chem. Bd. 22, p. 250.

separation of these metals, as they can be determined one after another with the same re-agent.

Determination of lead, copper, and zinc in one solution, according to this method.—These three metals are frequently associated in alloys, and are first conjointly precipitated from an acetic acid solution. The associated metallic sulphides are filtered off and decomposed with nitrate of silver, the silver sulphide formed by titrating with ammonium sulpho-cyanide, corresponds to the sulphides of the three metals dissolved. The filtrate from the silver sulphide again contains the three metals. After precipitating the excess of silver with HCl, and separating the lead with H_2SO_4, neutralize the slightly acid solution with soda, until somewhat turbid, then add acetic acid until the liquid becomes clear. Precipitate the copper and zinc with H_2S, filter, wash, and again decompose with nitrate of silver. The difference between the amounts of nitrate of silver used in the first and second titrations, corresponds to the amount of lead present. The excess of silver is separated, as before, with HCl, and after filtering it off the copper is precipitated with H_2S, and determined separately in the same manner. The amount used in the third titration corresponds to the copper, and by adding it to the difference first found, which corresponds to the lead, and subtracting the total from the total consumption of sulpho-cyanide used in the first titration, we have the amount of silver which corresponds to the zinc. Should silver be present in the substance assayed, it is determined in the beginning from the HNO_3 solution with ammonium sulpho-cyanide, the silver cyanide is filtered off, the other metals remaining in solution in the filtrate. In calculating the amount of lead, the number of c.c. of sulpho-cyanide used is multiplied by 0.95833, and for the copper the number of c.c. of sulpho-cyanide is multiplied by 0.29351.

3. *Schober's volumetric assay.*[1]—The zinc is precipitated with an alkaline sulphide solution. The excess of sulphide used is decomposed with silver solution, and finally the quantity of the solution of silver, which has been added in excess and remained undecomposed, is determined by ammonium sulpho-cyanide according to Volhard's method (p. 155).

---

[1] Oesterr. Ztschr. 1879, 5.

IX. CADMIUM.

48. ORES.

It occurs more seldom as an ore (*greenockite*, CdS, with 77.6 Cd) than as a constituent of calamine and zinc blende.

49. ELECTROLYTIC ASSAY.[1]

Cadmium sulphide is precipitated from an acid solution with sulphuretted hydrogen. This (and also cadmium oxide) is dissolved in nitric acid, the free acid is neutralized with caustic potassa, with an addition of potassium cyanide solution, until the precipitate is dissolved. It is then diluted with a sufficient quantity of water, so that about 0.2 gramme of cadmium is contained in 75 cubic centimeters of the liquid. The cadmium is then precipitated upon the platinum cone (p. 111) with 3 Bunsen elements, the glass containing the liquid being placed in a dish filled with cold water. The rate of precipitation should be from 80 to 90 milligrammes of metal per hour. The light gray cadmium is rinsed off with water, then with alcohol, and dried by placing it in a heated platinum dish (p. 112).

According to Clark,[2] from an ammoniacal solution cadmium is precipitated in a spongy porous condition containing impurities so that the results are too high. From neutral acetic acid solutions with a strong current cadmium is precipitated in a crystalline form suitable for weighing.[3]

According to Yver,[4] zinc and cadmium can be separated by electrolysis if the solution containing both metals as sulphates or acetates is mixed with 2 or 3 grammes of sodium acetate and a few drops of acetic acid. The cadmium alone is separated in a crystalline form on the negative pole. With two Daniell cells, 0.18 to 0.21 gramme of cadmium can be precipitated in 3 or 4 hours. The zinc remains behind in solution.

---

[1] Berichte der deutsch. Chem. Ges. 1879, No. vii. p. 759.
[2] Am. Journ. of Science and Arts, vol. 16, p. 200.
[3] Berichte d. deutsch. Chem. Ges. Bd. 11, p. 2048.
[4] Bulletin de la Société Chemique, vol. 34, p. 18.

According to Classen, cadmium may be precipitated completely from solutions of ammonium or potassium double oxalate, preferably the latter. The solution is treated with an excess of ammonium or potassium oxalate diluted to 200 c.c., and electrolyzed hot. Care must be taken that the volume of the solution is not reduced by evaporation. The end of the reaction is shown by testing with hydrogen sulphide.

According to Beilstein and Jawein, cadmium may be determined in the same way as zinc. If the solution contains free acid, it is neutralized with potassium hydroxide, and potassium cyanide added till the solution becomes clear. A current from three Bunsen cells is used, and the solution diluted so that 0.2 gramme of cadmium is contained in 75 c.c. The vessel in which the reaction takes place is cooled during the process.

X. TIN.

50. ORES.

Tinstone (cassiterite), SnO_2, with 78.7 per cent. tin.

51. DETERMINATION OF TINSTONE BY WASHING.

This method is used for testing borings from mines, in order to ascertain whether poor tin iron ores are worth working (Saxony); or in concentrating works, to determine the quantity of material worth smelting which may be obtained from an ore-heap (Cornwall). The specific gravity of tinstone = 6.8 to 7.0.

A. Saxon assay of tin.—A sample of dust, taken by volume, is washed in the vanning trough.

B. Determination of tin by washing in Cornwall.[1]—50 kilogrammes of samples are taken from different parts of the heap. The mass is comminuted and thoroughly mixed. From this, another sample is taken, sifted, and dried, and 55 to 56 grammes of it are weighed off. This is placed upon an iron shovel, and washed, by imparting to it, first a rotary motion, and then a decided upward and downward movement. By these operations,

---

[1] B. u. h. Ztg. 1859, p. 358. Muspratt's Chemie, vii. 1375.

the products free from tin will be washed away, while those yielding tin will, according to their specific gravity, be collected on different parts of the shovel. They are then removed, and, if necessary, roasted, and again washed. This manipulation requires considerable skill (see p. 28).

52. FIRE ASSAYS.

The object of these is to reduce the tin oxide (stannic acid) and slag off the admixtures of earths by solvent fluxes. The accuracy of the result is impaired, or the assay is made difficult, on account of the tin oxide being easily slagged off by acids and bases; by the difficulty of uniting the reduced particles of tin to a *single* button; and by the presence of many foreign metallic combinations and earths which promote the slagging off or the contamination of the tin.

The losses by the *German* method are less than by the *English* or *Cornish* method. The assay with potassium cyanide gives the highest yield. The tin buttons obtained by the fire assay must be tested in the wet way for the presence of copper, iron, etc., by treating them with nitric acid of 1.3 specific gravity, adding water, digesting, filtering, drying, igniting, and weighing the tin oxide.

A. German assay.—5 grammes of clean ore are intimately rubbed together with 0.75 to 1 gramme of powdered charcoal. The mixture is poured into a suitable crucible (Fig. 52, p. 65), and covered with 12.5 to 15 grammes of carbonaceous black flux or potash with 50 per cent. of flour, 1 to 1.25 grammes of borax glass, and finally with a cover of common salt and a small piece of coal. The charge is exposed for three-quarters to one hour to a very strong red heat in the reverberatory (p. 54) or muffle-furnace (p. 45), or for one-half to three-quarters of an hour in the blast furnace (p. 59). The crucible is then taken out and allowed to become entirely cold, as tin has a low fusing point. It is then freed from slag, and the result must be a single, ductile button of a tin-white color, which does not follow the magnet under water In case the tin is distributed in the slag, this must be washed off and the metal collected.

Other charges: 25 grammes of ore, 5 grammes of argol, 20 grammes of soda, and 3 grammes of lime, are intimately mixed together, and covered with a layer of soda and 10 grammes of borax. The charge is smelted at a strong red heat and kept in fusion for twenty minutes. Or for siliceous ores: 10 grammes of ore, and from 10 to 20 grammes of fluor-spar or cryolite are placed in a charcoal-lined crucible and covered with charcoal. A lid is luted on, and the charge is then very strongly heated for one hour. This assay gives a good yield.

Modifications become necessary—

1. When the ore contains many earthy admixtures. The ore, before it is reduced, must be washed in a vanning trough, in spitz-glasses (Fig. 4, p. 26), or in a beaker-glass (as silicic acid especially promotes slagging off of tin). This manipulation is effectual on account of the high specific gravity of tin, but metallic admixtures cannot (or can only partly) be removed by it.

Specific gravities: tinstone, 6.8 to 7; native bismuth, 9.6 to 9.8; tungsten, 7.2 to 7.5; arsenical pyrites, 6 to 6.4; copper glance, 5.5 to 5.8; iron pyrites, 4.9 to 5.1; copper pyrites, 4.1 to 4.3; molybdenite, 4.5 to 4.6; magnetic iron ore, 4.8 to 5.2; specular iron ore, 6 to 6.5; red hematite, 4.5 to 4.6; zinc blende, 3.9 to 4.2; quartz, 2.65 to 2.80; chlorite, 2.65 to 2.85; slate, 2.5.

2. When the ore contains *foreign metallic sulphides, arsenides,* and *antimonides.*

a. The unroasted ore is either digested with *aqua regia* for half an hour, and then washed by decantation, the tungstic acid, from tungsten ores, if any be present, removed by digesting with caustic ammonia for half an hour (the flask being frequently shaken), then washed by decantation and dried (*Levol*[1]), and then reduced:

b. Or, the dead-roasted ore is treated with hydrochloric acid as long as the acid, after the ore has been repeatedly decanted and washed, appears yellow, when a fresh addition is made at a boiling temperature; the ore is then washed by decantation, dried,

---

[1] Polyt. Ctrbl. 1857, p. 406.

and subjected to reducing and solvent fusion as above described (p. 224).

3. *On account of the ease with which tin oxide is slagged off.*— For these reasons—

a. The ore is intimately rubbed together with powdered charcoal (p. 225), or carbonaceous black-flux, but too large a percentage of carbon will render the charge more refractory.

b. The tin oxide is reduced, before the reducing and solvent fusion, by mixing the ore with $\frac{1}{4}$ part of powdered wood charcoal and igniting it in the crucible. It is then charged as above.

4. When *separate grains of tin are found.*—These must be collected with copper (bronze being formed), by mixing 5 grammes of ore with 5 grammes of pure copper oxide (with 79.14 Cu). The mixture is placed in a suitable crucible (Fig. 52, p. 65) when 15 grammes of black flux and 1.25 grammes of borax-glass are added, together with a cover of common salt and a fragment of coal. The charge is gradually heated to a high temperature, and after the "flaming" has ceased, is exposed for three-quarters to one hour to a white heat in the muffle or wind furnace, or for one-half to three-quarters of an hour in the blast furnace. It is then taken out, and, when it has become cold, the brittle bronze button is freed from slag and weighed. The weight of the copper contained in the copper oxide, which was added, is deducted. If the copper oxide is not entirely pure, 5 grammes of it are fused with the same additions as given above and the weight of the resulting copper button is deducted from that of the bronze button.

5. When tin oxide is combined with silicate (as, for instance, in tin-ore slags), 5 to 25 grammes of slag are pulverized as finely as possible, the metallic tin is sifted out and the fine substance is gradually added to 12 to 15 times the quantity of potassium bisulphate, which has been previously fused in an iron or porcelain crucible under the muffle. The mixture is then fused until no more gas bubbles are formed. The fused mass is then extracted with boiling water, washed with hot water, and the residue reduced as above. Or, 25 grammes of slag are mixed with 10 grammes of ferric oxide, 6 grammes of fluor-spar, and 100 grammes of charcoal powder. The mixture is placed in a

covered crucible and gradually heated to a strong red heat. This is kept up for half an hour, and finally kept at a white heat for half an hour longer.

B. Cornish assay of tin.—50 to 100 grammes of rich tinstone are mixed with one-fourth the quantity of anthracite and some fluor-spar. A large wind furnace, such as is used for assaying iron (for instance, 254 millimeters wide, 178 millimeters long, and 380 millimeters deep), is filled about two-thirds full with coke, and brought to a strong red heat. A few pieces of fresh coke are then added, and in these a graphite crucible is placed and made red hot. It is then taken out, the charge poured into it by means of an open mixing scoop, and the crucible covered. The charge is then fused for 20 minutes, when the white-hot crucible is taken out, and the contents are poured into an iron ingot mould. The ingot is freed from slag. The slag is pounded fine and sifted through a sieve of tin plate having meshes about as large as a pin-head. The tin remaining in the sieve is added to the ingot of tin, while the fine stuff that has passed through the sieve is washed and floated in a vanning shovel (Fig. 5, p. 27). The metallic residue from the washing is placed in a dish and dried. The three lots of tin are then weighed together. The ingot of tin is now refined by smelting it in an iron spoon. The film which is formed during this operation is removed until the liquid metal has a bright non-iridescent surface. It is then poured into a gutter in a marble plate, and the quality of the tin judged by the surface and ductility of the resulting rod of metal. When lead and copper are present, the surface will exhibit a play of colors and will be crystalline, especially towards the centre. This assay yields about 10 per cent. less than the true percentage.

The Cornish assay is less accurate than the German, but an experienced assayer obtains results which compare well with those obtained on a large scale, and it allows him to form a judgment of the quality of tin which an ore may be expected to yield, and to fix the price to be paid for it accordingly.[1]

C. Levol's assay with potassium cyanide.[2]—A sufficient quantity

---

[1] B. u. h. Ztg. 1862, p. 261.
[2] Journ. für prakt. Chemie, xcv. 503.

of powdered potassium cyanide is rammed into a capacious porcelain or fire-clay crucible to form a layer of from 12 to 15 millimeters thick, 5 grammes of the powdered ore, intimately mixed with 5 times the quantity of potassium cyanide, are added to that in the crucible, and the whole covered with a thin layer of the cyanide. The charge is then heated in a moderate fire until it fuses, and is kept in constant fusion for 10 minutes. The crucible is then taken out, and gently tapped to facilitate the formation of a single button, and allowed to cool. The button is then freed from adhering slag by water. In case copper or lead is present, the ore must be freed from them before the reduction, by treating it with acid. This is the most accurate method of assaying tin (to within one-half per cent.), and can be executed in a very short time. In case the ore is siliceous, a mixture of 10 grammes of ore, 3 to 8 grammes of ferric oxide, and 40 grammes of potassium cyanide, is placed in a crucible (Fig. 53, p. 65) lined with charcoal. The mixture is first covered with potassium cyanide and then with powdered charcoal. The cover is luted on and the charge heated at a high temperature for one-half to one hour.

In any case it is always better to mix a small quantity of powdered charcoal with the charge. One-half of the potassium cyanide may be mixed with the ore and balance used to cover the charge. Care must be taken not to allow the fire to become too hot or the charge to boil over. After pouring, the mould must not be moved until the slag has set, otherwise it is apt to penetrate into the button.

53. WET ASSAYS.

A. Gravimetric assays.

1. 1 gramme of tinstone is digested with diluted *aqua regia;* the residue is washed by decantation and dried. The dry mass is then fused with 3 parts sulphur and 3 parts sodium carbonate. The soluble double sulpho-salt of sodium and tin is lixiviated with water, and the tin sulphide precipitated with hydrochloric acid. Sulphuretted hydrogen is then introduced into the liquid, the resulting precipitate is filtered, dried, roasted, and finally weighed as stannic acid, containing 78 per cent. tin. By another

method the ore is digested in *aqua regia* for half an hour, then washed by decantation and dried. It is then introduced into a silver crucible (standing in a clay crucible), with 4 times the quantity of caustic potassa dissolved in water, the very finely powdered tinstone is stirred into it, and the whole brought to dryness. It is then fused at a low red heat for half an hour. The mass, when cold, is treated with diluted hydrochloric acid and evaporated to dryness. The dry mass is then taken up in some hydrochloric acid, gently heated, filtered, and precipitated with sulphuretted hydrogen. The sulphide is washed, dried, and roasted, some ammonium carbonate being added towards the end of the operation. The tin oxide is then weighed.

2. The ore is digested with *aqua regia*, the residue with some charcoal is placed in a porcelain crucible and heated to redness. The reduced tin is dissolved in hydrochloric acid, and precipitated from the solution with zinc, which, in the form of a flat button fastened on the end of a copper wire, is suspended in the fluid. According to the proportion of free acid present, the tin will appear in brilliant needles, in scales, mossy or spongy; the latter condition indicating the termination of precipitation. The zinc button is now taken from the liquor, free from the tin, and this is pressed together in an agate mortar. It is then dried and fused to a button with some sterine (Moissenet[1]). Or, the ore is digested with *aqua regia*, decomposed with potassium hydrate, and a solution of stannic chloride in hydrochloric acid is formed as above. The tin is precipitated by a rod of zinc. It is then washed and dried, treated with strong nitric acid, and evaporated to dryness. When cold, it is moistened with diluted nitric acid and filtered. The tin oxide is then dried, ignited, and weighed. Lead and copper are removed by the *aqua regia* at the commencement of the operation, or when the precipitated tin is dissolved in nitric acid.

Determination of tin in tin slags.—The following method is recommended by Fresenius and Hintz.[2] About 3 grammes of the finely powdered assay material are treated with *aqua regia*, after

---

[1] B. u. h. Ztg. 1861, p. 170.
[2] Ztschft. Anal. Chem., XXIV., p. 412.

heating the solution is diluted with water and the insoluble residue filtered off, and washed with water containing ammonium nitrate. Make filtrate alkaline with caustic soda and digest for a considerable time with an excess of sodium sulphide. The resulting precipitate is filtered off and repeatedly treated with sodium sulphide in the same manner. The insoluble residue is separated from filter and dried. The latter is incinerated at a low temperature, the resulting ash added to above residue, the whole fused with sulphur, taken up with water and filtered. The filtrate is mixed with the precipitate obtained with sodium sulphide, acidulated with HCl and the precipitate obtained treated with bromo-hydrochloric acid. It is then filtered, the filtrate mixed with potassium chloride, and evaporated to dryness. In case tungstic acid separates, the filtering must be repeated. Precipitate the tin in the filtrate with ammonium nitrate. The tungstic acid precipitated by the treatment with bromo-hydrochloric acid as well as that separated by evaporation, is fused with five times its weight of potassium cyanide, whereby any tin contained in it is obtained as metal, whilst the tungstic acid passes into the flux. The precipitate obtained with ammonium nitrate is fused in the same manner with KCy, the tin being obtained in metallic state. The insoluble residue, however, is repeatedly fused with KCy, the resulting globules of tin being each time washed and picked out. The gray powder which finally remains, as well as the impure tin obtained from the tungstic acid, is fused with sulphur, taken up with water, filtered, the filtrate acidulated with H_2SO_4, and the precipitated metallic sulphides dried. They are then heated in a porcelain dish in a current of hydrogen. The resulting residue is heated in air and fused with KCy. The tin thus obtained is added to the above, dried at 100° ·C., and weighed. As this tin is not entirely pure it is dissolved in HCl, the impurities partially remain undissolved and partially escape as hydrogen combinations. The latter are caught in a silver solution and determined as well as the portion remaining undissolved. The content of pure tin is known by deducting the amount of the impurities thus obtained.

B. Volumetric assays.

1. *Determination of tin by means of iodine.*—A few drops of potassium iodide of any desired concentration are added to the acid solution of stannous chloride, and then a few drops of diluted

starch paste. A solution of potassium bichromate of 0.02 or 0.01 gramme of the salt in 1 cubic centimeter is now added drop by drop under constant stirring, until the separated iodine does not again disappear, and the starch assumes a blue color when all the stannous chloride has been converted into stannic chloride ($3SnO + Cr_2O_6 = 3SnO_2 + Cr_2O_3$). The quantity of chromate decomposed by 100 parts of pure tin dissolved in hydrochloric acid is empirically determined (100 tin 83.2 chromate).

One to 2 grammes of tinstone are placed with four times the quantity of potassium cyanide in a porcelain dish, and heated for fifteen or twenty minutes. The mass is then poured upon an iron plate and treated with water. The metallic residue (tin and iron) is dissolved in hydrochloric acid, the tin precipitated with zinc, again dissolved in hydrochloric acid, and titrated with potassium bichromate in the presence of potassium iodide and starch. (*Hart*.)[1]

According to *Lenssen*,[2] more accurate results may be obtained by dissolving stannous salts (stannous chloride) with an addition of tartaric acid, or potassium-sodium tartrate, in sodium bicarbonate. Some starch paste is added to the clear solution, and it is then titrated with solution of iodine until the blue color appears ($SnO + 2I + Na_2O = SnO_2 + 2NaI$). The iodine solution is standarized by dissolving 12.7 grammes of pure iodine, and 20 to 30 grammes of potassium iodide in 1000 cubic centimeters of distilled water. A quantity of pure tin, accurately weighed, is dissolved in hydrochloric acid, tartaric acid is added, and the solution supersaturated with sodium bicarbonate. Solution of starch is added, and the iodine solution is gradually added from the burette until the liquid becomes blue. Two atoms of iodine (254) correspond to one atom of tin (118). This assay may be especially recommended for the detection of small quantities of tin.

2. *Determination of tin by means of potassium permanganate.*— 5 to 10 cubic centimeters of solution of stannous chloride are treated with a boiling solution of ferric chloride containing free hydrochloric acid ($SnCl_2 + Fe_2Cl_6 = SnCl_4 + 2FeCl_2$). The ferrous

---

[1] Dingler, ccx. 394.
[2] Journ. f. prakt. Chem. lxxviii. 200. Mohr, Titrirmethode, 1874, p. 311.

chloride which is formed is titrated, after dilution with water, with standard solution of potassium permanganate, until the liquid becomes reddish (p. 123). The value of the standard solution is determined by placing 0.2 gramme of freshly precipitated tin in a platinum crucible, and dissolving it in hydrochloric acid, in a current of carbonic acid. Ferric chloride in excess is added, and then the solution of potassium permanganate is added, until the last drop colors the fluid perceptibly. 2 equivalents of iron = 1 equivalent of tin. A correction becomes necessary, as experience has shown that more potassium permanganate is consumed in titrating ferrous chloride than stannous chloride.

3. *Determination of lead in tin.*[1]—According to Roux, the following method is used in the Paris Municipal Laboratory: 2.5 grammes of the metal are treated in a ¼ liter flask with 15 c.c. of HNO_3, and the nitrous vapors expelled by boiling; 40 c.c. of a saturated solution of sodium acetate are added, and the flask filled up to mark. When the precipitate has settled 100 c.c. of the clear supernatant liquid are taken out, 10 c.c. of a solution of potassium bichromate (7.13 grammes to the liter) are then added, and the lead chromate allowed to settle, another 10 c.c. of the potassium bichromate are added, and the same operation repeated until the supernatant liquid indicates an excess of lead chromate. Filter the precipitate, and determine the excess of lead chromate by a solution of 75 grammes of ammonium-ferric sulphate and 25 grammes of H_2SO_4 to the liter.

Detection of tin in presence of antimony.—According to Muir,[2] it is detected in the HCl solution by boiling with copper turnings for about ten minutes. This is reduced to stannous form and permits of being detected with mercuric chloride ($HgCl_2$).

C. Electrolytic determination of tin.—Tin can be very easily determined by this means; it separates completely from a HCl solution containing a little free acid, from the ammonium double oxalate, or from ammonium sulphide solution. Sodium and potassium sulphides cannot be used, as tin separates only partially from a dilute solution of such salts, and not at all from a concentrated solution. Potassium oxalate cannot be used; since, in this case, a basic

[1] Bull. Soc. Chim., XXXV., p. 596. Chemikerztg, 1881, 441.
[2] Chem. News, xlv. p. 69.

salt separates at the positive pole and cannot be reduced. If an acid solution is used, the current must not be interrupted during the washing; this precaution is unnecessary when ammonium oxalate or sulphide is used. When a solution of the double oxalate is used, it is freed as fully as possible from acid, treated with ammonium oxalate solution, then heated, and 3 or 4 grammes more of ammonium oxalate added. The hot solution is treated with a current of 9 to 10 c.c. oxyhydrogen gas per minute. The precipitation is complete in about five hours; the solution is then poured off, and the metal treated as usual. In the solution of the ammonium sulpho-salt, tin behaves like antimony. The tin solution (neutralized if necessary with ammonia) is treated with just sufficient ammonium sulphide to form the sulpho-salt, diluted to 200 c.c., and electrolyzed. The precipitation is complete in 5 or 6 hours. Should any sulphur adhere so strongly to the tin that it cannot be washed off, it can easily be removed, after washing with alcohol, by gentle rubbing with a linen cloth. In gravimetric analysis tin is often separated from other metals by sodium sulphide instead of ammonium sulphide. To determine tin electrolytically in such cases, the sodium sulphide must be converted into ammonium sulphide. To accomplish this, the solution is treated with about 25 grammes of pure ammonium sulphate free from iron, and heated very carefully, with the dish covered, till the H_2S has all escaped; the solution is then kept in gentle ebullition for about fifteen minutes. After it is completely cool, any sodium sulphate that may have separated is dissolved by the addition of water, and the solution electrolyzed with a current of 9 to 10 c.c. of oxyhydrogen gas per minute.[1]

XI. BISMUTH.

54. ORES.

Native bismuth, bismuth glace, Bi_2S_3, with 81.25 Bi; *cupriferous bismuth,* $CuBiS_2$, with 62 Bi and 18.9 Cu; *tetradymite,* Bi_2Te_3, with 51.94 Bi; *bismuth ochre,* Bi_2O_3, with 89.65 Bi, and others.

[1] Quant. Chem. Anal. by Electrolysis. Classen. [Trans.] New York. 1887.

55. FIRE ASSAYS.

These are inaccurate, as bismuth volatilizes, and in case the ore is impure, foreign metals collect in the brittle button.

1. *Ores and compounds free from sulphur* (native bismuth, tetradymite, bismuthic cupel ash, etc.). 5 grammes of ore with two and a half to three times the quantity of black flux, or potash and flour, and 2.5 to 5 grammes of borax-glass, are placed in a suitable crucible (Fig. 52, p. 65) and covered with a layer of common salt. It is then placed in a muffle, and, after the "flaming" has ceased, fused for twenty-five to thirty minutes at not too high a temperature. Or, 10 to 20 grammes of the substance are heated with two and a half to three times the quantity of borax-glass, an equal weight of soda, and 5 to 10 grammes of potassium cyanide, with a covering of common salt. Lead, tin, and copper pass partly into bismuth, and must be removed by the wet method.[1]

2. *Sulphurized bismuth ores.*—Five grammes of ore are placed in a crucible and covered with 1.25 to 1.5 grammes of thick iron wire, and 2.5 to 10 grammes of fine shreds of silver; upon this are placed two and a half to three times the quantity of black flux or potassium carbonate and flour, on this $1\frac{1}{2}$ to 2 grammes of borax, and a covering of salt. The charge is fused as in the lead assay, when the sufficiently ductile alloy of silver with bismuth can be separated from the iron. The percentage of bismuth is found by deducting the silver added. *Arsenic* must be removed by a previous ignition of the ore, with exclusion of air; but *antimony*, with an admittance of air. *Lead* passes into the bismuth, and *copper*, if not too much of it is present, is slagged off.

Joachimsthal: 5 grammes of ore are fused with 2 grammes of sodium carbonate and 1.25 grammes of iron turnings with a covering of common salt. The resulting plumbiferous button is dissolved in nitric acid, the lead is separated as lead chloride by *Patera's* method, which will be given later on, or the bismuth separated, as metal, from a weak acid solution by means of a strip of *lead. Tamm's process:* Ore free from copper is fused with a flux consisting of 2 parts potassium or sodium carbonate, and

1 part common salt, with an addition of some potassium cyanide. *Cupriferous ores:* 3 parts of ore with 5 parts potassium or sodium carbonate, 2 parts common salt, 1 part powdered wood-charcoal, and 2 parts flowers of sulphur. The bismuth is separated (with about 8 per cent. loss), containing a little copper, most of the latter passing into the slag as sulphide. The presence of iron induces more copper to unite with bismuth; antimony and arsenic and lead (partly) pass into the slag. *Rose* fuses the ore with 5 times the quantity of potassium cyanide, in a porcelain crucible, washes the resulting metal grains quickly with water, then with dilute alcohol, and weighs. In fusing, no black pulverulent residue of bismuth sulphide must remain.

56. WET ASSAYS.

While the *volumetric assays*[1] which have been recommended are of no practical importance, the gravimetric methods are mostly complicated, the available docimastic tests being, as a general rule, limited to the separation of bismuth from lead.

A. Assay of ore.—2 to 3 grammes of ore are dissolved in nitric acid, and evaporated to dryness. Some sulphuric acid is then added, the mass stirred, and again evaporated to dryness. The residue is now dissolved in water, and filtered. The filtrate is precipitated with ammonium carbonate in excess, filtered off, washed, dried, ignited, and then weighed as oxide, containing 89.65 per cent. bismuth.

B. Separation of bismuth from lead.

1. *According to Patera.*—The plumbiferous bismuth is dissolved in nitric acid. The solution is considerably diluted with water, upon which the liquid must not become turbid. The bismuth is then precipitated from the weak acid solution with a bright strip of lead. The pulverulent black bismuth is removed, and washed with water and alcohol. It is then dried at a temperature not over 120° C. (248° F.), and weighed. Or, hydrochloric acid in excess is added to the diluted nitric acid solution,

---

[1] Fleischer, Titrirmethode, 1876, p. 87. Muir. in Ber. d. deutsch. chem. Gesel. 1877, p. 2051. Buisson, in Fresenius's Ztschr. 1874, p. 61. Pearson, in Mitchell's Practical Assaying, 1888, p. 817.

and then some strong alcohol; the silver chloride and lead chloride are filtered off, and the bismuth is precipitated from the filtrate with ammonium carbonate. The carbonate is washed, dried, and ignited, and the bismuth determined as oxide (Bi_2O_3 with 89.65 Bi).

Determination of the percentage of lead and silver.—The combined metallic chlorides are weighed upon a weighed filter and then cupelled. The silver is calculated to silver chloride, the yield deducted from the weight of the combined chlorides, and the lead calculated from the lead chloride found by difference. *Copper* is determined by evaporating the ammoniacal filtrate to dryness with sulphuric acid. The dry mass is taken up in water, and the copper precipitated with iron or zinc (p. 106).

2. *According to Ullgreen.*—The solution of bismuth in nitric acid is precipitated with ammonium carbonate. The lead and bismuth carbonates are dissolved in acetic acid, and the bismuth is precipitated in a well-closed vessel with a bright strip of lead. It is then filtered, washed, and dissolved in nitric acid. The mass is then evaporated to dryness and heated, when bismuth oxide remains behind.

Tin oxide (also antimonious acid) remains behind in nitric acid in dissolving the alloy. The residue is then washed with alcohol, dried and weighed. *Lead* and *bismuth* are precipitated from the filtrate with ammonium carbonate (see above), and the copper from the filtrate as above.

Electrolytic assay.—Smith and Knerr[1] find that bismuth can be completely and rapidly precipitated by electrolysis from a sulphate solution containing free H_2SO_4. They add 1 c.c. of H_2SO_4 to 25 c.c. of solution, and use a current of 1 to 4 c.c. of oxyhydrogen gas per minute.

XII. MERCURY.

57. ORES.

Cinnabar, HgS, with 86.2 Hg; *native mercury, mercurial tetrahedrites*, with 0.5 to 17 per cent. Hg.

---

[1] Am. Chem. Journ., VIII., p. 207.

58. FIRE ASSAYS.

The object of these assays is to separate the sulphur, either by union with alkalies, iron, lime, etc., or to oxidize it by means of lead oxide, by heating the ores in retorts, tubes, or crucibles, and condensing the mercurial vapors liberated into liquid mercury, which is weighed either by itself, or, what is more accurate, in combination with gold. In smelting works, where many assays have to be made, small distilling furnaces[1] are generally on hand for this purpose. Notwithstanding the defects of the fire assays, it has not been possible to replace them by simple *wet* methods.[2]

A. Assays yielding free mercury.—140 to 1800 grammes, according to richness, of cinnabar ore, are heated with one half or equal parts of black flux, or 50 per cent. iron filings, or 30 per cent. lime, and 30 per cent. powdered wood charcoal, either in clay or iron retorts, in the latter case without iron filings (native mercury ore, amalgam, etc., it is best to heat in glass retorts), at a slowly increasing heat, in a suitable furnace (Figs. 44 to 46, pp. 61, 62). The vapors are condensed in the receivers represented by Figs. 44 to 46, or in a wet linen bag tied to the end of the stem of the retort, the lower part of the stem being at the same time kept cool by moist strips of linen or paper tied around it. The mercury adhering to the neck is removed by gently tapping and wiping out, and that from the receiver is dried with absorbent paper and caustic lime, and weighed in a watch-glass.

Idria:[3] 140 grammes of ore, with two or three spoonfuls of powdered lime, are placed in iron distilling tubes of about 52 millimeters diameter, lying in two rows, one above the other, on each side of a furnance. The receivers are luted on, and the process is finished when the tubes show a bright-red heat. The loss of mercury is considerable.[4] *Hungarian mercurial tetrahedrite:*[5] The ores are heated with the same quantity of iron

---

[1] B. u. h. Ztg. 1854, p. 394 (Idra).
[2] Muspratt's Chem. v. 1296. Mohr, Titrirmethode, 1874, pp. 236, 318, 441, 438.
[3] B. u. h. Ztg. 1854, p. 394. [4] B. u. h. Ztg. 1854, 357.
[5] B. u. h. Ztg. 1866, pp. 24, 262.

turnings—roasted ore at the same time with an equal quantity of lead oxide[1]—in glass retorts resting upon clay dishes. The neck of the retorts, when the operation is finished, are broken off by a blow, and the mercury is removed by means of a wiper of rabbit fur. It is then collected into a globule and weighed. *Rose's* method: A body of magnesite (or chalk with an equal quantity of sodium bicarbonate), 26 to 52 millimeters long is introduced into a glass tube, closed at one end, and measuring 314 to 470 millimeters in length, and 9 to 13 millimeters in width. Upon this is placed an intimate mixture of the ore and quicklime in excess, upon this more lime with which the mortar has been cleaned off, then more quicklime, and upon all a loose plug of asbestos. The open end of the glass tube is drawn out and bent to an obtuse angle, and introduced into a narrow-necked flask, so that its end just touches the water contained therein. The horizontal part of the tube is gradually heated from front to back in a *combustion furnace*, such as is used for organic analysis (Fig. 77,

Fig. 77.

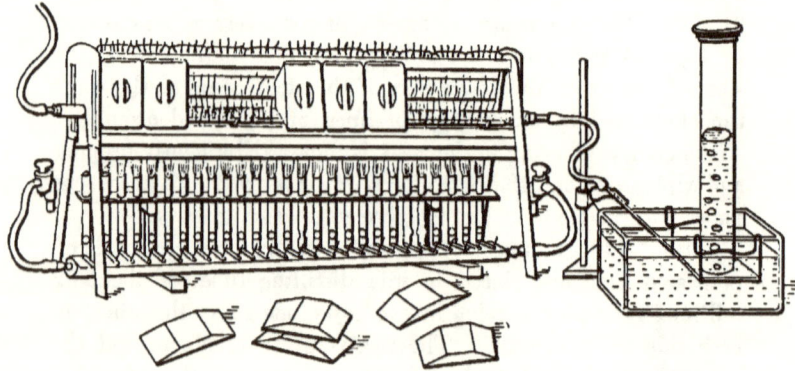

without the cylinder on the right). When the operation is finished, the bent end of the tube, in which the mercurial vapors have been condensed, is cut off, and the mercury, which has been protected from oxidation by the current of carbonic acid which is developed, is collected in the matrass. This is well shaken and

---

[1] Bergwerksfreund, v. 127 (Berthier). B. u. h. Ztg. 1879, p. 206 (Atwood).

allowed to settle. The clear water is then poured off, and the mercury placed in a previously weighed porcelain crucible. The water still adhering to it is removed with blotting paper. It is then dried under a bell-glass over sulphuric acid, or in an air-bath at 100° C. (212° F.), and weighed.

B. Assays in which the mercury is determined in combination with gold.—These are the most accurate assays.

1. *Eschka's process.*[1]—The quantity of ore taken for the assay varies according to its richness. If the ore carries as much as 1 per cent. 10 grammes, 1 to 10 per cent. 5 grammes, and over 10 per cent. 2 grammes are used. The sample is placed in a porcelain crucible, the edge of which has been ground smooth, and mixed with half its quantity of iron filings free from grease, covered with a layer of iron filings 5 to 10 millimeters thick. A well-fitting concave cover, made of fine gold, and previously accurately weighed, and the concavity of which is filled with distilled water, is now placed on the crucible. The lower part of the crucible is heated by a flame for about 10 minutes, during which time the mercury is volatilized, and deposits itself on the gold. The gold cover is now removed, the water in its concavity poured off, and the mirror of mercury washed with alcohol. The cover is then dried for about two or three minutes in the water-bath, placed upon a tared porcelain crucible and allowed to cool in the desiccator, and then both the mercury and crucible are weighed together. The most accurate results are obtained in the case of poor ores carrying up to 10 per cent. Hg.

The crucible should be heated very gradually, and the temperature only slightly raised towards the end of the operation. The water evaporated from the gold lid during the assay must be constantly replaced, and the lid, after the assay is finished, dried in an air-bath. The distillation of the mercury out of the lid must at first be done at a very moderate heat, and the temperature only raised when the mercury-mirror has disappeared, otherwise the gold is apt to volatilize with the mercury, and the lid to become spongy. If the assay material contains so much mercury that it is deposited

[1] Oestr. Ztschr. f. Berg. u. Hüttenwes, 1872, No. 9. B. u. h. Ztg. 1872, p. 173.

in drops, the quantity taken must be reduced. If the gold lid, after the distillation of the mercury, shows a decrease in weight as compared with that previously found, it is probably due to the volatilization of the gold.

In Idria (Austria) this assay is used as the prevailing working assay. The following compensative differences are allowed in the results:—[1]

Amount of mercury in the ore.	Differences allowed.
0 to 0.4 per cent.	0.04 per cent.
0.4 to 0.7 per cent.	0.06 "
0.7 " 1.0 "	0.08 "
1.0 " 3.0 "	0.15 "
3.0 " 5.0 "	0.20 "
5.0 " 10.0 "	0.25 "
10.0 " 20.0 "	0.35 "
20.0 " 30.0 "	0.45 "
30.0 " maximum	0.50 "

In testing for less than 0.001 gramme of mercury Teuber[2] recommends the following method: The dry and finely pulverized assay material is mixed in a porcelain crucible, shown in Fig. 78 ($\frac{3}{4}$ actual size), with thoroughly dried iron filings and minium. The charge is covered with a layer of minium, the lid luted on, and the crucible gradually and carefully heated over a lamp. The temperature is thus slowly raised until the bottom of the crucible shows a scarcely perceptible glow. The drops of moisture which condense on lid-tube are absorbed with blotting paper. The gold dish is then placed upon the tube a, b (which is about 1 millimeter in diameter), and filled with water. The mercury is deposited as a metallic mirror on the spot where the gold dish rests upon the tube. If the deposit of mercury is not too thin, its presence can be confirmed by moistening it with HNO_3, drying, and testing the nitrate of mercury found with a strip of filter paper moistened with

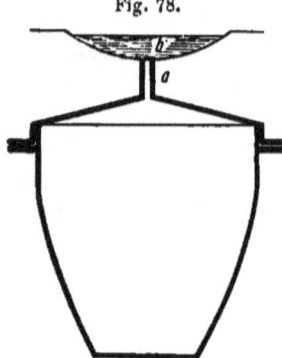

Fig. 78.

---

[1] Fortschritte im Probirwesen, Balling. Berlin, 1887, p. 142.
[2] Oesterr. Ztschft. f. Bg. u. Httuwsn, 1879, p. 423.

potassium iodide; the characteristic red color of mercuric iodide will immediately appear.

2. *Küstel's assay*.[1]—This is executed in a manner similar to the above, before the blowpipe, with the difference that the heating is done in a tube, the front end of which is provided with a gold spiral.

C. Assay of cinnabar.—10 grammes of cinnabar ore are introduced into a glass retort (Fig. 45, p. 62), and heated. The sublimate of mercury sulphide which deposits itself in the neck of the retort is collected and weighed. Some metallic mercury freed by organic substances, which may have been present, may be mixed with the sublimate. This is removed from the sublimate by nitric acid, and the quantity of mercury dissolved ascertained by the difference in weight. 86 parts of mercury correspond to 100 parts of cinnabar.

59. WET ASSAYS.

These may be—

A. Gravimetric assay.—1 gramme of cinnabar is heated with *aqua regia*, and repeatedly evaporated with hydrochloric acid, in order to expel the nitric acid. The solution of chloride is boiled with stannous chloride in excess, the clear fluid is poured off, and the beads of metal are collected into a coherent globule, by heating it again with some stannous chloride and a few drops of hydrochloric acid. The mercury is then washed by decantation, first with water containing hydrochloric acid, and then with pure water. It is now introduced into a small porcelain crucible, previously weighed, and the greater part of the adhering water removed by means of filtering paper. The mercury is then dried in a desiccator (Fig. 19, p. 40), with concentrated sulphuric acid, and weighed.

B. Volumetric assays.[2]—These are mostly complicated, and possess no advantage over the ordinary analytical determination

---

[1] B. u. h. Ztg. 1874, p. 70.
[2] Mohr, Titrirmethode, 1874, pp. 236, 318, 436, 438, 441. Fresenius's Ztschr. ii. 381.

by weight, or, are not generally available, as they require the absence of certain metals.

C. Electrolytic determination of mercury.—According to Escosura,[1] 0.5 gramme of the ore is digested with 10 to 15 c.c. of HCl and 20 c.c. of water in a porcelain dish; after boiling add from 0.5 to 1.0 gramme of potassium chlorate in small portions; when decomposition has taken place dilute with 50 c.c. of water and expel free chlorine by continued boiling. In order to separate selenium or tellurium, if present, add 20 to 30 c.c. of a saturated solution of ammonium sulphate and boil slightly; when the insoluble residue has settled, filter, and use filtrate as the electrolytic bath. The negative electrode should be a pure gold and the positive a platinum sheet. The solution is subjected to the galvanic current from 24 to 30 hours. The increased weight of the gold equals the content of mercury. Two Bunsen cells are generally sufficient. This assay may also be performed by treating the finely pulverized ore in a platinum dish with HCl, ammonium sulphate, and water. Of 10 per cent. ore only 0.2 gramme is used, and in 0.1 per cent. ore 10 grammes. The platinum dish serves as the negative pole and a disk of sheet gold, about 4 centimeters in diameter, as the positive pole. The current is supplied by 6 Meidinger cells, and the mercury is precipitated in 24 hours.

These determinations are said to be very accurate and exclusively used at Almaden.

According to Classen,[2] mercury is precipitated from a solution of its salt acidified with HNO_3 by a current of 0.2 to 0.5 c.c. oxyhydrogen gas, in the form of a mirror or of small globules, on the negative electrode; the metal adheres well and can be washed without loss. The washing must, however, be done without interrupting the current. Insoluble mercury compounds may easily be analyzed by suspending them in acidulated water, or in a dilute solution of common salt (1 : 10), and electrolyzing as usual.

Smith and Knerr[3] use a current of 4 c.c. oxyhydrogen gas per minute, and find precipitation complete in 30 to 45 minutes, the mercury forming a compact shining deposit.

[1] Journ. Pharm. Chem. 1886. p. 411. Chemikerztg. 1886. p. 100.
[2] Quan. Chem. Analysis, by Electrolysis. [Trans.] New York, 1887.
[3] Am. Chem. Journ., 8, p. 209.

XIII. ANTIMONY.

60. ORES.

Stibnite (*antimony sulphide*), Sb_2S_3, with 71.77 Sb; *valentinite* (*antimony oxide*), Sb_2O_3, with 83.56 Sb; *pyrostilbite* (*antimony oxysulphide*), $Sb_2O_3.Sb_2S_3$, with 77.21 Sb.

61. FIRE ASSAYS.

The assays of antimony sulphide, which is the principal ore, are inaccurate on account of the volatilization of antimony, incomplete decomposition by alkaline carbonates and even by potassium cyanide (4 parts), as well as on account of iron, which always passes partly into the antimony. There are likewise no simple docimastic tests by the *wet method*.[1] Sometimes the yield of antimony sulphide (*antimoniun crudum*) contained in the ore is ascertained by a liquation process assay on a small scale.

A. Liquation process for determining antimonium crudum.—1 to 1½ kilogrammes of ore comminuted to fragments the size of a hazel-nut or walnut are introduced into a covered crucible having a perforated bottom and fitting air-tight in another crucible in such a manner that sufficient space is left between the two to allow the fused antimony sulphide to collect in the lower crucible. The lid and the joint between the two crucibles should be luted with fire-clay and sand. The lower crucible is surrounded with some poor conductor of heat (ashes), and the upper ore with live coals, kept in a glow with a bellows, and heated to a moderate red heat. The antimony sulphide will melt and collect in the lower crucible.

B. Determination of antimony in antimony sulphide.

1. *Assay by precipitation.*—5 grammes of ore are fused in a crucible with the same or double the quantity of black flux, or

[1] Fresenius's Ztsch. xvii. 185 (Becker's gewichtsanalytische Probe); Mohr, Titrirmethode, 1874, p. 267, 309; Muspratt's Chemie, i. 820; Fleischer, Titrirmethode, 1876, p. 299.

potassium carbonate and flour, about 2 grammes of iron filings, 0.75 to 1.25 grammes of borax-glass, and a covering of common salt. The fusion is continued for three-quarters of an hour. When the operation is finished, the brittle regulus is freed from slag, and the adhering particles of slag are removed by washing. The yield by this assay is about 68 per cent. and if more iron is added, it will appear to be larger on account of the contamination of the antimony by iron.

Five grammes of antimony ore are mixed in a crucible (Fig. 49, p. 64) with 10 grammes of anhydrous potassium ferrocyanide. The mixture is covered with 2.5 grammes of potassium cyanide and heated to a cherry-red heat. The yield is 72 per cent.[1]—*Or:* 10 grammes of ore, and 40 to 50 grammes of potassium cyanide, with a covering of common salt, are fused at a high temperature. —*Or:* 100 parts of antimony sulphide, 42 parts of iron filings, 10 parts of sodium sulphate, and 2 parts of powdered wood charcoal are fused. Yield 62 per cent. antimony.—*Or:* 100 parts of antimony sulphide, 80 parts of iron slag, 50 parts of sodium carbonate, and 10 parts wood charcoal. Yield 60 per cent. Impure buttons are comminuted and treated with concentrated nitric acid, filtered, the precipitate of antimonic acid washed, dried, and ignited in a porcelain crucible. It is then weighed, and the weight found multiplied by 0.7922 gives the metallic antimony.

2. *Roasting and reducing assay.*—The ore, which is very fusible, is carefully roasted at a very gradually rising temperature until a yellowish-white powder has been formed. This is fused for about three-quarters of an hour with the same or double the quantity of black flux, or potassium carbonate and flour, with a cover of common salt. If necessary, some borax is added. The yield will be at the utmost from 64 to 65 per cent. Sb. *Oxidized ores* do not require roasting, and are fused with 3 parts of black flux (with 1 part of argol), 1 part of sodium carbonate, and 15 per cent. of powdered wood charcoal, at not too high a temperature: or, 10 grammes of ore with 25 grammes of black flux and 1 part of argol with a cover of salt.

[1] B. u. h. Ztg. 1856, p. 319.

62. WET ASSAYS.

Namely :—

1. *Gravimetric assay.*—0.5 gramme of antimony sulphide is dissolved in *aqua regia*, some tartaric acid added, and the solution filtered. The filtrate is saturated with ammonia and an excess of yellow ammonium sulphide, and digested for some time on the water-bath. It is then filtered, but without bringing the precipitate upon the filter. This is again digested with ammonium sulphide and then filtered. The antimony sulphide is thrown down out of the filtrate by the addition of diluted hydrochloric acid. The sulphuretted hydrogen is driven off on the water-bath, and the liquid then filtered upon a filter, previously dried at 100° C. (212° F.), weighed, washed with sulphuretted hydrogen water, and the precipitate dried at 100° C. (212° F.) until the weight remains constant. A part of the contents of the filter is then weighed off in a small porcelain boat, placed in a glass tube, and heated in a current of carbonic acid gas (Fig. 77, p. 238), in order to remove the sulphur in excess, until no more of the latter escapes. The Sb_2S_3, containing 71.8 per cent. Sb, is weighed, and from this the antimony contained in the entire mass in the filter is calculated.

Becker's method.[1]—Fuse the assay material with 3 parts of sodium or potassium carbonate and 3 parts of sulphur. Strongly sulphuretted combinations of these metals are thus obtained. They easily give off sulphur to the air, and when decomposed with HCl large quantities of sulphur are precipitated; this latter circumstance interferes greatly with the subsequent operations. Donath[2] observes that sodium hyposulphite completely freed from water by careful melting in a dish and finely pulverized is much better adapted for this purpose. In this case disintegration proceeds readily, and the solution lixiviated from the fused mass shows but a slight color. From this solution HCl precipitates the respective sulphides mixed with but little sulphur. It is best to weigh the antimonious sulphide as such, as it has been shown by Bunsen that

---

[1] Ztschft. f. Anal. Chem. Bd. 17, p. 185.
[2] Ztschft. f. Anal. Chem. Bd. 19, p. 23.

an accurate determination cannot be made by converting it into antimonic oxide.

2. *Volumetric method.*—The antimony sulphide is dissolved in hydrochloric acid, and heated until the odor of sulphuretted hydrogen can no longer be detected. Tartaric acid, or potassium-sodium tartrate is added, and the solution supersaturated with a cold saturated solution of sodium bicarbonate in the proportion of about 20 cubic centimeters to 0.1 gramme of Sb_2O_3. The result will be a solution, which, on the addition of iodine, will be converted into antimonic acid and hydriodic acid, $SbHO_2 + 2I + H_2O = SbHO_3 + 2HI$. Starch solution is now added, and a sufficient quantity of titrated solution of iodine to color the fluid blue is allowed to drop into the solution from a burette. A decinormal solution of iodine is prepared by dissolving 12.7 grammes of iodine in potassium iodide, and diluting this to the bulk of 1 liter. 2 atoms of consumed iodine correspond to 1 molecule of antimony oxide, or 1 cubic centimeter of solution of iodine to 0.0061 gramme of antimony. The standard of the solution of iodine is determined by titration with tartar emetic, observing the same conditions regarding the concentration and quantity of the reagents (tartaric acid, sodium hydrocarbonate, etc.) in the subsequent titration.

See "XIX., Sulphur," concerning the indirect determination of antimony in Sb_2S_3 by the quantity of sulphuretted hydrogen evolved therefrom.

Weil's method of determining antimony with protochloride of tin.—If nothing but the antimony is to be determined in the assay, treat, according to the richness of the assay material, from 2 to 5 grammes in *aqua regia* and excess of HCl; add potassium permanganate until the solution retains a rose color, boil until the rose color disappears and potassium iodide starch-paper is no longer colored blue by the escaping vapors. The solution thus obtained contains the antimony, as antimonic acid. Dilute to $\frac{1}{4}$ liter with an aqueous solution of tartaric acid. Take 25 c.c. of this solution for the assay, and add a measured quantity of cupric sulphate solution containing a known amount of copper. Heat mixture to boiling and add 25 c.c. of concentrated HCl, then titrate the boiling solution with protochloride of tin until discoloration takes place. The number c.c. of protochloride of tin which corresponds to the amount of the copper salt added must be known; the number of c.c of protochloride of tin used over this amount corresponds to the amount of antimony. By calculating the amount of copper corresponding to this volume and multiplying the result by 0.96214, we get the content of antimony in the assay material

$$SbCl_5 + SnCl_2 = SnCl_4 + SbCl_3$$

and $2CuCl + SnCl_2 = SnCl_4 + 2CuCl$, hence 1 equivalent of antimony corresponds to 2 equivalents of copper, also

$$\frac{Sb}{2Cu} = \frac{122}{126.4} = 0.96214$$

the results can be more quickly obtained by the use of the following table:—

ASSAYING.

Titer of the protochloride of tin solution, in cubic centimeters, for 0.1 gramme of copper.	Cubic centimeters of protochloride of tin solution used for every 25 cubic centimeters of assay solution (2 grammes = 250 cubic centimeters) correspond to *per cent. of antimony.*									
	1	2	3	4	5	6	7	8	9	10
15.0	3.20	6.40	9.61	12.81	16.01	19.22	22.42	25.63	28.83	32.03
15.1	18	36	55	73	15.92	10	29	47	66	31.84
15.2	15	31	46	62	77	18.93	09	24	40	55
15.3	13	27	40	54	68	81	21.95	09	22	36
15.4	11	23	35	46	58	70	82	24.93	05	17
15.5	09	19	29	39	49	58	68	78	27.88	30.98
15.6	07	15	23	31	39	47	55	63	70	78
15.7	05	11	17	23	29	35	41	47	53	59
15.8	04	08	12	16	20	24	28	32	36	40
15.9	02	04	06	09	11	13	16	18	20	23
16.0	00	00	00	00	00	01	01	01	01	01
16.1	2.98	5 96	8.94	11.93	14.91	17.89	20 92	23.86	26.84	29.82
16.2	96	92	88	85	81	77	74	70	66	63
16.3	94	88	83	77	72	66	60	55	49	44
16.4	92	84	77	69	62	54	47	39	32	24
16.5	91	83	74	66	57	49	30	32	23	15
16.6	89	79	68	58	48	37	29	16	06	28.96
16.7	87	75	62	50	38	25	13	01	25.88	76
16.8	85	71	57	43	28	14	00	22.86	71	57
16.9	83	67	51	35	19	02	19.86	70	54	38
17.0	82	65	48	31	14	16.97	79	62	45	28
17.1	80	61	42	23	04	85	66	47	28	09
17.2	79	58	37	16	13.95	74	53	32	11	27.90
17.3	78	56	34	12	90	68	46	24	02	80
17.4	76	52	28	04	80	56	32	09	24.85	61
17.5	74	48	22	10.96	70	45	19	21.93	67	41
17.6	73	46	19	92	66	39	12	85	59	32
17.7	71	42	13	85	56	27	18.99	70	41	13
17.8	70	40	11	81	51	22	92	62	33	03
17.9	68	36	05	73	42	10	79	47	15	26.84
18.0	66	33	7.99	66	32	15.99	65	32	23.98	65
18.1	65	31	96	62	27	93	58	24	89	55
18.2	62	25	87	50	13	75	38	00	73	26
18.3	61	23	85	47	08	70	32	20.94	55	17
18.4	60	21	82	42	03	64	25	85	46	07
18.5	59	19	79	39	12.98	58	18	78	37	25.97
18.6	58	17	76	35	94	52	11	70	29	88
18.7	56	13	70	27	84	41	17.98	55	12	68
18.8	55	11	67	22	79	35	91	47	03	59
18.9	54	08	62	16	70	24	78	32	22.86	40
19.0	52	06	59	12	65	18	71	24	77	30
19.1	51	02	53	04	55	06	57	08	54	11
19.2	50	00	50	00	50	00	51	01	51	01
19.3	49	4.98	47	9.96	45	14.95	44	19.93	42	24.91
19.4	47	94	41	89	36	83	30	78	25	72
19.5	46	92	38	85	31	77	24	70	16	63
19.6	45	90	35	81	26	72	17	62	08	53
19.7	43	86	30	73	17	60	03	47	21.90	34
19.8	42	84	27	69	12	54	16.97	39	82	24
19.9	41	82	24	66	07	49	90	32	73	14
20.0	40	81	21	62	02	43	83	24	64	05

The simultaneous presence of iron, copper, and antimony can also be very quickly and sharply determined by means of protochloride of tin. Dissolve in *aqua regia*, pass H_2S through solution, filter, oxidize in the filtrate with potassium chlorate, add HCl, and after filling up to the $\frac{1}{4}$ liter mark titrate the iron. Dissolve the precipitated metallic sulphides in *aqua regia*, evaporate considerably, and after adding HCl and filling up to $\frac{1}{4}$ liter mark, simultaneously titrate the copper and antimony, allow the titrated solution to stand over night, whereby the copper alone becomes completely reoxidized, and the next day titrate the copper alone. The difference in the quantity consumed corresponds to the antimony. The number of c.c. used in titrating the iron multiplied by 0.883 give its content.

Volumetric estimation of antimony in the presence of tin.[1]— This method is applicable to alloys such as Britannia metal or type metal, and depends upon the fact that antimonic chloride is reduced to antimonious chloride by hydriodic acid, with the constant liberation of iodine, whilst the stannic chloride is unreduced. Dissolve the finely-divided alloy in strong HCl, and add from time to time small quantities of potassium chlorate. After all the metal is dissolved and the antimonious chloride converted into antimonic chloride, the solution is gently boiled until all the oxides of chlorine have been expelled. Cool, and add a slight excess of a strong solution of potassium iodide. The free iodine is then estimated by a standard solution of sodium hyposulphite. Since 122 parts of antimony liberate 254 parts of iodine, the amount of iodine found multiplied by 0.48031 will give the amount of antimony present.

3. *Electrolytic determination of antimony.*—According to Parodi and Mascazzini,[2] antimony can be separated in a compact form from a solution of the chloride in ammonium tartrate, or from its sulphosalt. According to Luckow,[3] this precipitate from the trichloride is of a pale-gray color and has a metallic lustre. It dissolves with difficulty in HCl, but readily in HNO_3, especially after moistening with HCl.

According to Classen, antimony can be precipitated from a HCl solution, but the metal does not adhere with sufficient tenacity to

---

[1] Chem. News, xlv., p. 101.
[2] Gazz. Chimikal, 8.
[3] Ztschft. f. Anal. Chem., Bd. 18, p. 588.

the electrode, even if potassium oxalate has been added. An adherent metallic deposit can be obtained by adding potassium tartrate, but the separation is then too slow. The precipitation of antimony from its sulpho-salts is complete and satisfactory. If ammonium sulphide is used to produce a double salt, it must contain neither free ammonia nor polysulphides. When a sulpho-salt is used 0.2 gramme of the assay material is melted at a moderate heat, with four times its weight of a mixture of soda and sulphur. Lixiviate with water, filter, pour hot yellow ammonium sulphide repeatedly over the residue upon the filter, wash with water, and subject the cold filtrate to the action of the electric current. When a solution of antimony containing ammonium sulphide is electrolyzed, there is formed over the metal a coating of sulphur which cannot be washed off with water. After the metal is washed with alcohol, however, the thin coating of sulphur can be rubbed off with the finger, or a handkerchief moistened with alcohol, without danger of loss. A separation of antimony sulphide will be observed in the positive electrode, but it will again be reduced if an excess of ammonium sulphide is present. Alkaline polysulphides should not be present, as they impair the accuracy of the results. Classen and Ludwig[1] have found that polysulphides are converted by hydrogen dioxide into sulphates in the same manner as monosulphides. They therefore recommend fusing the precipitate, containing the antimony and lead, in the form of sulphides, with ten times its weight of dehydrated sodium hyposulphide, lixiviating with water, and heating solution in a carefully cleaned platinum dish with ammoniacal hydrogen dioxide until the solution has become colorless, or a precipitate of antimonious sulphide has formed. Add 10 c.c. of a concentrated solution of sodium monosulphide and the *cold* solution electrolyzed with a current of 1.5 to 2 c.c. of oxyhydrogen gas per minute. A battery of 5 or 6 Meidinger cells is best adapted for the purpose. Precipitation is complete in 10 or 12 hours, the precipitate is washed with water and alcohol, dried for a short time at 80° or 100° C., and weighed.

[1] Berichte der deutsch. chem. Gesellschaft, 1885, Bd. 18, p. 1104. Chemikerztg, 1885, p. 891.

XIV. ARSENIC.

63. ORES.

Native arsenic, *mispickel*, FeAsS, with 46 As; *leucopyrite*, Fe_2As_3, with 66.8 As; nickel and cobalt ores, etc.

64. FIRE ASSAYS.

The object of these is to determine the quantity of arsenic, or of its compounds, which may be obtained from the ores.

A. Native arsenic.—300 to 500 grammes of mispickel or leucopyrite (when arsenious acid is present the ores are mixed with 16 to 20 per cent. powdered charcoal, and with potassium carbonate in case metallic sulphides should be contained in the ore) are introduced into a clay tube, one end of which is closed. A spiral of sheet iron is placed in the open end, and the tube is closed by a sheet-metal cap, which is loosely luted on. The tube is then brought into the furnace, and the charge is gradually heated for one to one and a half hours at a red heat, the end with the spiral projecting from the furnace. When the operation is finished, the tube is allowed to cool, and is taken out. The sheet-iron spiral is then removed and unrolled, and, if the operation has been conducted at the proper temperatures, scales of white flaky arsenic and some gray powder, both allotropic modifications of arsenic, will fall off. The product is then weighed (FeAsS = FeS + As and Fe_2As_3 = 2FeAs + As).

White metallic scales will be the result, when the receiver is small and has nearly the same temperature as that of the arsenical vapors; and *gray* powder, when the receiver is large or has a decidedly lower temperature than that of the arsenical vapor, and when it is evolved along with other heated gases (as in the reduction of arsenious acid by charcoal).

B. Arsenious acid.—2 to 5 grammes of ore are placed in the open end of a refractory glass tube, the other end of which, somewhat drawn out and bent in a right angle, projects into a large *Woulff* bottle. The end of the tube containing the ore is laid so as to incline somewhat upwards, in a combustion furnace

(Fig. 77, p. 238), and heated. The *Woulff* bottle is connected with an aspirator, when the arsenious acid will deposit itself in the straight and curved part of the tube, and in the *Woulff* bottle. It is driven by heating, from the horizontal to the bent part of the tube; this is cut off, the arsenious acid is wiped out with a feather, and then weighed together with that contained in the bottle.

C. Realgar (red orpiment), with 70.03 As, and *yellow orpiment*, As_2S_3, with 60.9 As.—The object of the docimastic tests is to determine the quantity of orpiments which can be prepared from the raw materials in question, or to ascertain the proportions of these to be added in certain technical processes.

1. *Assays for the determination of realgar.*—Iron pyrites ($7FeS_2 = FeS_2 + 6FeS + 6S = 23$ per cent. S) and arsenical pyrites ($FeAsS = FeS + As = 46$ per cent. As) in different proportions (20 to 30 grammes, or more) are heated in a glass tube closed at one end, or in a glass retort, whereby realgar is sublimed. This is fused in a porcelain crucible, and its color examined. Lighter or darker shades can be given to the color by adding, respectively, sulphur or arsenic.

2. *Assays for the determination of yellow orpiment.*—A mixture of arsenious acid with sulphur (together 10 to 20 grammes) in different proportions is heated (generally with 6 to 12 per cent. of sulphur, or less) in a glass retort at a gradually rising temperature, until the sublimation of the yellow product is complete.

While the *native yellow orpiment* is a combination of As_2S_3, difficult to dissolve in acids, the *artificial* product consists principally of arsenious acid colored by a few per cent. of arsenic sulphide, and is consequently very poisonous.

65. WET ASSAYS.

These may be—
A. Gravimetric assays.

1. *Wet assay.*—$\frac{1}{2}$ to 1 gramme of ore is ignited in a porcelain crucible at a strong red heat, with 4 to 5 times the quantity of saltpetre, and $1\frac{1}{2}$ times the quantity of calcined sodium carbonate, with a thick covering of these fluxes. When the crucible is

cold, the potassium arseniate is extracted by lixiviation with hot water, and evaporated to dryness with nitric acid. The dry mass is treated with water, the silicic acid filtered off; the filtrate is treated with ammonia in excess, then with a solution of magnesium sulphate (or at once with magnesia mixture prepared from 110 parts of crystallized magnesium chloride, 700 parts concentrated ammonia, 140 parts ammonium chloride, and 1300 parts water). The liquid is allowed to stand for 12 hours, the ammonium-magnesium arseniate, $2(MgNH_4.AsO_4)+H_2O$, is filtered off upon a filter previously dried at 100° C. (212° F.), and weighed; 100 parts of salt dried at 100° C. (212° F.) and weighed, will contain 60.51 parts of arsenic acid, corresponding to 65.21 per cent. of arsenic and 86.08 per cent. of arsenious acid. By igniting *very carefully* and not too quickly, the magnesium salt passes into $Mg_2As_2O_7$. Or the sample is digested in strong nitric acid with the addition of a few crystals of potassium chlorate. The solution is diluted with water and filtered. Some lead nitrate in solution is added to the acid liquid, when lead sulphate will be separated, lead arseniate remaining in solution. The precipitate of lead sulphate is filtered off, the filtrate saturated with soda, when the lead arseniate will be separated. This is filtered off, washed, dried, and weighed. 100 parts arseniate = 22.2 parts of metallic arsenic, or 29 parts of arsenious acid.

2. *Wet method combined with the dry.*[1]—1 to 1½ grammes of the substance is fused in the manner indicated under 1, with saltpetre and sodium carbonate, in order to obtain a solution of alkaline arseniate. The filtrate is saturated with nitric acid, and strongly diluted in case sulphuric acid is present. Silver nitrate in excess is then added, and sufficient ammonia to cause the precipitate to disappear. The excess of ammonia is evaporated without boiling until its odor has disappeared. The silver arseniate is then filtered off, dried, and smelted with lead, and the arsenic calculated from the quantity of silver found, 1 atom of arsenic being precipitated to 3 atoms of silver, or 100 silver corresponds to 23.15 arsenic = 35.5 arsenic acid.

[1] Plattner-Ritcher's Löthrohprobirkunst, 1878, p. 651.

B. *Volumetric assays.*—The method most used is *T. Mohr's*[1] *estimation of arsenious acid;* according to which arsenious acid combined with soda is completely converted into arsenic acid by iodine ($As_2O_3 + 2I_2 + 2Na_2O = 4NaI + As_2O_5$).

A solution containing 500 cubic centimeters 2.5 grammes of iodine and 4 grammes potassium iodide, or in 1 c.c. 0.005 gramme of iodine, is added to a solution containing sodium arsenite, sodium bicarbonate, and some starch paste. Fixing of the standard: 4.95 grammes of arsenious acid are placed in a flask and dissolved in sodium bicarbonate, in about 200 cubic centimeters of water. The clear solution is poured off, and sodium salt and water are added in small portions until solution is complete. It is then transferred to a liter flask, 20 to 25 grammes more of sodium bicarbonate are then added, and the flask is filled up to the mark. 10 c.cm. of this solution are taken, some fresh starch solution and sodium bicarbonate added, the liquid diluted to about 150 c.c., and then titrated with the solution of iodine until the blue color appears, which must not disappear even when sodium bicarbonate is added. *Arsenic acid* must be reduced to arsenious acid by introducing sulphurous acid gas into the hot acid solution, or by the addition of sulphites.

Pearce's method of determining arsenic.[2]—The finely powdered assay material is mixed in a porcelain crucible with six to ten times its weight of a mixture of equal parts of carbonate of sodium and nitrate of potassium. The mass is then heated with a gradually increasing temperature to fusion, and kept for a few minutes in that state. Lixiviate with water, filter, and wash with hot water. The arsenic is in the filtrate as alkaline arseniate. Acidify the filtrate with HNO_3, and boil to expel excess of carbonic acid and nitrous fumes. Exactly neutralize by alternate additions of ammonia and HNO_3. If the neutralization has caused a precipitate, it is best to filter off at once. A solution of nitrate of silver is now added in slight excess and stir vigorously. Filter and wash thoroughly with cold water. Dissolve on filter the arseniate of silver with dilute HNO_3. Cool and add about 5 c.c. of ferric sulphate. Titrate with

---

[1] Mohr, Titrirmethode, 1874.
[2] Pro. Colorado Scientific Society, vol. i. Engineering and Mining Journ. May 5, 1883, p. 256.

a standardized solution of ammonium sulpho-cyanide until a faint red tinge is obtained which remains after considerable shaking. From the formula $3Ag_2O.As_2O_5$ we deduct 648 parts $Ag = 150$ parts As, or $Ag : As = 108 : 25$. Canby[1] simplifies this method by neutralizing with oxide of zinc instead of ammonia, obtaining quite as satisfactory results and a great saving of time. The oxide of zinc is added in excess, and as it seems to hold the gelatinous silica and alumina, which are usually precipitated upon neutralizing, the washing of the precipitated arseniate of silver is usually much more rapid.

Arsenic in glass.[2]—Fresenius finds that many samples of glass contain arsenic which under some circumstances may be set free and cause errors. When an alkaline flux is melted in contact with an arseniferous glass, either in a current of carbonic acid or hydrogen, the arsenic will be set free. By the reducing action of hydrogen alone the amount set free is small. If a solution of an alkaline carbonate be boiled in a vessel of such glass, a very perceptible amount of arsenic will be taken into solution, while, on the other hand, HCl solutions have no solvent effect. The glass used in the above experiment contained 0.20 per cent. of arsenic.

XV. URANIUM.

66. ORES.

Pitch blende, Ur_3O_8, with 84.9 Ur.

67. WET ASSAYS.

Namely :—

A. Gravimetric assays.

1. *More accurate analytical process.*—1 to 2 grammes of ore, etc., are decomposed with concentrated nitric acid. The solution is diluted and the residue and precipitate (silicic acid, lead sulphate, basic antimony, and bismuth salts) are filtered off. The arsenic acid in the filtrate is reduced by means of sulphurous acid to arsenious acid, and the sulphurous acid removed by boiling. It is then precipitated with sulphuretted

[1] Trans. Am. Inst. M. E. vol. XVII., p. 77.
[2] Ztschft. f. Anal. Chem. Bd. 22, p. 397.

hydrogen (As, Sb, Pb, Bi, Cu), filtered, and the sulphuretted hydrogen removed by boiling. The iron is then oxidized by potassium chlorate, ammonium carbonate in excess added, and the liquid filtered. Ammonium sulphide is now cautiously added in the cold in order to precipitate Mn, Zn, Ni, Co, leaving the uranium in solution. The liquid is then filtered, and the filtrate heated with nitric acid to separate sulphur. It is now again filtered, and when the solution has become cold the uranium is precipitated as brownish-yellow ammonia-uranic oxide, by ammonia. This is filtered off, washed, dried, and ignited to green uranoso-uranic oxide ($UrO_2.2UrO_3 = Ur_3O_8$), with 84.8 uranium, corresponding to 96.22 uranous oxide (UrO_2) and 101.9 uranic oxide (UrO_3).

2. *Patera's technical test.*[1]—5 grammes of ore are dissolved in nitric acid, not in excess. The unfiltered solution, which has been freed from the excess of nitric acid by boiling, is super-saturated with sodium carbonate, boiled for a short time, and the solution containing the sodic-uranic carbonate is filtered into a golden dish. The solution is evaporated to dryness, the residue ignited and extracted with hot water, and the insoluble acid sodium uranate ($Ur_4Na_2O_7$) is filtered off, ignited, and weighed. 100 parts of the weight obtained equal 88.3 parts of uranoso-uranic oxide.

In case a golden capsule should not be at hand, the solution of the uranic oxide in soda is treated with sodium hydrate in order to precipitate hydrated acid sodium uranate. It is then filtered, washed, and dried, and the precipitate is removed as much as possible from the filter and ignited together with the ash of the filter. It is then again washed upon the filter, dried, and ignited. In case a considerable quantity of copper is present, a small quantity of it passes into the alkaline solution.

According to Alibegoff,[2] uranium cannot be separated from calcium by means of ammonium sulphide, but, if present in the solution as chloride, it is completely precipitated by mercurous oxide. To completely precipitate uranium in the presence of alkaline earths

[1] Dingler, clxxx. 242 (Patera). Fresenius's Ztschr. v. 229 (Fresenius); viii. 387 (Winkler).
[2] Liebig's Anal. d. Chem. 1886, Bd. 233, p. 143.

it is recommended to add sufficient ammonium chloride to the solution containing the chlorides, and, after maintaining it at a boiling temperature for some time, to add *pure* mercurous oxide. The solution is then agitated, and after boiling once more decanted. In the presence of many alkaline earths it is several times boiled with water containing ammonium chloride, filtered and washed with cold water also containing ammonium chloride. Small particles of alkaline earths which may still adhere to the olive green uranic oxide after ignition are recognized by their yellow color. The precipitate and filter must be carefully ignited in a closed platinum crucible, then in an open crucible, and finally over a blast lamp. The separation of the uranium from barite ($BaSO_4$) being incomplete it is advisable, if the latter be present, previously to separate it with $H\,SO_4$.

B. Volumetric assay.[1]—1 to 2 grammes of ore is dissolved in concentrated sulphuric acid. The sulphuric acid (not hydrochloric acid) solution is diluted according to the richness of the ore, to $\frac{1}{2}$ or $\frac{1}{4}$ liter. 50 cubic centimeters of this are taken, placed in a suitable flask, and diluted with 100 cubic centimeters of water. The liquid is boiled for half an hour with zinc until the yellow solution of uranic oxide has assumed the sea-green color of uranous oxide. All the zinc is dissolved, and the uranous oxide is titrated with potassium permanganate (p. 123). Uranous oxide requires the same quantity of potassium permanganate to become oxidized, as ferrous oxide.

According to Zimmermann,[2] the oxysalts of uranium cannot well be determined in HCl solution on account of their being converted into uranous subchloride (Ur_4Cl_3), which is a very unstable compound. The determination of uranium in HCl solution is therefore made by destroying the injurious influence of the HCl by the addition of manganous sulphate and excluding the air in titrating. It is consequently best to reduce the uranium compound in a flask with zinc and HCl, the air being excluded. At first the solution becomes green, but later a dirty green, then brown, and finally a

---

[1] Journ. f. prakt. Chem. xcix. 231 (Belohoubeck). Fresenius's Ztschr. xi. 179 ; xvi. 104. Gouyard's Probe in Chem. Centr. 1864, p. 339.—Analyse der Uranoxydalkalien in Fresenius's Ztschr. iii. 71 (Stolba).

[2] Liebig's Anal. d. Chem. Bd. 213, p. 285.

17

beautiful red. When the latter color remains constant the reaction is finished. In the meanwhile a known excess of permanganate is placed in a porcelain dish, strongly acidulated with H_2SO_4, and manganous sulphate added; the hot solution of uranous subchloride is then quickly added. The excess of permanganate is then removed by a solution of·ferrous oxide (whose titer with permanganate is known) and the excess of it added measured back into the same permanganate until the appearance of a slight rose color.

Method of determining uranium with potassium dichromate and iodine.—According to Zimmermann,[1] this method is executed by acidulating the reduced oxysalt with HCl and adding an excess of potassium dichromate solution of known strength, whereby the uranous oxide is converted into uranic oxide. Dilute with water to 150 c.c. and add, with constant stirring, a solution of potassium iodide whereby iodine is separated from the bichromate remaining in excess. After adding starch paste this iodine is determined with a solution of sodium hyposulphate of known strength. The titrated iodine solution is finally added, drop by drop, until the liquid has acquired a very slight blue color. If uranous subchloride is to be determined, the reduced solution is allowed to run into an excess of potassium chromate solution mixed with HCl; the process otherwise is the same as above.

XVI. TUNGSTEN.

The only important tungsten mineral is wolfram, which is essentially $FeWo_4$, but always containing more or less manganese. In the pure state it contains 76.4 per cent. tungstic acid or 60.5 per cent. metallic tungsten. Wolfram is brownish-black, gives a reddish-brown or blackish-brown streak. It fuses before the blowpipe to a magnetic globule; with borax it gives the iron reaction; with soda upon platinum foil the manganese reaction. Next to gold and platinum metallic tungsten is specifically the heaviest metal (17.0 to 17.6). It is very hard and brittle, has a color and lustre of iron and remains unchanged in the air, but ignites when heated in a finely divided state and burns to tungstic acid. It is very refractory and has never as yet been fused by itself.[2]

[1] Ztschft. f. Anal. Chem. Bd. 23, p. 65.
[2] Fortschritte im Probirwesen. Balling, Berlin, 1887, p. 162.

68. FIRE ASSAY.

By melting wolfram in crucibles lined with charcoal at a very high temperature, a hard, brittle, foliated regulus resembling white pig iron is obtained, which contains tungsten, manganese, and iron, and whose content of tungsten can only be determined in the wet way.

69. WET ASSAYS.

A. Gravimetric assays.

1. Sheele's method.—The very finely pulverized wolfram is several times evaporated to dryness with an excess of HCl to which some HNO_3 is finally added. The lumps that form are carefully crushed before each addition of HCl and the dish heated to 120°C. in an air bath. The final residue is treated with water acidulated with HCl and the separated tungstic acid filtered off, ignited, and weighed. The digestion must be continued until the brown tungsten powder is converted into yellow tungstic acid. If the ore is not entirely pure, the tungstic acid contains silica and undecomposed silicates. In this case wash residue into a beaker and dissolve out the tungstic acid with ammonia. The ammonium tungstate is filtered off, the residue thoroughly washed, ignited, weighed, and weight subtracted from that previously obtained. As a check on this operation, evaporate the ammonium tungstate, forming an acid salt in lustrous scales which dissolves with difficulty. After evaporating to dryness, the pure tungstic acid is ignited and weighed.

2. Berzelius's method.—Fuse ore with double its weight of sodium carbonate, lixiviate with hot water, neutralize the free alkali so far that the solution is scarcely acid, and add mercurous nitrate in excess; wash the precipitate formed with water containing sodium chloride, and, after drying, ignite, whereby the tungstic acid remains behind in a pure state. According to Schiebler,[1] it is advisable to add, after precipitation, a few drops of ammonia until the white precipitate appears brownish-black.

3. Marguerite's method.— The solution obtained after fusion with soda is evaporated in water bath with an excess of H_2SO_4, the residue slightly heated to expel free H_2SO_4, taken up with water and the alkaline bisulphate filtered off. The tungstic acid remaining

---
[1] Ztschft. f. Anal. Chem. Bd. 1, p. 71.

upon the filter, is dried, detached, the filter incinerated, and placed together with the tungstic acid in a small crucible. After moistening the tungstic acid with a few drops of HNO_3 and allowing the latter to evaporate, the crucible and contents are heated. In heating the acid readily acquires by deoxidation a greenish-blue color, which, however, does not affect the result.

4. Cobenzl's method.[1] — The assay material previously bolted through fine linen is placed in a small flask and treated with concentrated HNO_3. Heated upon a water bath, HCl added from time to time until disintegration is effected. After standing 5 or 6 days nothing but the yellow tungstic acid can be seen. Now evaporate to dusty dryness, take up with very dilute HNO_3, evaporate again, and repeat these operations three times or more. Finally, take up with dilute HNO_3 and some tartaric acid (on account of the possible presence of some antimony). Heat to 100° C., filter, wash the tungstic acid together with the undecomposed silicates and silica several times by decantation, and then with hot water upon the filter; dissolve the tungstic acid from off the filter with very dilute ammonia, leaving the silica and undecomposed silicates behind. The ammonium tungstate is then evaporated to dryness in a porcelain crucible, and after heating, the straw yellow tungstic acid is weighed together with the crucible.

5. Separation of tungsten from tin. — Metallic sulphides and arsenides are frequently associated with wolfram. They are generally removed by digestion with *aqua regia*. Tungsten and tin are, however, most frequently associated, and their separation has often to be effected.

According to Talbott, this separation is made as follows: The material containing the two metals is intimately mixed with potassium cyanide, previously fused and powdered. The mixture is fused in a porcelain crucible, lixiviated with hot water, and the tungstic acid in the potassium tungstate formed, determined by one of the methods already given. In order to neutralize the excess of potassium cyanide, the solution must be boiled with HNO_3, and the tungstic acid again brought into solution by an alkali. Mercurous nitrate is recommended for precipitating the tungstic acid.[2]

---

[1] Monatshefte für Chemie, etc. Sitzungsber. der Kaiserl. Akad. d. Wissenschaften zu Wien, April, 1881, p. 259.

[2] Fortschritte im Probirwesen. Balling, Berlin, 1887, p. 166.

B. Volumetric assay. Zettner's method.[1]—Fuse 0.5 to 1.0 gramme of the finely powdered assay material with 4 parts of sodium or potassium carbonate, or with 3 parts sodium carbonate and 1 part saltpetre in a platinum crucible over a Bunsen burner. Lixiviate with hot water, filter, neutralize the excess of alkali in the heated filtrate with acetic acid; acidulate slightly with the same acid, then heat to boiling, and add a solution of acetate of lead from a burette until no more precipitate is formed. The end of the precipitation is indicated by the more rapid clearing of the solution. Now add a few drops more of acetate of lead, and filter small portions of the tungsten solution into test-tubes near at hand in order to observe the precipitation more readily. Continue testing in this manner until precipitation is complete. The number of c.c. of acetate of lead multiplied by 0.0116 gives the amount of tungstic acid in the solution.

Preparation of the titrating solution.—Make a decimal lead solution by dissolving 18.950 grammes of extremely pure crystallized acetate of lead, acidulate slightly with acetic acid, and dilute to one liter. The effective strength of this solution is tested by fusing 11.6 grammes of pure, ignited tungstic acid with sodium carbonate, taking up the fused mass with water, acidulating slightly with acetic acid, and then determining the lead solution, which is equivalent to it. If they are not equal, they must be made so by repeated experiment.

XVII. CHROMIUM.

70. ORES.

Chrome iron ore, Cr_2FeO_4, with 30 to 65 Cr_2O_3; *crocoisite* (*red lead ore*), $Pb.CrO_4$, with 30.96 CrO_3.

71. WET ASSAYS.

Volumetric assays are less frequently made use of than gravimetric assays, and the latter vary, especially in the manner in which the very difficultly decomposable chrome iron-ore is decomposed.

---

[1] Poggendorff, Anal. Bd. 130, p. 16. Ztschft. f. Anal. Chem. Bd. 6, p. 229.

A. Gravimetric assays.[1]

1. *Direct assay.*—According to *Pourcel's* method : 2 grammes of the ore in coarse powder are highly heated in order to facilitate its comminution. It is then ground in an agate mortar, or on a porphyry plate, to an impalpable powder, which should show no glistening particles. This powder is heated to 120° C. (248° F.) until it loses no more in weight, and then a sample of 0.5 gramme is quickly weighed off. 5 grammes of sodium carbonate are introduced into a platinum crucible and heated in the mouth of the muffle in order to dry it, but not sufficiently to fuse it. The ore is then intimately mixed in an agate mortar, with 0.5 gramme of saltpetre and the warm sodium carbonate until the mixture assumes a uniform color. It is then placed in a platinum crucible and strongly heated, first at the front of the muffle and then at the back, at a white heat for three hours.

The crucible is then taken out, wiped off after it has become cold, and placed in a porcelain dish containing 0.5 liter of distilled water which completely impregnates the contents of the crucible. This is allowed to stand for about twelve hours on a sand-bath, at a temperature somewhat less than 100° C. (212° F.). It is then filtered on a very small filter, and the crucible and dish are washed out with hot distilled water. The filtrate containing the potassium chromate and sodium-potassium aluminate is weakly acidulated with sulphuric acid, ammonia in excess is added, and it is then heated to nearly 100° C. (212° F.) for at least three hours. The alumina is then filtered off, the filtrate is introduced into a capacious flask, hydrochloric acid strongly in excess and 100 cubic centimeters of pure alcohol of 40° are added, the contents of the flask then agitated and its neck closed by inserting the neck of a smaller flask into it. The liquid is now heated to nearly 100° C. (212° F.) for about forty-eight hours, until the reduction of chromic acid to chromic oxide is complete (whereby

---

[1] Journ f. prakt. Chem. lvii. 256 (Calvert). Fresenius's Ztschr. i. 497 (O'Neill, Oudesluys, Genth) ; iv. 63 (Souchay) ; 1861, p. 34 (Mitscherlich) ; 1870, p. 71 (Storer). Polyt. Centr. 1856, p. 701 (Hart). Dingler, cxciii. 33 (Clouet) ; cxcvii. 503 (Britton) ; ccxxi. 450 (Dittmar) ; ccxxiv. 86 (Fels). Bullet. de la soc. de l'industr. minér., St. Etienne, 1878, livr. iv. p. 867 (Pourcel).

the liquid assumes an emerald-green color), and the odor of alcohol has disappeared. Ammonia in excess is then added, the liquid is allowed to stand for about twelve hours at a temperature of nearly 100° C. (212° F.), so that the bubbles are formed on the sides of the flask. The precipitate of hydrated chromic oxide is filtered on a small filter, washed with boiling, slightly ammoniacal water, and dried. The chromic oxide is then ignited and weighed.

According to *Fels*, *Calvert's*, *Britton's* and *Dittmar's* are the best methods for decomposing chrome iron ores.

The following additional methods are recommended for decomposing chrome iron ores. By Hager.—Mix the ore with 3 parts of sodium fluoride, place in a small graphite crucible, cover with 12 parts of powdered potassium bisulphate and fuse. In five minutes the charge should be liquid, and in ten minutes viscous, then disintegration is finished. Fels recommends allowing this fused green mass, containing the chromium in the form of a fluoride, to cool and then remelt it with the addition of potassium chlorate. In a few minutes the mass becomes yellow and disintegration is complete.

By F. Clarke.—Mix the finely pulverized ore with 3 parts of sodium fluoride or cryolite, covering with 12 parts of potassium bisulphate and fusing in a covered platinum crucible.

2. *Indirect assay.*[1]—The ore is fused with potassium nitrate and sodium carbonate (in the same manner as arsenic, p. 252), the alkaline chromates are lixiviated. The solution is saturated with acetic acid, and boiled, in order to remove the carbonic acid. It is then diluted with water (to prevent a separation of silver acetate), and a sufficient quantity of silver nitrate is added. The precipitate of silver chromate will then contain to one atom of chromium, one-half atom of silver (100 Ag = 48.69 Cr = 70.92 Cr_2O_3 = 93.15 CrO_3). The precipitate, together with the filter, is boiled in strongly diluted hydrochloric acid, the silver chloride formed is filtered off, smelted with lead, and cupelled, and the chromium calculated from the resulting quantity of silver.

B. Volumetric assay.—1 gramme of the ore is powdered and

[1] Plattner-Richner's Löthrohrprobirkunst, 1878, p. 651.

ground as fine as possible. It is then fused with soda and saltpetre and converted into alkaline chromate (p. 262). It is then supersaturated with sulphuric acid and a weighed quantity of pure ferrous sulphate or ammonio-ferrous sulphate, when, the ferrous oxide becoming oxidized at the cost of the chromic acid, the latter is transformed into chromic oxide, and the color of the reddish-yellow solution becomes distinctly green ($2CrO_3 + 6FeSO_4 + 6SO_3 = Cr_2S_3O_{12} + 3Fe_2S_3O_{12}$); therefore, 6 equivalents of FeO correspond to 2 equivalents $CrO_3 = 3 : 1$. The residue of ferrous oxide which has not been decomposed is titrated with solution of potassium permanganate, when the final reaction will be more distinct the stronger the liquid has been acidulated with sulphuric acid, which causes the green color of the chromic oxide to become pale.

XVIII. MANGANESE.

72. ORES.

Pyrolusite, MnO_2, with 62.8 Mn and 37.2 O; *braunite*, Mn_2O_3, with 69.23 Mn and 30.77 O; *hausmannite*, Mn_3O_4, with 71.7 Mn and 28.3 O; *manganite*, $Mn_2O_3 + H_2O$, with 89.9 Mn_2O_3 and 10.1 H_2O; *varvicite*, $(Mn_2O_3 + H_2O) + 2MnO_2$, with 14.23 MnO, 80.79 O, and 4.98 H_2O; *psilomelane*, $(Mn.Ba.K_2.Li_2) O + 4 MnO_2$, with 20 to 60 MnO_2; *wad*, $MnO.2Mn_2O_3 + 3H_2O$.

73. ASSAYS OF PYROLUSITE.[1]

The percentage of manganese in an ore, which is of interest to the iron manufacturer, is generally determined by chemical analysis (*Tamm*[2] has given a process of preparing *manganese carbide*, from pyrolusite), while the value of an ore for other technical purposes (manufacture of chlorine and of chloride of ime, the preparation of oxygen) is judged—

1. By the *quantity of chlorine* which the ore yields on treatment

---

[1] Muspratt's Chemie, iv. 1111.
[2] Dingler, ccvi. 136. B. u. h. Ztg. 1873, p. 55.

with hydrochloric acid ($MnO_2 + 4ClH = MnCl_2 + 2H_2O + Cl_2 =$ 81.2 per cent. Cl), or, with common salt and sulphuric acid ($MnO_2 + 2NaCl + 2H_2SO_4 = MnSO_4 + Na_2SO_4 + 2H_2O + 2Cl =$ 81.2 per cent.).

If an ore contains ferrous oxide, for instance in the form of spathic or magnetic iron, a part of the chlorine evolved from the hydrochloric acid is consumed in the higher oxidation of the ferrous oxide, and is therefore lost for technical use. Only so much of the chlorine as is actually obtained from pyrolusite is of value to the buyer, and the assays which give this (for instance, *Bunsen's* and *Gay-Lussac's*) are to be preferred to those which give the total amount of chlorine without taking into consideration that, when ferrous oxide is present, the chlorine is not all available for practical purposes.

2. By the *quantity of oxygen* the pyrolusite yields when ignited or treated with sulphuric acid ($MnO_2 + SO_3 = MnSO_4 + O =$ 18.3 per cent.).

3. By the *foreign admixtures* according to their quality and quantity. *Substances soluble in acids* (for instance, calcium and iron carbonates) are especially injurious. They increase unnecessarily the cost of manufacturing chlorine, and, by evolving considerable quantities of carbonic acid, exert a disturbing influence on the preparation of chloride of lime. For this reason, the quantity allowed is sometimes limited by contract to one per cent.

The quantity of acid required for the decomposition is determined by finding the quantity of pure marble dissolved by a given quantity of hydrochloric acid; then allowing the same quantity of acid to act on a known quantity of the ore; then when chlorine ceases to be evolved, introducing the marble, and when the evolution of carbonic acid has ceased, removing the marble and weighing. The quantity of acid is ascertained from the difference in the loss of marble in the two cases. 100 parts of marble saturate 70.5 parts of dry and 205 parts of aqueous hydrochloric acid of 1.17 specific gravity; and when the acid is of 1.09 specific gravity, corresponding to 18.2 per cent. of dry acid, 7.3 milligrammes of dry acid correspond to 1 gramme of dissolved marble.

4. *By the decomposability and the physical condition of the ore,* as they require different quantities of acid, for instance, *Spanish* ore more than *Nassau* ore.

Hydrochloric acid dissolves the ferrous oxide contained in manganese ores more readily than sulphuric acid, and differences may, therefore, arise in the yield of available chlorine from an ore, according as one or the other acid has been used, since the more ferrous oxide dissolved, the more chlorine is kept back (p. 265).

5. By the *constitution* of the ore, which requires also different quantities of acid.

1 atom pyrolusite requires 4 atoms hydrochloric acid to yield 2 atoms of chlorine ($MnO_2 + 4ClH = MnCl_2 + Cl_2 + 2H_2O$); while braunite requires 6 atoms of the acid ($Mn_2O_3 + 6ClH = 2MnCl_2 + Cl_2 + 3H_2O$).

6. By the *amount of hygroscopic water*, which is sometimes considerable, and must be removed by drying, for instance, upon *Fresenius's* disk (Fig. 2, p. 25), at 100° or 110° to 115° C. (212° or 230° to 239° F.).

It is customary to represent the commercial value of manganese ores, whether the available chlorine or oxygen is to be ascertained, in terms of manganese peroxide equivalent to the chlorine or oxygen yielded (even if the ore does not contain it, as, for instance, braunite); namely, 2 atoms of chlorine (17), or 1 atom oxygen (16) = 1 atom peroxide. Frequently 60 per cent. of peroxide is taken as the standard in commerce, and from a fixed price for this, the value of the ore is determined according to the higher or lower percentage of peroxide it contains.

The following Table gives the theoretical yield of oxygen, chlorine, and peroxide of the different ores:—

	Oxygen per cent.	Chlorine per cent.	Peroxide per cent.
Pyrolusite	18	81.2	100
Braunite	10	45.1	55.5
Hausmannite	6.8	30.6	37.7
Manganite	9	45.6	50.
Vårvicite	13.8	62.2	76.6

A. *Gravimetric assays.*—These require but simple utensils, and can be easily executed by not very experienced operators, but the results, if certain substances are present, are inaccurate, or require check assays.

1. *Method of Fresenius-Will.*[1]—2 to 5 grammes of very finely powdered ore are weighed out and placed in the flask A (Fig. 79), made of thin glass, and capable of holding about 120 cubic centimeters. To this are added two and a half times the quantity of powdered neutral potassium oxalate (5 to 12.5 grammes), and the flask is then filled to about one-third with water. The flask

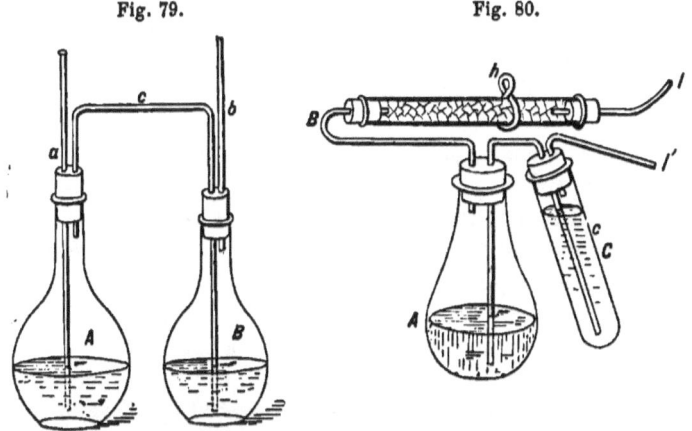

Fig. 79.   Fig. 80.

A is hermetically connected by the tube c (corks saturated with wax or paraffine, or caoutchouc stoppers should be used) with the flask B half filled with English sulphuric acid. The tube a is closed with a plug of wax, or a caoutchouc tube with a glass rod, and the whole apparatus is weighed. The air is then aspirated at b, in order to create a partial vacuum in the flask B, by which the air above the liquid in flask A will likewise be somewhat rarefied, in consequence of which the acid from B will pass through c into A, after the suction ceases. Carbonic acid is developed by the action of the sulphuric acid upon the ore and oxalate ($MnO_2 + C_2O_3 + SO_3 = MnSO_4 + 2CO_2$), which passes through c into B, is dried in passing through the sulphuric acid, and escapes through b. The drawing over of acid by suction is repeated

[1] Ann. der Chemie u. Pharm. xlix. 137. Fresenius's Ztschr. i. 48 (Röhr); i. 110 Kolbe); i. 81, 110 (Kolbe). Dingler, clxxxvi. 210 (Lunge).

until no more carbonic acid is developed, and no black residuum is observed in A, gentle heating being employed towards the last, the plug or stopper is now removed from a, suction is applied at b, in order to remove the carbonic acid contained in the apparatus. When this is *entirely* cold the apparatus is weighed, and the percentage of peroxide calculated from the loss of carbonic acid; 2 atoms CO_2 (88) being equal to 1 atom MnO_2 (87.14, or, in round numbers, 87).

If carbonates are present, the result will be affected, and the following process becomes necessary to neutralize their influence. A little sulphuric acid is drawn over from the flask B into A, the latter containing only ore and water. The carbonic acid evolved is removed by aspiration, and the apparatus is weighed, the required quantity of potassium oxalate being weighed along with the apparatus upon the same scale pan. The oxalate is then quickly poured into the flask A, and the further operation conducted as above.—According to *Mohr*, a percentage of ferrous oxide[1] in the ore (for instance, magnetic and spathic iron) does not affect the result of the assay, but gives too high a percentage of chlorine for commercial purposes, since a part of the chlorine is consumed in the higher oxidation of the iron.—Lighter apparatus than that of *Will* and *Fresenius* is described by *Mohr*, *Kolbe*, and *Rose* (Fig. 80); A, flask for the reception of the ore, potassium oxalate, and water; C, glass tube for sulphuric acid, diluted with an equal volume of water; B, drying-tube with calcium chloride, supported at h. The apparatus is weighed, and then tilted in order to transfer acid from C through the tube c to A (or by aspirating on the calcium chloride tube at l). The carbonic acid developed is allowed to escape through the drying-tube B, and finally the carbonic acid is expelled from the apparatus by suction at l'.

2. *Method of Fikentscher-Nolte.*[2]—1 to 5 grammes of ore, 8 times the quantity of ferrous sulphate free from ferric oxide, an

---

[1] Fresenius's Ztschr. 1869, p. 314 (Mohr); 1871, p. 310 (Luck); Dingler, cxcvii. 422 (Pattinson). B. u. h. Ztg. 1871, p. 312 (Sherer).

[2] Journ. f. prakt. Chem. xviii. 160, 173 (Fikentscher). B. u. h. Ztg. 1859, p. 149; 1864, p. 374 (Nolte).

accurately weighed strip of bright sheet-copper 4 to 5 times the weight of the ore, 30 to 35 cubic centimeters of hydrochloric acid of 1.12 specific gravity, and distilled water are placed in a suitable flask provided with a rubber valve (Fig. 11, p. 36). The contents are kept at a boiling heat (at least two hours) until the solution, at first brown, becomes nearly colorless, when the ferric chloride, which was first formed, is reduced to ferrous chloride by the copper, the weight of which correspondingly decreases ($MnO_2 + 2FeCl_2 + 2HCl = MnCl_2 + Fe_2Cl_6 + 2H_2O$; then, $Fe_2Cl_6 + 2Cu = 2FeCl_2 + Cu_2Cl_2$). The flask is then quickly filled with boiling water free from air. The liquid is then poured off, the strip of copper quickly thrown into water, rinsed, and dried with absorbent paper, but without rubbing it; then completely dried at 100° C. (212° F.), allowed to become cold in the desiccator, and then weighed. The percentage of manganese peroxide is calculated from the loss of copper, as, according to the above equation, 2 atoms of dissolved copper (126.8) correspond to 1 atom of peroxide (87.14).

The oxidizing action of the air upon copper must be avoided, as otherwise more would be dissolved than the required quantity. For this reason a flask provided with a rubber-valve should be used, ferrous sulphate free from ferric oxide. The copper should further be kept constantly covered by the hydrochloric acid, thoroughly boiled water used, and the copper quickly rinsed off. In case the ore contains ferric oxide, a check assay must be made, as it forms ferric chloride with hydrochloric acid, and this contributes to the solution of the copper. This must be calculated for by treating a quantity of ore equal to that of the principal assay, with hydrochloric acid, but first *without* copper. This is now heated until all the chlorine has been expelled, the weighed copper is now added, and the liquid boiled until it becomes colorless. The loss in weight of copper by ferric chloride alone is now determined, and this is deducted from the loss which takes place in the principal assay.

B. *Volumetric assays.*—Those methods which give only the actually available chlorine, as *Bunsen's* and *Gay-Lussac's*, deserve the preference, as other methods indicate too large a percentage of chlorine, which, in operations on the large scale where the ore contains ferrous oxide, cannot be depended on.

Perry[1] found that an ore which gave 100 MnO_2 according to Fresenius's method yielded 99.1 according to *Mohr's*, 100.6 according to *Hempel's*, 97.8 according to *Gay-Lussac's*, and 98.4 per cent. MnO_2 according to *Bunsen's*. The differences are partly explained by the fact that, in making the assays, the action of the ferrous oxide present is taken into consideration in some, while in others it is not; and that, when sulphuric acid is used (*Fresenius, Hempel*), or hydrochloric acid (*Bunsen, Gay-Lussac*), more or less ferrous oxide is dissolved, as it is not equally soluble in these acids.

1. *Bunsen's method with iodine.*[2]—This is based upon the principle that, when chlorine evolved from manganese ore by hydrochloric acid ($MnO_2 + 4ClH = MnCl_2 + 2Cl + 2H_2O$) is introduced into a solution of potassium iodide, it separates from this a corresponding quantity of iodine, indicated by the appearance of a brown color. After the addition of starch, which colors the solution blue, the quantity of separated iodine is determined by titrated solution of sodium hyposulphite, which is added until the color has disappeared ($2Na_2S_2O_3 + 2I = Na_2S_4O_6 + 2NaI$); whence 2 atoms of chlorine correspond to 2 atoms of iodine, and these to 1 atom of manganese peroxide. This assay is very accurate, and the results obtained are not affected by any admixtures, but it requires somewhat skilful manipulation.

The retort d (Fig. 81), the neck of which is provided with two bulbs in order to lessen the danger of loss by overflow, is filled one-third full with a freshly prepared solution of one part potassium iodide in ten parts of water. The potassium iodide must not become colored on the addition of acid, as this would indicate the presence of potassium iodate, from which more iodine would be separated by chlorine. The retort is then inverted (see Fig. 81), and a glass tube provided with a bulb c is introduced

---

[1] Dingler, ccxxvi. 194.

[2] Mohr, Titrirmethode, 1874, p. 625. Muspratt's Chemie, iv. 1118. Fresenius's Ztschr. 1869, p. 314; 1870, p. 410. Journ. f. prakt. Chem., Neue Folge, xviii. 101 (Marawski und Stingl).

into the neck of the retort. The other somewhat wider end of the glass tube is furnished with a caoutchouc tube b, previously boiled in a solution of potassium hydrate. 0.1 to 0.5 gramme of very finely powdered ore is poured into the flask a. This is then

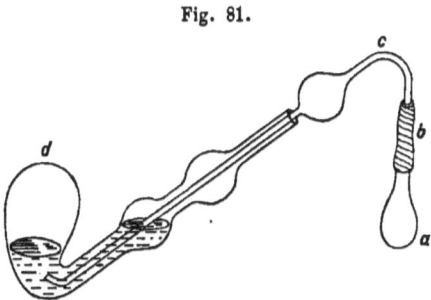

Fig. 81.

about two-thirds filled with fuming hydrochloric acid, the neck of the flask is quickly pushed into the moistened caoutchouc tube b, so that glass touches glass. The flask is now slowly heated over a spirit-lamp, so as to avoid the passage of liquid from d back into a. The heating is continued until the greenish color of the chlorine in the bulb of the connecting tube disappears, the ore is entirely decomposed, and a peculiar crackling noise is heard in the flask. The heat is somewhat increased for about half a minute after the crackling noise has been perceived, to prevent the liquid, containing about 5 grammes of solid potassium iodide and colored brown by the separated iodine, from passing over into the flask. The gas-conducting tube c is then withdrawn from the retort with the left hand, the heating of the flask a being continued all the while by holding the spirit-lamp below it with the right hand (if these manipulations are not conducted with proper skill, there is danger that the liquid in the retort will pass over). The tube is then washed off in a beaker-glass. The retort, containing the brown iodine solution, is corked, and carefully shaken, so that the solution may take up all the free iodine, during which operation the liquid must not come in contact with the cork. After the fluid has become entirely cold, it is poured out in the beaker-glass. Should the liquid contain any undissolved iodine, a few crystals of potassium iodide must be added before the contents of the retort are emptied into the beaker-glass. It is then diluted to the bulk of one-half liter. 100 cubic centimeters of this are taken, introduced into a separate vessel, and titrated sodium hyposulphite, which is added to it as long as a red color is still *distinctly* perceptible. About 2 grammes of starch liquor are added, and then more sodium hyposulphite, drop by drop, until the blue color which has been formed just commences to disappear. If necessary, the assay may be controlled by titrating back with a normal solution of iodine, until the blue color appears again. The standard solution of sodium hyposulphite is prepared by dissolving 24.8 grammes of sodium hyposulphite in water, and diluting the solution to 1 liter. 0.1 to 0.2 gramme of pure iodine is then dissolved in 18 grammes of potassium iodide, free from iodic acid, and this solution is also diluted to 1 liter, when 1 cubic centimeter of normal solution of sodium hyposulphite will correspond to 0.0127 gramme of iodine contained in 1 cubic centimeter of solution, so that equal volumes correspond.

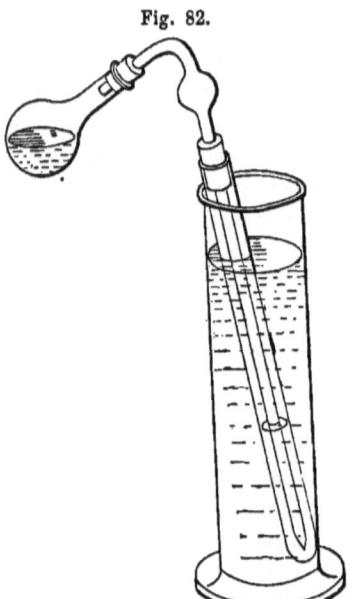

Fig. 82.

Fig. 82 represents a modified form of this apparatus, in which the bulbed-tube attached to the dissolving flask passes into a long and narrow tube, which is cooled by immersion in cold water, and contains a higher column of potassium iodide solution.

2. *Levol's method with iron.*[1] —0.5 to 0.6 gramme of manganese ore, and 0.8 to 1 gramme of piano wire (containing on an average 99.6 per cent. of pure iron), are treated with exclusion of air with hydrochloric acid, in a flask

---

[1] Dingler, lxxxv. 299 (Levol) ; cxcvii. 422 (Pattinson). Fresenius's Ztschr. 1869, p. 509 (Teschemacher u. Smith).

provided with a rubber valve (Fig. 11, p. 36); when, from the resulting ferrous chloride, a quantity of ferric chloride corresponding to the chlorine developed will be formed ($MnO_2 + 4HCl = MnCl_2 + 2Cl + 2H_2O$ and $2Cl + 2FeCl_2 = Fe_2Cl_4$). The liquid is diluted to 0.5 liter. 100 cubic centimeters of this are taken, and, after having become entirely cold, the non-oxidized quantity of ferrous oxide is determined by titration with potassium permanganate (p. 123). The quantity of iron oxidized by the chlorine is determined from the difference; and 2 atoms of iron (112) correspond to 1 atom of manganese peroxide (87.14).

Ammonio-ferrous sulphate, $FeSO_4 + (NH_4)_2SO_4 + 6H_2O$, may be used instead of metallic iron, in the proportion of about 7 grammes to 1 gramme of manganese ore with 70 per cent. of peroxide, and 8 to 9 grammes if the percentage is higher; when 2 atoms of the salt (784) correspond to 1 atom of manganese peroxide (87.14). The assay solution to be titrated should be cold, and strongly diluted, to prevent the hydrochloric acid from being decomposed by the potassium permanganate; or, potassium bichromate[1] may be used instead of potassium permanganate. In this assay (iron test), which is much used in England, a part of the chlorine is consumed for the higher oxidation of the ferrous oxide, if any be contained in the ore.

Volhard's method.[2]—This method is based upon the fact that only in the presence of salts of calcium, magnesium, barium, and tin monoxide, manganese is completely precipitated as dioxide by potassium permanganate.

$$3MnO + Mn_2O_7 = 5MnO_2.$$

In the absence of these salts a combination of dioxide with varying quantities of monoxide is precipitated, and consequently a portion of the manganese is withdrawn from the titering solution. Heat and concentration aid in the formation and separation of the dioxide, while the addition of an acid retards the reaction; HCl and H_2SO_4 should therefore not be present. A drop of HNO_3, however, added

---

[1] Mohr, Titrirmethode, 1874, p. 632. Polyt. Centrbl. 1871, p. 1117; Oxalsäureprobe in Fresenius's Ztschr. 1870, p. 410.

[2] Annal. d. Chem. t. 198, p. 318. Dingler's Journ. Bd. 235, p. 387. Bg. u. Httnmsch. Ztg. 1880, p. 150. Oesterr. Ztschft. f. Bg. u. Httnwsn. 1880, p. 168.

before the beginning of the titration, facilitates the precipitation of MnO_2 in case there is a small amount of *organic matter* present, which has the effect of producing a turbidity which renders the determination difficult; while small portions of ferric oxysalts exert no disturbing influence in the acidulated solution, they prevent precipitation in a neutral solution and render it difficult to recognize the color of the solution. The ferric salts are removed by adding pure zinc oxide, whereby all the ferric oxide is precipitated as zinc oxide containing hydroxide.

The assay is executed as follows.—Dissolve 2 grammes of the finely powdered ore in a mixture of 3 parts H_2SO_4 (1.13 sp. gr.) and 1 part HNO_3 (1.4 sp. gr.) and evaporate to dryness. Digest the residue with HNO_3, dilute with water and rinse the solution into a half liter flask, neutralize excess of acid with caustic soda and precipitate the ferric oxide with pure oxide of zinc, adding the latter while shaking until the supernatant liquid assumes a milky turbidity; when the precipitation is complete, dilute up to mark, and after allowing to settle, filter. Place 100 c.c. of the filtrate in a boiling flask, add *one drop* of HNO_3, heat nearly to boiling and titrate with permanganate solution until the fluid assumes a pale-red color.

Determining the strength of the permanganate.—Volhard[1] recommends the following method. Dilute 10 c.c. of a 5 per cent. solution of potassium iodide with 150 to 200 c.c. of water, add 4 or 5 drops of HCl and run in, with constant stirring, 20 c.c. of permanganate. The liberated iodine is titrated with sodium hyposulphite; this sodium salt is measured for comparison with potassium bichromate, which also separates iodine in acid solutions of potassium iodide.

Hampe's method.[2]—1. Decompose one gramme of the assay material with 20 c.c. of HNO_3 (1.2 sp. gr.). From this concentrated solution, containing free acid, precipitate the manganese as dioxide by potassium chlorate or bromate by adding, in two or three portions, 8 to 10 grammes of one of these salts in the solid state. Boil solution for 1½ hours, dilute with hot water, filter through asbestus and place the precipitate with the asbestus and a few grains of sodium bicarbonate into the same flask in which the solution and precipitation were made. Now add dilute H_2SO_4 and a

---

[1] Ztschft. f. Anal. Chem. Bd. 20, p. 274.
[2] Chemikerztg, 1883, p. 1106. Bg. u. Httnmsch. Ztg. 1883, p. 537.

measured quantity of ammonium ferric sulphate solution, and after closing flask with cork and rubber valve, heat in sand bath until the dark-brown precipitate disappears. Allow to cool and titer back the excess of ferric sulphate with permanganate. . . . It is best to use for the titrating solution 14.2817 grammes of double iron salt and 1.15095 grammes of permanganate, each dissolved in one liter of water. Both solutions are equivalent, and 1 c.c. of either corresponds to 0.001 gramme of manganese.

2. Heat 0.25 to 1.0 gramme of ferro-manganese in a covered porcelain dish with 20 c.c of HNO_3 (sp. gr. 1.12). Add 2 grammes of solid ammonium nitrate, evaporate to dryness, heat strongly until no more vapors escape, in order to decompose the nitrates, and add 20 to 40 c.c. of phosphoric acid of 1.7 sp. gr. Stir with a glass rod, cover dish, and heat in an air bath to 140° C. In from 6 to 10 hours beautiful violet manganese phosphate and colorless ferric phosphate are formed. The titrating solutions used in this determination are made only one-half the strength of those used in the previous method. The reaction is as follows: $Mn_2O_3 + 2FeO = 2MnO + Fe_2O_3$. In the previous method it is $MnO_2 + 2FeO = MnO + Fe_2O_3$.

Belani's method.[1]—Dissolve 0.5 to 2.0 grammes of the ore in ½ liter Erlenmeyer flask with 25 to 40 c.c. of HCl (1.19 sp. gr.); if ferrous oxide or manganous oxide be present in the ore, the solution is again boiled with 10 c.c. of HNO_3 (1.4 sp. gr.) for each gramme of ore. Rinse solution and residue into a graduated liter flask and neutralize with zinc oxide. The rest of the method is the same as Volhard's. With a very small content of iron the solution is turbid, but with a considerable quantity precipitation takes place suddenly. An excess of zinc oxide is not injurious. When the precipitation is finished the flask is filled up to the liter mark, and, after filtering off, 250 c.c. of the filtrate are taken for titration.

C. Electrolytic assays.—Manganese is one of the metals which are oxidized by the current to the peroxide, which separates at the positive electrode. The separation of the dioxide is complete from a solution of potassium manganese oxalate (it is not complete from a solution of the ammonium double salt), and also from solutions containing free HNO_3 or H_2SO_4. By the former method the solution of manganese salt is treated with a slight excess of potassium oxalate, diluted and precipitated with a current of 9 to 12 c.c. of

[1] Stahl und Eisen, 1886, p. 152.

oxyhydrogen gas per minute. If the quantity is small, the dioxide adheres to the positive electrode firmly enough to be converted, after washing, into mangano-manganic oxide by ignition of the previously weighed electrode. With larger quantities it is best to make the platinum dish the positive electrode, and when the reaction is ended to endeavor to convert the dioxide into mangano-manganic oxide by ignition without filtration. If this is impossible, the precipitate is filtered off, washed with hot water, and converted into either mangano-manganic oxide or sulphate. As ammonium sulphide is unsuited to show the end of the reaction, owing to the formation of oxalic acid, which hinders the precipitation of the sulphide, the best test is to evaporate a small portion of the solution in platinum foil, and fuse with sodium carbonate.[1]

According to Luckow,[2] in order to make the dioxide adhere firmly to the positive electrode, the solution must be neutral or only very slightly acid, and the current not too strong. In very dilute solutions acidulated with much HNO_3 or a mixture of HNO_3 and H_2SO_4 permanganic acid is formed, which is readily recognized by its red color. The presence of organic acids as well as ferrous oxide, chromic oxide, or of ammoniacal salts prevents, according to Schucht,[3] the formation of permanganic acid.

In the separation of manganese as dioxide, from an acid solution, it is better to use a few drops of H_2SO_4 instead of HNO_3, as the latter is converted by the current into ammonia, which may precipitate some of the other metals present. Manganese may be separated from the acid solution of a number of metals (copper, nickel, cobalt, zinc, etc.) without determining the other metals (excepting copper) at the same time.[4]

XIX. SULPHUR.

74. ORES.

Native sulphur (sulphur earths); *iron pyrites*, FeS_2, with 53.33 S and 46.47 Fe; *magnetic iron pyrites*, $5FeS.Fe_2S_3$, with 39.5 S; *copper pyrites*, $CuFeS_2$ with 34.89 S.

---

[1] Classen's Quan. Chem. Anal. by Electrolysis.
[2] Ztschft. f. Anal. Chem. Bd. 19, p. 17.
[3] Ztschft. f. Anal. Chem. Bd. 22, p. 494.   [4] Classen.

75. ASSAYS BY DISTILLATION FOR THE DETERMINATION OF THE AMOUNT OF SULPHUR WHICH AN ORE MAY YIELD.

a. Sulphur earths.—0.5 to 1 gramme of the ore is heated to a strong red heat in an impervious clay retort, on the neck of which is luted a porcelain tube (Figs. 45, 46, p. 62), when the sulphur vapors will deposit themselves in the porcelain tube, the end of which just dips in water. The tube is then removed and the sulphur collected, dried, and weighed.

Gerlach[1] conducts superheated steam into a glass retort containing the ore, from the neck of which the sulphur, which passes over, drops into a dish containing water.

b. Iron pyrites.—2 to 5 grammes, preferably mixed with the same volume of quartz or powdered charcoal to prevent caking, are placed in a glass tube 30 to 40 centimeters long and 13 to 15 millemeters wide, closed at one end. The other end is introduced into another glass tube, also closed at one end, and the substance is then heated in a combustion furnace (Fig. 77, p. 238), or by another source of heat (p. 61). The end of the tube containing the sublimed sulphur is cut off and weighed. The sulphur is then volatilized by heat, and the tube again weighed. Pure iron pyrites gives on a large scale at the utmost 23 per cent. of sulphur ($7FeS_2 = 6FeS.FeS_2 + 6S$); copper pyrites not more than 9 per cent. ($Cu_2S + Fe_2S_3 = Cu_2S + 2FeS + S$).

76. ASSAYS OF SULPHUR FOR THE DETERMINATION OF THE QUANTITY OF SULPHUR CONTAINED IN A SUBSTANCE.

These assays may be executed in order to determine the yield of an ore in sulphurous acid, for the manufacture of sulphuric acid, or its yield of sulphur for the formation of raw matt, for controlling roasting, etc. For this the wet method is more frequently used than the dry method.

A. Dry assay (raw matt assay).—The object of this assay is to determine the quantity of metallic sulphides, especially iron sul-

---

[1] Dingler, ccxxx. 66.

phide, contained in an ore, after the oxidized and earthy, etc., substances mixed with it have been separated by solvent agents.

A mixture of 5 grammes of ore and 0.5 gramme of resin is introduced into a crucible (Fig. 52, p. 65). Upon this are placed 10 to 15 grammes of borax, 5 to 10 grammes of glass free from heavy metals, a cover of common salt, and a fragment of coal. The charge is fused at a bright red heat in the muffle or wind furnace for 30 to 45 minutes after the "flaming" has ceased. The resulting button of iron sulphide which is brittle, oxidizes and disintegrates quickly, is carefully freed from the slag which should be well fused. It is then weighed and broken up in order to recognize the presence of foreign metallic sulphides by the appearance of the fracture. When only iron pyrites are present, this has a fine grain and speiss yellow color; with copper pyrites, brass yellow; with lead and sulphide, grayish and foliated; with zinc blende, radiated or foliated, of a sub-metallic lustre, and blackish-gray; with metallic antimony and arsenic, a fine grain and light gray.

Hungary :[1] If the ores are easily fusible, 5 grammes are charged with a flux composed of 2 parts of calcined borax and 1 part of glass free from iron, by placing a mixture of ore and 11.5 grammes of flux in the bottom of the crucible, upon this 23 grammes of flux, and on the top 8 to 10 grammes of common salt. This is fused in the muffle furnace. When the ores are refractory, 1.25 grammes are fused with the same flux, and 3.2 grammes of a pure easily fusible concentrated pyrite ore, the percentage of matt contained in this being afterwards deducted; or, with 0.25 to 3.5 grammes of copper as a collecting agent.—*Pribram:* 5 grammes are mixed with $\frac{2}{3}$ the quantity of a flux consisting of 10 grammes of borax, 2 grammes of glass, and 0.4 gramme of coal dust. The remaining $\frac{1}{3}$ part of the flux is strewn over this, and on top a cover of common salt and fragment of coal. 0.05 gramme of copper may be added to the roasted, and as much as 0.2 gramme to unroasted ores. The crucible with luted cover is gradually heated to a moderate red heat in an anthracite furnace (p. 51), and then fused for from 30 to 35

[1] B. u. h. Ztg. 1871, p. 255.

minutes in a bright red heat.—*Slags* with mechanical inclosures of matt are charged in the following manner: 30 grammes of slag, 20 grammes of borax, and 50 grammes of glass with a covering of common salt and a fragment of coal. The charge is fused for $\frac{1}{2}$ to $\frac{3}{4}$ of an hour at a bright red heat.

B. *Wet assays.*[1]

1. *Gravimetric assays.*—1 gramme of ore is decomposed by fuming nitric acid, then nearly all the nitric acid removed by boiling. Hydrochloric acid is now added and heat is applied, when the sulphur will be quickly dissolved (the heating should not be continued too long, or loss will ensue by the escape of sulphuric acid). The solution is then evaporated to dryness, the residue is heated with hydrochloric acid in order to expel the nitric acid (as otherwise the results would be too high). The liquid is now diluted and filtered, the filtrate precipitated with barium chloride, and the barium sulphate filtered through impervious Swedish paper, dried, ignited, and weighed ($BaSO_4$, with 34.356 per cent. SO_3 and 13.73 per cent. S). The sulphur in *pyrites*, roasted pyritous ores, and the same *lixiviated*,[2] is determined by heating 0.5 gramme of the substance in a platinum crucible, together with 10 parts of a mixture of 2 parts of sodium carbonate and 1 part of saltpetre. This is lixiviated, etc., and finally precipitated by barium chloride, etc., as above.—Process for determining small quantities of *sulphur in materials* and *products of the iron-works* at *Creuzot*.[3] The substance is heated in a porcelain tube, and a mixture of $\frac{3}{4}$ of hydrogen and $\frac{1}{4}$ carbonic acid is conducted over it. The sulphuretted hydrogen, which is formed, is led into an acid silver solution, and the quantity of sulphur calculated from the weight of the silver sulphide.

Gravimetric determination of sulphur in pyrites, according to Lunge.[4]—Pour over the finely-powdered and sifted mineral about 50 times its weight of *aqua regia*; if reaction does not immediately appear, heat upon water bath until a vigorous effect is produced.

---

[1] Muspratt's Chem. vi. 8.
[2] Fresenius's Ztschr. 1877, p. 335. B. u. h. Ztg. 1877, p. 241.
[3] Dingler, ccxxxiii. (Rollet).
[4] "Handbuch d. Sodaindustrie," Braunschweig, 1879, p. 92.

Then at once remove from water bath and replace only when reaction becomes weaker. When the mineral is powdered sufficiently fine disintegration is complete in 10 or 15 minutes. If complete disintegration has not been effected, add more *aqua regia* and heat. Now add an excess of HCl and evaporate to dryness in order to expel all the HNO_3, as in the presence of the latter the results obtained in precipitating the H_2SO_4 with barium chloride are too high. Take up with water acidulated with HCl, heat to boiling, and add a hot solution of barium chloride, just a little more than sufficient to precipitate the H_2SO_4 present. The precipitate settles quickly and can be filtered at once, wash with boiling water acidulated with a few drops of HCl, boil filtrate, and after settling filter again. Repeat these operations several times, but without the addition of any more HCl. Wash the precipitate of $BaSO_4$ very thoroughly and when wash water shows no further traces of $BaCl_2$ treat precipitate and filter in usual manner and weigh as $BaSO_4$. The latter should not be baked together, show an alkaline reaction, or yield a barium salt when treated with hot dilute HCl.

If the $BaSO_4$ is colored yellowish from the presence of iron, it is digested in moderately strong HCl, filtered, washed, and again ignited. Or else it can be fused with soda, fused mass taken up with water, filtered, washed, and the H_2SO_4 again precipitated with $BaCl_2$.

Fresenius[1] observes that all methods, like the above for determining sulphur in pyrites, have two sources of error, mainly the $BaSO_4$ contains ferric oxide and is sensibly soluble in the acid solution containing ferric chloride. Hence by dissolving pyrites in *aqua regia* and precipitating the H_2SO_4 with $BaCl_2$ the results are always inaccurate.

Deutocom[2] mixes 1 gramme of pyrites with 8 grammes of a mixture of equal parts potassium chlorate, sodium carbonate, and sodium chloride, and very slowly heats the mixture in a large covered porcelain crucible until it is thoroughly dry, then fuses with a strong heat. After cooling the mass is treated with boiling water and placed together with the insoluble residue into a graduated 250 c.c. flask. After filling up and filtering, the H_2SO_4 is determined in 50 c.c. of the filtrate. The insoluble residue should retain no H_2SO_4.

Bodewig's method.[3]—One-half gramme of pyrites is treated in a

---

[1] Ztschft. f. Anal. Chem. Bd. 19, p. 57.
[2] Ztschft. f. Anal. Chem. Bd. 19, p. 313.
[3] Ztschft. f. Anal. Chem. Bd. 22, p. 571.

glass-stoppered vessel, of about 100 c.c. capacity, with 30 c.c. of water and 4 c.c. of bromine. The stopper is quickly inserted, and the vessel shaken for five minutes. When oxidation is ended, which is known by the disappearance of all pulverulent sulphur adhering to the sides of the glass, the solution is emptied into a casserole and most of the bromine allowed to evaporate in the cold. The solution is almost neutralized with ammonia, then poured into an excess of hot ammonia contained in a platinum dish and digested from 10 to 15 minutes at a gentle heat. The H_2SO_4 is determined in the filtrate from this in the usual manner. The whole amount of bromine employed must not be added all at once, as otherwise there may be some loss of sulphur in the form of H_2S. Some iron volatilizes as a bromide with the excess of bromine, hence it cannot be determined in the precipitate made by the ammonia.

Determination of sulphur in pyrites waste.—According to Bockmann,[1] in sulphuric acid works the pyrites waste is tested in regard to its content of sulphur as follows: Intimately mix in a platinum dish 2 grammes of finely pulverized waste with 25 grammes of a mixture of 6 parts sodium carbonate and 1 part potassium chlorate. Fuse over blast lamp, take up with water, and filter into a tall beaker containing HCl in excess. After thoroughly washing the residue upon filter, the H_2SO_4 in the heated filtrate is precipitated with $BaCl_2$. The $BaSo_4$ is then treated and weighed in the usual manner.

Working test for determining the residue of sulphur in the roasting charge.[2]—In the upper Silesian zinc works the process of roasting is controlled in the following manner: The workman heats a shovel in the roasting furnace, and when red hot it is removed, and about 2 grammes of potassium chlorate strewed upon it. This quickly melts, and upon the fused mass some of the roasting charge to be tested is strewn. If no flaming or burning of separate grains takes place, the blende is completely roasted; if only a grain here and there burns, the roasting is considered complete enough, for according to experience the roasted charge then contains only about 1 per cent. of sulphur, which is practically no detriment to the subsequent distillation process. The roasted charge, however, should not contain more than 1 per cent. sulphur.

---

[1] Ztschft. f. Anal. Chem. Bd. 21, p. 90.
[2] Bg. u. Httnmsch. Ztg. 1883, p. 443.

2. *Volumetric assays.*[1]

a. 1 gramme of ore is intimately rubbed together with 2 grammes of pure saltpetre, or with 3 grammes of sodium carbonate and the same quantity of saltpetre. The mixture is placed in a small dish of sheet iron, 55 millimeters wide and 25 millimeters deep, which is placed in a scorifier (Fig. 47, p. 64). It is then gradually heated in a red-hot muffle. After fusing quietly from five to eight minutes, the small dish is taken out and allowed to cool off. The mass contained in it is then lixiviated with hot water, and filtered into a small beaker-glass. The residue is washed out with as little water as possible, and hydrochloric acid in excess then gradually added. It is now heated in a sand-bath, in order to expel the nitrous compounds. A titrated solution of barium chloride is then added drop by drop from a burette to the hot solution until no further white turbidity is formed in the supernatant liquid. 1 cubic centimeter of a normal solution, with 0.152 gramme of barium chloride, precipitates 0.050 gramme of sulphuric acid, corresponding to 0.020 gramme of sulphur; or 5 per cent. of sulphuric acid and 2 per cent. of sulphur. Some experience is required to detect the final reaction. Indicators for this have been proposed, as, for instance, potassium chromate, by *Wildenstein*, or filtering off a few drops and adding barium chloride.

According to *Wildenstein*, the solution containing sulphuric acid is diluted to a bulk of 45 to 55 cubic centimeters, to which is added a slight excess of a titrated solution of barium chloride. It is then boiled for a half to one minute, a slight excess of ammonia free from carbonic acid having first been added. A titrated solution of neutral potassium chromate is now added in small quantities, which should not exceed $\frac{1}{2}$ cubic centimeter at a time, in order to precipitate the excess of barium monoxide, until the liquid, after having been shaken and allowed to become clear, shows a distinct yellow color. It is then titrated back with a few drops of barium chloride until the fluid becomes colorless. During this operation the precipitate must be allowed to settle every time,

---

[1] Muspratt's Chem. vi. 9. Fresenius's Ztschr. i. 323 (Wildenstein).

or a few drops should be filtered off, and tested. *Normal solutions:* 1 cubic centimeter of barium chloride = 0.015 gramme of sulphuric acid, and 1 cubic centimeter of a solution of potassium chromate = 0.01 gramme of sulphuric acid. This assay gives accurate results to within $\frac{1}{5}$ per cent. of sulphur.

b. Metallic sulphides decomposable by hydrochloric acid are treated with it in a flask connected with a retort (Fig. 81, p. 271). The sulphuretted hydrogen developed is introduced into a titrated solution of iodine in potassium iodide ($H_2S + I_2 = 2HI + S$), and the unchanged iodine titrated with sodium hyposulphite[1] (p. 270).

Volumetric determination of sulphur in ores which contain either sulphur alone or also sulphates.— Weil gives the following method:[2] Place 1 to 2 grammes of the finely powdered ore in a flask provided with a cork through which passes a bent tube. The outside end of this tube dips into an ammoniacal solution of copper of known strength. A few small pieces of granulated zinc are placed in the flask and 75 c.c. of HCl poured over its contents; quickly close the flask and heat. The H_2S developed precipitates an equivalent portion of copper, and when precipitation ceases, the sulphide is allowed to settle, is filtered and washed. The object of adding the zinc is to dilute the H_2S with hydrogen and to carry off the last traces of H_2S which may remain in the flask and glass tube. The amount of filtrate from the precipitated copper sulphide is carefully measured and 10 to 20 c.c. of it saturated with from 25 to 50 c.c. of HCl. Heat to boiling, and when boiling titrate with a standardized solution of stannous chloride. The amount of copper found must of course be calculated for the entire volume of the filtrate. The amount of copper being known from the volume of the copper solution used for the precipitation of the H_2S, the weight of the precipitated copper sulphide is found by deducting the amount of copper found remaining in the solution on titrating with stannous chloride. This quantity multiplied by 0.50393 gives the content of sulphur sought. The zinc added facilitates the action of the acid upon the ore, and if any galena is present it is also more readily decomposed, the chloride of lead formed being reduced to metallic lead by the zinc.

[1] Dingler, ccx. p. 184.
[2] Compt. Rend. June 22, 1886. Genie. civ. tom. ix. No. 16, August 14, 1886.

Estimation of sulphur in metallic lead.[1]—Treat 20 to 30 grammes of very fine chips of the lead with a considerable excess of concentrated HCl; sulphur will be set free as H_2S. Pass this H_2S, by means of an aspirator, into bromine water, in which it is decomposed, H_2SO_4 being formed. Determine the sulphur in the latter with $BaCl_2$, as usual. The solution of the lead is aided by gentle heat. A large excess of HCl prevents the separation of lead chloride.

XX. FUELS.

77. FUELS.[2]

These may be in either of the following forms : solid (raw or natural, carbonized, or artificial, agglomerated, or patent fuel briquetts); or liquid (petroleum, tar-oils); and gaseous (natural gas, waste and generator gases, illuminating gas).

The different varieties of raw fuel are, approximately, composed as follows :—

	C	H	O
Woody fibre (cellulose = $C_6H_{10}O_5$)	44.44	6.17	49.39
Peat	60.44	5.96	33.60
Lignite	66.96	5.27	27.76
Earthy brown coal	74.20	5.89	19.90
Bituminous coal, recent	76.18	5.64	18.07
" " ancient	90.50	5.05	4.40
Anthracite coal, recent	92.85	3.46	3.19
" " ancient	94.20	2.50	3.30

In order to remove *earthy admixtures* from fossil fuel before subjecting it to docimastic test, it is comminuted and stirred into sulphuric acid of 1.4 specific gravity (in soda manufactories, in a solution of sodium sulphate).[3] The heavier earths will fall to the bottom, while the coal rising to the surface is removed with a spoon, thoroughly washed and dried.

---

[1] School of Mines, Quarterly, vol. ii. p 211.

[2] Kerl, Grundr. der allgemeinen Huttenkunde, 2 Aufl. 1879, p. 64. Muck in B. u. h. Ztg. 1876, p. 286 (Steinkohlen).

[3] Dingler, cxc. 76.

78. ASSAYS OF FUEL.

The examination extends to the following points, on which the value of fuel chiefly depends:—

1. *Determination of the amount of hydroscopic water.*—5 grammes of the powdered sample are placed in a watch-glass and heated on a water-bath (raw fuel), or (wood-charcoal, coke) at a high temperature (120° to 150° C., 248° to 302° F.) in an air-bath, or on a drying disk (Fig. 2, p. 25). It is allowed to become cold in the desiccator (Fig. 19, p. 40), and then weighed: is again dried and weighed until two weighings agree.

Air-dried wood and peat contain 15 to 20 per cent. of water; lignite, 10 to 15 per cent.; brown coal with a conchoidal fracture, 10 to 5 per cent.; earthy coal, as much as 25 per cent.; bituminous and anthracite coal, fresh from the pit, 1 to 10 per cent.; wood-charcoal, 10 to 12 per cent.; coke, 5 to 10 per cent.

2. *Yield of carbon.*—5 to 10 grammes of the material, either in small fragments, or in the form of powder, are placed in a covered crucible (Fig. 52, p. 65), and gradually heated to a red heat in the muffle furnace, until the flame which shows itself at the lid of the crucible disappears. The residue, upon cooling, is weighed, and (in tests of coal) the physical condition of the coke is observed at the same time. This may be more accurately ascertained by heating 1 gramme of coal in a platinum crucible 40 millimeters high with a bottom diameter of 24 millimeters, keeping the crucible at a distance of 3 centimeters over the flame of a gas-burner.[1]

The yield of carbon will vary according as the heat is raised more or less quickly, and with the degree of temperature, decreasing as the latter is more intense. Therefore, if several varieties of fuel are to be compared, the carbonization must be conducted at the same temperatures. The yield is generally less than that indicated by assay on the large scale. The average yield from wood charcoal in heaps is 21 to 22 per cent. by weight; from pine wood, 55 per cent., and from hard wood, 48 per cent. by volume; and respectively 25 to 27 and 60 to 65 per

---

[1] B. u. h. Ztg. 1876, p. 287.

cent. in furnaces; from more recent bituminous coal as high as 60 per cent.; from semi-bituminous coal, 78 to 83 per cent.; from anthracite coal, 84 to 87 per cent., and from real anthracite, 88 to 93 per cent. by weight.

Determination of the coking quality of coal according to *Richter*.[1] 1 gramme of coal in a finely powdered condition is mixed either with 0.1, 0.2, 0.3, etc., that is to say, as many times 0.1 gramme of powdered quartz as may be necessary to just crush the cake of coke remaining in the covered porcelain crucible after ignition, when weighed with a 0.5 kilogramme weight carefully placed upon it. If 0.5 gramme of powdered quartz has been used, the coking quality of the coal would be represented by 5, etc.

3. *Volatile products* are determined from the difference in weight between the coke and raw fuel, after deducting the percentage of water. The amount of gas a coal will yield is ascertained by heating 5 grammes of the sample in a glass retort or tube. The gas evolved (after passing through two wash-bottles filled respectively with baryta water and lead acetate to absorb carbonic acid and sulphuretted hydrogen) is collected over mercury in a graduated cylinder (Fig. 77, p. 238).

4. *Determination of the ash.*—The residue from the assay for carbon (containing carbon and ash) is pulverized as fine as possible, placed in a roasting-dish (Fig. 8, p. 32) and heated in the muffle, which should not be exposed to too strong a draught of air, until the black particles have entirely disappeared. The ash is then weighed, and its physical properties (color, whether caked or pulverulent, etc.) are at the same time examined.

Gypsum and iron pyrites undergo alteration during this operation, thus impairing the result of the assay. This must be especially taken into consideration when a contract for the purchase of coal is based upon the minimum amount of ash.[2]

In the production of pig iron it is important to know the composition of the ash of the fuel used, as in calculating the charges the amount of silica, etc., in it is an important consideration. A coal containing pyrites is apt to contain small amounts of copper, which exerts a very injurious effect upon the pig iron. A small quantity

---

[1] Dingler, cxcv. 71.   [2] B. u. h. Ztg. 1878, p. 61 (Muck).

of copper can be readily detected in the ash by moistening it with HCl and heating it in the non-luminous part of a gas flame or before a blow-pipe, whereby the flame will acquire an azure color. A simpler method is to throw a few spoonfuls of common salt on the burning coals and stir. There will immediately appear small azure blue flames of burning carbonic oxide containing cuprous chloride.[1]

Amount of ash in different kinds of fuel: Wood, 0.15 to 2 per cent., an average 1 per cent. (composed of about 70 per cent. of calcium carbonate and 20 per cent. of alkaline carbonates); wood charcoal, 3 to 4 per cent.; peat, 0.5 to 50, on an average from 6 to 12 per cent. (composed of, approximately, 35 per cent. of argillaceous sand, as much as 40 per cent of magnesian gypsum, about 30 per cent. of ferric oxide, and 3 per cent. of alkalies, as well as some phosphoric acid and chlorine); brown coal, as high as 50 per cent., on an average from 5, to 15 per cent. (chiefly silicic acid, alumina, ferric oxide, lime, sulphuric acid, lesser quantities of magnesia, alkalies, chlorine, rich in sulphur in the form of gypsum and iron pyrites, poor in phosphorus); hard coal, 0.5 to 30 per cent.; the best coal, at an average from 4 to 7; medium quality 8 to 14; and poorer qualities over 14 per cent. Ash, mostly bisilicate of alumina with lime (1 to 20 per cent.), ferric oxide (1 to 75 per cent.), alkalies (0 to 3 per cent.), sulphur (0.5 to 2 per cent.). Coke, 1 to 30 per cent. (good coke about 10 per cent.), phosphorus 0.0025 to 0.05 per cent.—*The amount of sulphur*[2] *contained in a coal or its ash* is determined by fusing 1 gramme in a platinum crucible with 8 grammes of saltpetre, 4 grammes of potassium carbonate, and 16 grammes of common salt. The fused mass is lixiviated with water, hydrochloric acid is added, and the mass then filtered. The filtrate is precipitated with barium chloride, and the resulting barium sulphate containing 13.8 per cent. of sulphur is weighed (see also assays of sulphur).

5. *Determination of heating power.*—In estimating the availability of a fuel for a given purpose, it is of the greatest importance

---

[1] Sitzungsterichte der königl. böhm. Gesellsch. d. Wissensch. p. 9. Chemikerzeritung, 1880, p. 276.

[2] Schwefelbestimmung in Steinkohlen, etc. Oestr. Ztschr. f. Berg. u. Hüttenwesen, 1874, p. 11 (Eschka). Fresenius's Ztschr. xii. 32, 178; xiv. 16 (Sauer). B. u. h. Ztg. 1875, p. 228 (Hayes).

to determine how much heat, equal parts by weight *(absolute heating effect)*, or equal parts by volume *(specific heating effect)*, of different fuels will produce. The intensity of the heat produced *(pyrometric heating effect)* may also be a point of investigation.

The latter may be determined by calculation or by the pyrometer,[1] either *Fischer's*[2] calorimeter or *Siemens's* electric pyrometer;[3] or by means of the fusing point of alloys by a modification of *Prinsep's* principle.[4] The specific heating effect is found by multiplying the absolute heating effect by the specific gravity of the fuel in question. *Berthier's* method of determining the absolute heating effect may be especially recommended for docimastic purposes.

Berthier's method of determining the absolute heating power is based upon *Welter's* law, according to which the absolute heating power of different combustible substances is proportional to the amounts of oxygen required for their complete combustion. The oxygen is taken from oxides (lead oxide), and the quantity of metal (lead) set free represents the quantity of liberated oxygen, and therefore the absolute heating power.

Welter's law is based upon the fact that, according to former experiments by *Rumford, Despretz,* and others, the absolute heating power of carbon to hydrogen is in the proportion of 1 : 3.03, and the respective amounts of oxygen required for the combustion of 1 part of these elements are in nearly the same proportion (1 : 3). But, according to recent investigations by *Favre, Silbermann,* and others, the absolute heating powers of carbon and hydrogen are, respectively, as 1 : 4.3 ; and, therefore, *Welter's* law has become obsolete, and the results obtained by *Berthier's* method are only approximate. They approach more closely to the truth, the richer in carbon and poorer in hydrogen the fuel is, while the results from combustible substances rich in hydrogen are from $\frac{1}{5}$ to $\frac{1}{2}$ too low compared with more accurate calorimetric determinations. His method is, nevertheless, frequently used

---

[1] Kerl, Grundr. der alleg. Hüttenkunde, 1879, p. 85.

[2] Dingler. ccxxv. 468. Ber. d. deutsch. chem. Ges. 1879, p. 1694.

[3] Dingler. ccxvii. 291. B. u. h. Ztg. 1871, p. 450; 1873, p. 231, 396; 1874, p. 463; 1876, p. 156; 1877, p. 109.

[4] Freiberger Jahrb. 1879, p. 154. B. u. h. Ztg. 1879, p. 126 (see Appendix).

ASSAYS OF FUEL. 289

in practice, it being very convenient and quickly executed, and under the above-mentioned conditions gives practically available results, especially in the examination of different varieties of the same kind of fuel.

Exactly 1 gramme of the finely divided fuel is weighed out and intimately mixed with 40 to 50 grammes of litharge finely sifted and free from globules of lead, organic substances, and minium; or, still better, with 70 to 90 grammes of white lead. The mixture is covered with 20 to 25 grammes of litharge (or, 30 to 40 grammes of white lead), a lid is placed upon the crucible, and the charge gradually heated in the muffle furnace (Fig. 27, p. 47) until it is completely fused. The heat is then increased for a short time, when the oxidizable constituents of the fuel will be consumed at the expense of lead oxide and will separate the more lead the richer they are in such constituents. After the charge has been fused, which requires from $\frac{1}{2}$ to $\frac{3}{4}$ of an hour, the crucible is taken out and allowed to cool off. The lead button is then freed from slag, brushed off, and weighed. The resulting weight is divided by the quantity of the sample used in order to learn how much lead has been reduced by it. The thermal value of *graphite*[1] may also be determined by this assay.

In the case of combustible materials *decomposable at a lower temperature*, lead oxychloride, which is more easily fusible, should be used. It is obtained by fusing 3 parts by weight of red litharge together with 1 part of lead chloride in a Hessian crucible. 1 gramme of the fuel is mixed with 40 grammes of the oxychloride, and the mixture covered with 30 grammes of the latter.

One part of pure carbon reduces 34 (more accurately 34.52) times the quantity of lead; wood, 12 to 15, on an average 13.95 parts (= 3200 heat units = 0.41 per cent. carbon); peat 8 to 18, Irish varieties as much as 27 parts; brown coal 14 to 26 parts; bituminous coal 23 to 31 parts, and anthracite 26 to 33 parts; wood charcoal 28 to 33.7 parts; coke 22 to 30 parts of lead. If the fuel contains iron pyrites, the quantity of lead reduced increases (1 part FeS_2 reduces 8.72 parts, and 1 part FeS, 7.18 parts of

---

[1] Kerl, Thonwaarenindustrie, 1879, p. 91.

lead from litharge). Suppose p to be the weight of the lead button, the heating power in calories or heat units is expressed by $x = \frac{8080}{34.5} p = 234 p$. In case the fuel contains a large amount of hydrogen, this value must be multiplied with a co-efficient lying between 1 and ⅞.

Suppose 100 kilogrammes of coal must be replaced by wood, in a reverberatory furnace, in some smelting process. How much of the latter must be taken? The absolute heating power of both must be determined according to *Berthier's* method, by which they are capable of reducing, respectively, 24 and 14 parts of lead. Then $14 : 24 = 100 : x$ and $x = 170$ kilogrammes (374 pounds) of wood supply the place of 100 kilogrammes (220 pounds) of coal. If the quantities of fuel are to be determined by the *volume*, it is only necessary to multiply the above numbers, 24 and 14, by the specific gravity of the fuel in question, and to formulate the resulting products into a similar proportion.

6. *Determination of the sulphur.*—According to Drown,[1] a solution of sodium hydrate, of 1.25 sp. gr., is saturated with bromine. If an excess of bromine is used, it must be neutralized by the addition of a little more sodium hydrate. The finely powdered coal is moistened with 25 c.c. of this solution and heated, then HCl is added cautiously to just acid reaction. This operation is repeated with another 25 c.c. of the alkaline solution and again made acid. The mixture should be kept hot. The contents of the beaker or dish are then evaporated to dryness to separate silica and taken up with dilute HCl. The H_2SO_4 is precipitated with the usual precautions by barium chloride. Instead of using pure bromine a saturated solution of bromine in potassium bromide may be used with equally good effect. According to Atkinson,[2] the content of sulphur is determined as follows: Dry the finely powdered fuel at 100° C., mix one gramme of it with 5 grammes of anhydrous sodium carbonate, introduced into a flat platinum dish and spread uniformly over bottom. Heat very slowly in muffle, for not over a half hour, to clear cherry-redness. Allow the air to enter muffle freely, the heat must not be sufficient to sinter or fuse the sodium carbonate. At

[1] Trans. Am. Inst. M. E., vol. viii. p. 569, and vol. ix. p. 656.
[2] Journ. Soc. Chem. Ind. 1886, p. 154.

the end of the half hour the sulphur will be oxidized and absorbed by the sodium carbonate, the carbon will also be all burnt off. Lixiviate with 150 c.c. of hot water. When the insoluble portion has settled, filter, wash with hot water containing some pure sodium chloride, rinse precipitate with the same solution acidulated with 10 c.c. of pure HCl, boil, and precipitate in usual manner with barium chloride. It is claimed that the addition of the sodium chloride solution prevents the passage of the finely divided ash through the filter.

7. *Physical and chemical behavior.*—The following points must be considered as exerting an influence upon the action of combustible substances in the fire, during transportation, etc.: Structure, density (compactness), form and size of the lumps, specific gravity, behavior when thrown into the glowing muffle, or in the furnace (whether they kindle easily or with difficulty, burn quietly or fly into pieces, whether the flame is short or long, or more or less smoking, the liberation of odors, brittleness, etc.), chemical composition of the fuel, and of the pulverulent, sintered, or clinkered ash, etc.

79. EXAMINATION OF FURNACE GASES.

To be able to judge the processes taking place during combustion,[1] the velocity of the flue gases, and the amount of air passing through the furnace are determined by an *anemometer*,[2] the strength of the draught by a *draught meter*,[3] the intensity of the heat by a *pyrometer*,[4] and the amount of *carbonic acid, carbonic oxide*, and *free oxygen* in the furnace gases are ascertained—

1. By means of *Orsat's apparatus*.[5]—With some experience

[1] Kerl, Thonwaarenindustrie, 1879, p. 301.
[2] Töpfer- u. Ziegler-Ztg. 1878, No. 1.
[3] Dingler, clxxi. 43 (List). Notizbl. der deutsch. Ver. f. Fabrikation von Ziegeln u. s. w. ix. 96; xi. 191; xiii. 40, 42. Töpfer- u. Ziegler-Ztg. 1877, No. 46.
[4] Kerl, Grundr. d. allg. Hüttenkunde, 1879, p. 85. Mitchell, Practical Assaying, 1888, p. 167.
[5] Fichet-Ramdohr, Gasfeuerung, Halle, 1875. Ann. des mines, vol. viii. livr. 6 de 1875. B. u. h. Ztg. 1874, p. 232; 1875, p. 143; 1876, p. 72; 1877, p. 147. Dingler, ccxix. 420 (Weinhold). Fresenius's Ztschr. 1877, p. 343

and intelligence this apparatus gives results available for practical purposes, even in unscientific hands. It is based upon the principle, that a measured volume of gas is conducted through agents for the absorption of its principal constituents (caustic potassa for carbonic acid, potassium pyrogallate for oxygen, and solution of cuprous chloride for carbonic oxide), the volume of gas remaining after each absorption being measured, when the amount of each will be ascertained from the difference. The ap-

Figs. 83 and 84.

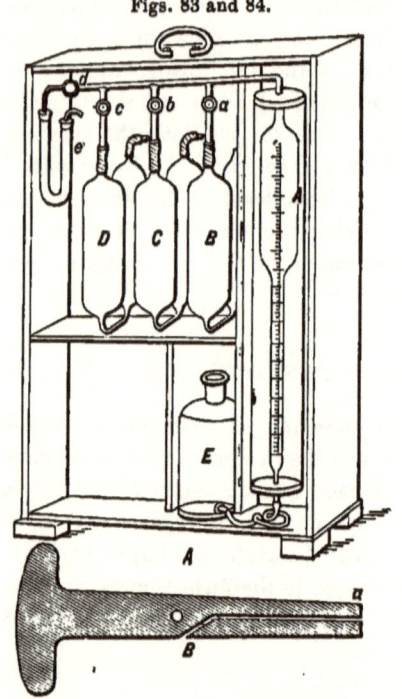

paratus is placed in a portable wooden case, and, according to *Fischer's* latest construction, is arranged as follows, (Fig. 83):—

A is a burette inclosed in a glass cylinder. Its lower end is

(Seyberth). Oest. Ztschr. 1877, No. 11, 13, 16. Ztschr. de Ver. deutsch. Ing. xx. 318. Dingler, ccxxvii. 258; ccxxix. 262 (Fischer.) Winkler, Anleitung z. Chem. Untersuchung der Industriegase, 2 Abth., 1877, p. 1859.

connected with the water flask E by means of a rubber tube. The burette is capable of holding 100 cubic centimeters. Its lower part holding 40 cubic centimeters is graduated to one-fifth cubic centimeter, and the upper part in whole cubic centimeters. B C D are the absorption vessels (B for caustic potassa, C for potassium pyrogallate, and D for solution of cuprous chloride, or fluid obtained by shaking copper hammer scale with a mixture of equal volumes of ammonia and cold saturated solution of sal ammoniac). The vessels are filled with fine glass tubes and connected with the burette by means of a system of thick-walled capillary tubes. a b c are plain cocks; d is a *Winkler* cock (Fig. 84, B), which besides having a simple perforation is also cut lengthwise. The outer end, a, is connected with an aspirator by means of a rubber tube. When the cock is properly set, the tube e (which is provided with a little water, and loosely filled with cotton, in order to saturate the gas with water vapor and to retain dust), and the gas-conducting tube connected with it, can then be filled with the gas to be examined.

The operation is conducted as follows: The cock d is set so that it communicates with the outer air. The flask E, filled with water, is raised so that A will become completely filled with water, the air escaping from d. d is then closed towards A. The cock a is now opened and the flask E lowered, whereby the absorption vessel B is filled with the absorbing liquid (potassium hydrate) to the mark immediately below the cock a, whereupon this is closed. C is filled in a similar manner with potassium pyrogallate, and D with a solution of cuprous chloride from vessels of equal size communicating with and placed behind them. The furnace gas to be examined is aspirated through the aspirator, and the connection between e and A is established by the cock d, after the burette A has been completely filled by raising the flask E. The latter is then lowered when A will be become filled with the gas. d being properly set, the gas is allowed to escape by again raising E, in order to expel any small quantities of air which may still be contained in the capillary tubes. After A has in this manner been filled with gas, this is successively forced by the same manipulation through B, C, and

294 ASSAYING.

D, and each time returned to A in order to measure the volume of gas which has been absorbed respectively in B, C, and D.

Orsat has further enlarged his apparatus so that hydrogen and carburetted hydrogen can be also determined.[1] An apparatus for examining, by Schwackhöfer,[2] has recently been recommended, it being claimed that it is less easily broken and safer to handle than Orsat's.

2. *By means of Bunte's burette*[3] (Fig. 85).—An unlimited number of absorbing agents can be used in this apparatus, as it allows of the removal of the absorbing liquids from the burette without a loss of gas every time after they have been used, and further permits the gas inclosed in the burette to be brought, after each absorption, to the same pressure. A is a burette divided from the *Winkler* cock a to the common cock b into somewhat more than 110 cubic centimeters and fractions. t is a funnel forming the upper part of the burette, having a capacity of 25 cubic centimeters to the mark m. The burette A is filled with gas by connecting a with the gas-conductor by means of a rubber tube, and aspirating the gas through b until all the air has been expelled from A. a and b are then closed, and a rub-

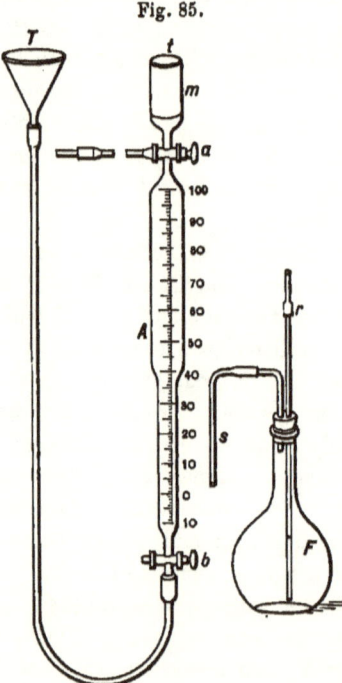

Fig. 85.

---

[1] Ann. d. min. 1875, t. viii. p. 501. Oest. Ztschr. 1877, No. 13. B. u. h. Ztg. 1878, 123. Dingler, ccxxi. 284; ccxxvii. 171 (Fischer). Winkler, Anl. d. chem, Untersuchung der Industriegase, 2 Thl. p. 198.

[2] Ztschr. der berg. u. hüttenm Ver. für. Steyermark u. Karnthen, 1878, Nos. 3-6, p. 78.

[3] Dingler, ccxxvii. 167; ccxxviii. 46.

ber tube, plugged at one end with a small glass rod, is pushed over the point of a. In order to bring the volume of gas in the burette to 100 cubic centimeters under a known pressure, water is forced into the burette up to the 0 point by means of a rubber hose filled completely with water, and connected with the funnel T and the point of the burette. b is then closed and a opened towards m, whereby a part of the gas escapes, and the inclosed remaining part stands under the pressure of the atmosphere and a column of water a few cubic centimeters high. The gas can in a similar manner be brought under equal conditions of pressure at any desired position of the liquid in the burette, and a correction for the pressure, which is made the same at every reading, is not required in the customary statement of the results of the experiments in per cents. of the total volume.

The following is the process of manipulating when furnace gases in the burette contain carbonic acid, carbonic oxide, and oxygen. The minutely perforated point b of the burette is connected with the flask F by the rubber tube r. The cock b is then opened and air drawn out through the tube s by suction, whereby water is drawn from the burette A to F, and then b is immediately closed. The flask F is removed from the burette, and the point of this is dipped into a dish containing solution of potassium hydrate. When the cock b is opened this enters A and replaces the water drawn out. The burette is held by its extremities and gently shaken for the absorption of the carbonic acid. As soon as this has been accomplished t is connected with A by the cock a, whereby water enters into A until the pressure is completely equalized. The volume of *carbonic acid*, which has disappeared, is read off on the burette. The *oxygen* is determined in a similar manner by withdrawing a part of the potassium hydrate by suction, and allowing potassium pyrogallate to enter; while for the determination of the *carbonic oxide* the potassium pyrogallate must be entirely removed from the burette by repeatedly aspirating the liquid from A through b, and allowing water to flow in through a from t until the absorbing agent has been entirely removed, when solution of cuprous chloride is introduced in the manner indicated.

APPENDIX.

A. Tabular Synopses.

1. *Atomic Weights.*

	Old equivalent weights.	New atomic weights.		Old equivalent weights.	New atomic weights.
Aluminium	13.7	27.4	Mercury	100	200
Antimony	122	122	Molybdenum	46	96
Arsenic	75	75	Nickel	29.3	58.6
Barium	68.5	137	Niobium	47	94
Beryllium	4.7	9.3	Nitrogen	14	14
Bismuth	210	210	Osmium	99.2	199.4
Boron	11	11	Oxygen	8	16
Bromine	80	80	Palladium	53.3	106.6
Cadmium	56	112	Phosphorus	31	31
Cæsium	133	133	Platinum	99	198
Calcium	20	40	Potassium	39	39.1
Carbon	6	12	Rhodium	52.2	104.4
Cerium	—	91.2	Rubidium	85.5	85.5
Chlorine	35.5	35.5	Ruthenium	52.2	104.4
Chromium	26.1	52.2	Selenium	39.7	79.4
Cobalt	29.5	59	Silicium	14	28
Copper	31.7	63.4	Silver	108	108
Didymium	48	95	Sodium	23	23
Erbium	56.3	112.6	Strontium	43.8	87.6
Fluorine	19	19	Sulphur	16	32
Gallium	—	68.8	Tantalium	91	182
Gold	197	197	Tellurium	64	128
Hydrogen	1	1	Thallium	204	204
Indium	56.7	113.4	Thorium	—	232.4
Iodine	127	127	Tin	59	118
Iridium	99	198	Titanium	25	50
Iron	28	56	Uranium	60	120
Lanthanium	46	93	Vanadium	—	51.3
Lead	103.5	207	Wolfram	92	184
Lithium	7	7	Yttrium	30.85	61.7
Magnesium	12	24	Zinc	32.6	65.2
Manganese	27.5	55	Zirconium	44.8	89.6

2. *Fusing Points of Metals and Furnace Products,*[1] *Glowing Temperatures.*

	Fusing point (Celsius.)	Fusing point (Fahr.).	Glowing temperature (Celsius).	Glowing temperature (Fahr.).
Tin	228°	442.4°		
Bismuth	264	507.2		
Thallium	290	554		
Cadmium (455° C., 851° F.) (boiling point 891° C., 1635.8° F.)	320	608		
Lead	335	635		
Zinc (boiling point according to Becquerell 891° C., 1635.8° F., according to Deville 1040° C., 1904° F.)	412	773.6		
Antimony	432	809.6		
Incipient redness	—	—	525°	977°
Dark redness	—	—	700	1292
Aluminium	700	1292		
Incipient cherry-redness	—	—	800	1472
Strong cherry-redness	—	—	900	1652
Bronze	900	1652		
Litharge	954	1749.2		
Complete cherry-redness	—	—	1000	1832
Silver (according to Becquerell 916° C., 1680.8° F.)	1000	1832		
Copper matt	1002	1835.6		
Brass	1015	1859		
Lead matt	1027	1880.6		
Black copper	1027	1880.6		
Raw matt	1047	1916.6		
Lead speiss	1062	1943.6		
Copper	1090	1994		
Gold (according to Becquerell 1037° C., 1898.6° F.)	1200	2192		
Bright redness	—	—	1200	2192
White heat	—	—	1300	2372
Lead and lead matt slag	1315–1330	2399–2426		
Raw slag	1330–1360	2426–2480		
Black copper slag	1345	2453		
Blast-furnace slag	1390–1430	2534–2606		
Cobalt (1400° C., 2552° F.).				
Strong white heat	—	—	1400	2552
Dazzling white heat	—	—	1500–1600	2732–2912
Cast-iron (according to Becquerell 1050° to 1200° C., 1922° to 2192° F.)	1500–1700	2732–3092		
White crystalline pig-iron according to v. Tunner	1600	2912		
Gray charcoal pig-iron according to v. Tunner	1700	3092		

[1] The older fusing points are mostly according to Plattner; those inclosed in brackets are newer, according to Becquerell.

	Fusing point (Celsius.)	Fusing point (Fahr.).	Glowing temperature (Celsius.)	Glowing temperature (Fahr.).
Palladium (according to Becquerell 1360° to 1380° C., 2480° to 2516° F.)	1600°	2912°		
Nickel (1600° C., 2912° F.)				
Wolfram (1700° C., 3092° F.)				
Manganese (according to John 1500° C., 2732° F., according to Becquerell 1600° C., 2912° F.)				
Uranium and molybdenum (1600° C., 2912° F.)				
Chromium (1700° C., 3092° F.)				
Steel (according to Becquerell 1300° to 1400° C., according to v. Tunner 1850° C., 3362° F.)	1700–1900	3092–3452		
Malleable iron (according to Becquerell 1600° C., 2912° F.)	1900–2100	3452–3812		
Platinum (according to Debray 2000° C., 3632° F., according to Becquerell 1460° to 1480° C., 2660° to 2696° F.)	2534	4593.2		
Iridium (2400° C., 4352° F.)				

Plattner's method is based upon Prinsep's principle (determination of temperatures by the fusing points of alloys), but his deductions are untenable, although probably approximate for the average of commencing fusion of substances. The determination of the fusing point of the old furnace products of the Freiberg smelting works is also of but little importance at the present time, as these products have lately been materially changed by new processes.

Erhard and Schertel have recently made experiments with the aid of their improved pyrometer to determine the fusing points of metals, alloys, furnace products, silicates, minerals, and rocks. As these may be considered at the present as the most reliable, we give them in the following table:—

3. *Fusing Points of Metals, Alloys, Furnace Products, Rocks, and Silicates, according to Erhard and Schertel.*[1]

a. Metals and Alloys.

				Degrees Celsius.	Degrees Fahrenheit.
	Ag			954	1749.2
80	Ag	20	Au	975	1787
60	"	40	"	995	1823
40	"	60	"	1020	1868
20	"	80	"	1045	1913
	Au			1075	1967
95	Au	5	Pt	1100	2012
90	"	10	"	1130	2066
85	"	15	"	1160	2120
80	"	20	"	1190	2174
75	"	25	"	1220	2228
70	"	30	"	1255	2291
65	"	35	"	1285	2345
60	"	40	"	1320	2408
55	"	45	"	1350	2462
50	"	50	"	1385	2525
45	"	55	"	1420	2588
40	"	60	"	1460	2660
35	"	65	"	1495	2723
30	"	70	"	1535	2795
25	"	75	"	1570	2858
20	"	80	"	1610	2930
15	"	85	"	1650	3002
10	"	90	"	1690	3074
5	"	95	"	1730	3146
	Pt			1775	3227

[1] Freiberger Jahrb. 1879, p. 154. B. u. h. Ztg. 1879, p. 126.

b. *Furnace Products and Gangues.*

	Corresponding alloy.			Temperature degrees Celsius.	Temperature degrees Fahrenheit.
Slag from working the ores in the Muldner smelting works (porous)	70 Au		30 Ag	1030	1886
The same rich in zinc (porous)					
The same two slags entirely fused	90 Au		10 Pt	1130	2066
Concentration slag of copper matt	80 "		20 Ag	1045	1913
Melaphyre from Mulatto	94 "		6 Pt	1106	2022.8
Pitchstone from Arran					
Asbestos	60 "		40 "	1300	2372
Temperature in the porcelain kiln	50 "		50 "	1400	2552
Blast-furnace slag (50SiO$_2$, 17Al$_2$O$_3$, 3FeO, 30CaO)	49 "		51 "	1392 formation temp.	2537.6
Fusing point 77Au 23Pt = 1208° C. (2206.4° F.).					
Freiberg raw slag (48SiO$_2$, 9Al$_2$O$_3$, 37FeO, 4.5CaO, 1.5MgO)	59 "		41 "	1326	2418.8
Freiberg lead slag (36.5SiO$_2$, 40.5FeO, 8.5Al$_2$O$_3$, 4CaO, 3MgO, 7.5 BaO)	75 "		25 "	1220	2228
Fusing point 85Au 15Pt = 1160° C. (2120° F.).					
Freiberg black copper slag (32.7SiO$_2$, 60.3FeO, 7Al$_2$O$_3$)	67 "		33 "	1273	2323.4
Fusing point 83Au 17Pt = 1172° C. (2141.6° F.).					
Raw slag from 3.45 dry ore and 5.25 lead slag	68 "		32 "	1267	2312.6
2.4 raw matt, fusing point 83Au 17Pt = 1172° C. (2141.6° F.).					
Lead slag from 2.34 roasted lead ore, 1.80 roasted raw matt, 5.00 lead slag, 0.30 powdered coke	77 "		23 "	1208	2206.4

B. Lower Harz Working Assays.

According to *Bräuning*,[1] the following methods are used in the *Oker* assay laboratory for assaying the *Rammelsberg* ores and the furnace products obtained from them, after 25,000 to 75,000 kilogrammes have been broken into pieces the size of a fist. A few grammes of samples are taken by crossing; these are comminuted in a stamping-mill, powdered fine in a mortar, and sifted.

a. Lead (p. 88).—3.75 to 7.5 grammes of the ore are decomposed with *aqua regia* and evaporated with addition of sulphuric acid. The mass is then digested with diluted sulphuric acid, and

---

[1] Zeitschr. f. Berg-, Hütten-, u. Salinenwesen im Preuss. Staate, Bd. 25.

filtered. The filtrate is placed on the roasting-dish and dried under the muffle. The mass is then mixed with three times the quantity of black flux (1 part of saltpetre and 2 parts of argol), and placed in a crucible (Fig. 49, p. 64). 0.75 to 1.13 grammes of iron wire and a thin covering of common salt are added, and the charge is fused under the muffle for from fifteen to twenty minutes.

b. Copper.—The filtrate from the test with sulphuric acid (*a*) is diluted to the bulk of 1 liter. 250 cubic centimeters of this are taken and treated with 10 to 15 cubic centimeters of nitric acid of 1.2 specific gravity. The solution is then precipitated by sulphuretted hydrogen, and the precipitate, in case its color indicates the presence of considerable quantities of antimony and arsenic, is treated with sodium sulphide. The residue of copper sulphide is dissolved in 20 to 30 cubic centimeters of moderately diluted nitric acid. This is diluted with water, filtered, and the copper precipitated by electrolysis (p. 110).—*Swedish assay* for the determination of copper in intermediate products. The object is attained more quickly by this method, but it is done at the expense of accuracy. 3.75 grammes of the assay sample are decomposed with *aqua regia*, and evaporated to dryness with sulphuric acid. The dry mass is then taken up with some diluted sulphuric acid, filtered, and the copper precipitated with zinc. The precipitated copper is then ignited under the muffle.—*Heine's colorimetric assay* (p. 125) for products poor in copper (slags, lixiviation residues). 3.75 grammes of the substance are decomposed by means of *aqua regia*. The solution is supersaturated with ammonia and filtered. It is then diluted to a known volume and compared with standard solutions.

c. Iron.—The filtrate from the precipitation with sulphuretted hydrogen (*b*) is oxidized with nitric acid. It is then evaporated to a small volume, and precipitated with ammonia. The precipitate is dissolved in hydrochloric acid, reduced with stannous chloride, and the excess of stannous chloride is titrated back with solution of iodine (*Kerl's Eisenprobirkunst*, 1875, p. 16).

d. Zinc (p. 218).—The filtrate from the iron precipitate is slightly acidulated with hydrochloric acid, and then tested for zinc with potassium ferrocyanide by the volumetric method.

e. Silver.—3.75 grammes of ore, etc., are placed in a scorifier with a mixture of 37.5 grammes of granulated lead and 0.55 to 0.75 grammes of borax, and with a cover of 0.37 gramme of borax, and the charge fused. The lead buttons are cupelled in cupels made of 3 parts of wood-ash and 1 part bone-meal. The buttons of silver are weighed.—*Black copper* and *crude copper* are fused with twenty times the quantity of granulated lead. The assay of fine silver is executed according to *Valhard's* method, it being simpler and more expeditious than that of *Gay-Lussac*, and equals the last-named in accuracy.

f. Gold.—The buttons obtained from fusing with lead (see *e*) are dissolved in nitric acid (free from chlorine) of 1.2 specific gravity. The solution is carefully heated to the boiling point, and, when the action of the acid can no longer be perceived, the argentiferous solution is separated from the gold. The latter is carefully washed, placed in a tared porcelain crucible, gently ignited, and then weighed. In fine silver assays the percentage of gold is determined in the manner indicated on p. 155.

g. Sulphur (p. 279).—1.87 grammes of raw or roasted ore are digested for several hours, in the cold, with concentrated fuming nitric acid. An equal quantity of concentrated hydrochloric acid is added, and the liquid heated on the sand-bath, until the nitric acid has been driven off. It is then diluted with water, filtered, and the sulphuric acid precipitated with barium chloride. Or, by another method, which is more easily executed and gives sufficiently accurate results, the ore is fused with alkaline nitrates and carbonates in iron dishes under the muffle, etc. (p. 282).

C. Schaffner's Assay of Zinc (p. 214) as modified by Brunnlechner.[1]

Crystals of sodium sulphide are dissolved in water with the application of heat until the solution is supersaturated. The solution is then allowed to cool off and settle. The clear solution is poured off and diluted with 10 to 11 times the quantity by volume of water. It is then poured into a flask having the

---

[1] Oesterr. Zeitschr. f. Berg-, u. Hüttenwesen, 1879, No. 37.

capacity of 4 to 5 liters and provided with a double perforated cork. Into one of them is fitted a glass siphon with a rubber tube and pinch-cock, and into the other a short, small, glass tube which may be closed by a cock or cork. The contents of the flask are thoroughly shaken, and, the flask being closed, the siphon is filled by blowing into the small tube and opening the pinch-cock at the same time. The liquid is then allowed to stand for at least twelve hours before it is used. Its strength will be diminished within twenty-four hours, to the extent of 2.5 to 3-thousandths by the oxidation of the sodium sulphide. As much chemically pure zinc as is approximately contained in the sample ore is then weighed off. If, for instance, the ore contains 40 per cent. and 0.5 gramme have been weighed off for the assay, about 0.2 gramme of pure zinc will be required. This is placed in a flask capable of holding half a liter and dissolved in 10 cubic centimeters of concentrated hydrochloric acid. The solution is diluted with 100 cubic centimeters of water, and treated with 50 cubic centimeters of ammonia. It is then thoroughly shaken and allowed to stand for some time, as, otherwise, the indicator would be affected too quickly and the titer of the standard solution would be too high. 1 cubic centimeter of the solution should precipitate at least 8, and at the utmost, 10 milligrammes of zinc from the assay.

When the ore contains over 20 per cent., 0.5 gramme, and, if less, 1 gramme of zinc carbonate and roasted zinc blende is dissolved in concentrated hydrochloric acid, to which a few drops of nitric acid have been added, but raw zinc blende and calamine are dissolved in *aqua regia*. In case gelatinous (ferruginous) silica should be separated, the solution should be diluted, the liquid poured off from the residue, the sediment on the bottom detached by means of a glass rod, again heated with acid, and the two liquids then united. In case sulphur should be separated from silicious zinc ores, fuming nitric acid or potassium chlorate is added. The solution is then evaporated to the consistency of syrup, in order to remove the excess of acid. The residue is then moistened with a few drops of hydrochloric acid and diluted with 20 cubic centimeters of water. To this are added 30 cubic centimeters of ammonia and 15 cubic centimeters of ammonium carbonate. The solution is allowed to settle and then filtered into a flask of 500 cubic centimeters capacity. The filter is washed with 30 cubic centimeters of warm ammonia, and finally with warm ammoniacal water. When much ferric hydrate is present the precipitate is again dissolved and precipitated with ammonia and ammonium carbonate.—*Plumbiferous ores*: The assay sample is dissolved in nitric acid, and the solution evaporated with sulphuric acid. The lead and calcium sulphates are then filtered off. If the ore contains but little calcium, it may be treated with nitric acid, and the lead precipitated with 30 cubic centimeters of ammonia, and 15 cubic centimeters of sodium phosphate. When much calcium is present, the precipitate is again dissolved and precipitated.

A stand with three shelves is used. The uppermost shelf serves for the reception of the flask containing the sodium sulphide, the second smaller one, for a small flask with a pipette containing the indicator solution, and upon the lowest shelf, which projects somewhat, the flask containing the solution of zinc is placed. Over this hangs a burette with a pinch-cock fastened to arms on the vertical wall, its mouth being directly under the orifice of the pinch-cock on the flask containing the standard solution. Duplicate titrations are then made, between which only decimal differences are allowable. Suppose E is the weight of the assay sample, Q the quantity of standard solution in c.c. used for the precipitation, M the total quantity of the assay solution in c.c.; the titer will be—

$$T = \frac{100\,E}{[Q - (M \times 0.007)]}$$

if hydrated peroxide of iron has been used as the indicator, or

$$T = \frac{100\,E}{[Q - (M \times 0.005)]}$$

if paper saturated with ferric chloride (p. 215) has been used.

In titrating the samples, one-third to one-half of the approximate quantity of the precipitating agent required should be added every time to the solution, and this well shaken. A drop of solution of ferric chloride is added to the zinc solution from a small pipette. The mass of hydrated peroxide of iron which is formed is broken up into as equally sized flakes of 1 to 1.5 millimeters diameter as possible by vigorously swinging the flask to and fro. This swinging is constantly continued while solution of sodium sulphide is added by cubic centimeters, until the color of the flakes commences to change. A minute is then allowed for the reaction of the particles of liquid which have remained ineffective, and the precipitation is then finished by adding sodium sulphide drop by drop. If V is the quantity of the precipitating agent consumed in c.c., T the titer, M the total quantity of the solution in c.c. after the assay is finished, the percentage of zinc when ferric hydrate has been used as indicator will be—

$$Z = \frac{T}{100}[V-(M\times 0.007)]$$

and

$$Z = \frac{T}{100}[V-(M\times 0.005)]$$

when slips of paper saturated with ferric chloride have been used. When the first plan is employed the titration must be continued, in order to ascertain the titer, until the flakes become entirely black, but in titrating solutions of ore, only until the reddish-brown color has passed into a greenish tint. The following conditions are required for obtaining a sharp reaction: The flakes must be nearly of the same size, the flask should be carefully swung to and fro to prevent the flakes from being broken up any further; not too much of the precipitating agent must be added at one time, and it should not fall directly upon the flakes, but run down the sides of the flask; the titration should be done at the ordinary temperature, whereby the reaction appears more gradually and uniformly, but the precaution must at the same time be observed of adding the standard solution at longer intervals; uniformity of quantity and time in treating the assays; judging the tone of color by reflected light; and finally the flake reaction should be controlled by a drop test.

If it is required to determine the amount of over 0.5 per cent. of lead in zinc ore, 2 grammes of the sample are dissolved in nitric acid and evaporated to dryness. Some diluted sulphuric acid is added to the dry mass, and this is again evaporated until white vapors appear. It is allowed to become cold, diluted with 20 cubic centimeters of water, filtered and washed until the wash water shows no reaction with ammonium sulphide. The precipitate (lead sulphate, calcium sulphate, gangue) is rinsed off into a beaker and digested with a mixture of ammonium tartrate and ammonia in excess. The lead solution is filtered off, the lead precipitated with sulphuric acid, and the lead sulphate dried and weighed; or, it is detached from the filter, the latter incinerated, and the precipitate ignited (see also p. 93).

D. Experiments on a Heat-Regulator at the United States Assay-Office, New York.[1]

Whoever has had much practical experience in cupellation must have noticed that the variation of light depending upon the clearness or cloudiness of the day deceives the operator in his judgment as to the degree of heat in the muffle. There is very perceptible difference even between the light of the morning and of the afternoon during a perfectly clear day, so that, if reliance be placed upon the eye alone, as is now done, there will be misjudgments resulting in very considerable error.

It often happens that errors arising from this cause are so serious as to necessitate repetitions of the assay, which would have been needless could the furnace have been maintained at the proper temperature. Very often there are annoying delays and embarrassments at the assay office on this account, which, in the old way of working, are unavoidable.

With the purpose of obviating this difficulty, a pyrometer, or, more properly speaking, a heat regulator, which has now been perfected so that a perfectly uniform temperature may be automatically maintained at any desired degree, has been designed. By this apparatus the furnace may be kept at a "copper," "silver," or "gold" heat, or in a few minutes changed from one to the other with perfect precision.

The value of such an appliance to the bullion assayer is manifest, and will at once be appreciated.

This automatic heat-regulator or pyrometer is based upon the expansion and contraction of a single metallic bar, so mounted as to be independent of the expansions of the furnace to which it is attached, thus giving a greater change in length for any given variation of temperature than those composed of two metals whose

[1] H. G. Torrey, Engineering and Mining Journal, August 28, 1886, p. 147.

co-efficients of expansion are different, the difference being the measuring co-efficient.

For assay furnaces, platinum has been found to be the most suitable metal, on account of its quality of remaining unoxidized upon exposure to high temperatures, in the presence of currents of air, and on account of its infusibility.

Fig. 87.

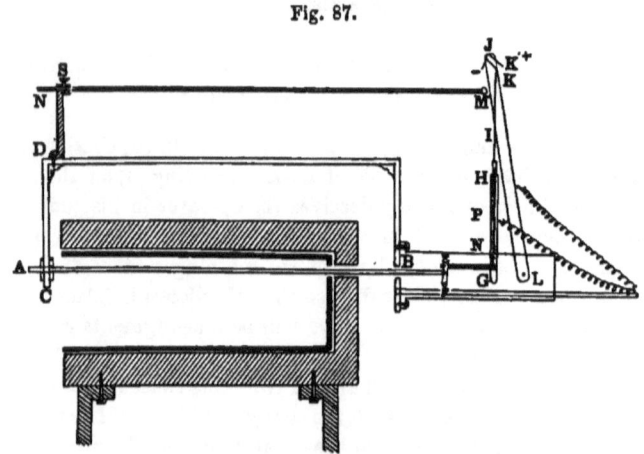

A rod $A B$ (Fig. 87), No. 12 wire gauge, extends along the interior of the muffle close to the roof, being suspended at the fore end by an iron bar $C D$, about half an inch wide, passing freely through it, and extending through the rear end of the muffle and furnace.

The rear end of the rod is connected by a link to the centre of a short lever whose fulcrum is at one end. To the other end of the lever a second link connects with a second lever $G H$, which serves to further magnify the expansion of the platinum rod. This second lever is extended into a pointer I, and has a total length of six inches.

The levers are so proportioned that the extremity of the pointer passes through an arc of six inches from a cold furnace to a cupelling heat for gold bullion assays.

The pointer plays between two platinum contacts $K K'$, which are one-tenth of an inch apart. These contacts are electrically connected with an apparatus controlling the gas supply of the furnace; and the pointer is also connected electrically with the apparatus in such a manner that, when it rests against the contact nearer the furnace, it causes the formation of a circuit operating machinery that opens the gas-cock. When the pointer, leaving the contact on account of the increased heat, reaches the other contact, the operation is reversed.

The temperature is therefore kept within the limits afforded by the movement of the pointer through one-sixteenth of an inch, which represents a variation of temperature of less than 25 degrees Fahrenheit.

The contact points are fastened to a rod JL, which is pivoted at its lower end close to the pivot of the pointer. A rod MN is also pivoted to its upper end, which passes to the front of the furnace, and is clamped by a thumb-screw S in any desired position to the frame that supports the whole apparatus.

If it becomes necessary to cool the furnace for silver bullion assays, the operator has only to draw the rod MN toward him, thus bringing the contact-bar toward the zero from which the pointer started when the furnace was cold. This movement causes the pointer to bear against the outer contact, shutting off the gas and cooling the furnace to the required point, and no lower.

The pointer is in two sections, pivoted in the centre, and is kept in a straight line by a spring PN, which allows, without straining the apparatus, the deflections that are made when the position of the contact-bar is changed. Should the platinum bar stretch, or should there be any lost motion in the levers, causing the pointer, when the furnace is cold, to rest at some point other than a fixed zero, the error may be compensated for by two nuts that are placed one on each side of the supporting bar in front. The platinum bar is kept in tension by a spiral spring fastened to the pointer, and to a point beyond it.

Mr. Charles Taylor, of the U. S. Assay Office in New York, writes under the date of April 8, 1889, as follows:—

"Since the above article appeared in the Engineering and Mining Journal, I have found that good results can be obtained by bolting the apparatus direct to the furnace, thus obtaining the use of an independent frame. The index has also been simplified so that but one spring is necessary to take up the expansion, lost motion, etc., and also to allow for the deflection of the index."

E. Dry Assay of Iron Ores.

While it is perfectly true that the results obtained by assaying iron ores in the dry way are less accurate than most of those obtained in the wet way, on the other hand they approximate much more nearly to the economic value of an iron ore when smelted on a large scale. The percentage of metallic iron is not only obtained, but obtained associated or combined with carbon, phosphorus, silica, sulphur, and manganese, according to the nature of the ore, these elements being present relatively in about the same proportions as when the ore is smelted in a blast furnace. The assay consequently gives a very fair and accurate *idea* of what may be expected of any particular iron ore in a blast furnace, not only as regards the yield of iron, but also the composition and action of the fluxes. From a few *systematic* assays one is not only enabled to form a good idea of the quality of the iron which would be produced on a large scale, but the fluxes giving the best results with the ore can readily be determined. This cannot always be done on the basis of a chemical analysis alone, no matter how complete and accurate it may be. It will therefore be observed that this system of dry assays particularly commends itself for use with new and untried ores. It is usually best to make a preliminary assay in the wet way in order to determine the earthy matter present. Berthier recommends the following process :[1]—

A weighed quantity of the finely powdered ore is heated to redness in a porcelain crucible; the loss of weight gives the amount of water, carbonic acid, and other volatile matters present. Another weighed portion is heated with dilute nitric acid, which dissolves out carbonates of lime and magnesium; the residue contains, under ordinary circumstances, only oxide of iron, clay, and quartz, the difference giving the amount of the earthy carbonates. A third portion is treated with strong hydrochloric acid, whereby the carbonates of lime and magnesium, and the oxides of iron ore dissolved out, leaving an insoluble residue of quartz and clay. This is weighed, the oxides of iron being determined from the difference of weight by deducting that of the carbonates previously found. From these results the proportion of fluxes necessary to be added can be approximately determined with a view to produce an easily fusible slag. In many instances, however, this preliminary assay can be dispensed with, as a sufficiently good idea of the nature of the fluxes

---

[1] Metallurgy of Iron. H. Bauerman. London, 1882, p. 104.

DRY ASSAY OF IRON ORES.

to be added can be determined by the appearance of the ore alone. The dry assay of iron may be performed either in plain clay crucibles, or, preferably, in basqued, or charcoal lined crucibles or some other carbonaceous material. In the former case the ore must be mixed with from 20 to 30 per cent. of powdered charcoal.

The basqued crucibles are prepared as follows:[1] The charcoal powder is mixed with just sufficient gum-water or molasses to make it cohere readily. The crucible is gently rammed full of this charcoal, and a cylindrical cavity of sufficient size to contain the charge is made in it with a spatula or some other kind of boring instrument.

The crucibles best adapted for this assay are made of a mixture composed of two parts unburnt and one of burnt clay. They are about $1\frac{3}{4}$ inches high and $1\frac{1}{4}$ inches in diameter at the top. As the amount of ore in the charges never exceeds from 10 to 15 grains (0.64 to 0.96 grammes), four crucibles are placed in the furnace at a time. The charges in each are exactly alike, and if there is but a slight variation in the weight of the resulting button, the assay is probably correct; the mean weight of the four should be taken as the result.[2] It is generally best to stand or lute the four crucibles to a half brick, so that when the fusion is finished all the crucibles can be removed from the furnace at one time. Although lids are sometimes used to cover the crucibles it is better, after the charge has been introduced, to stop the cavity with a charcoal plug and to cover the entire top of the crucible with a clay luting as shown in Fig. 86.

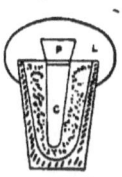

Fig. 86.

c. Cavity containing charge.
P. Charcoal plug.
L. Clay luting.

The half brick holding the crucibles should be placed near or directly upon the fire bars at the bottom of the furnace, and the anthracite or coke fire built up around and over them. When the crucibles have been in the fire about $1\frac{1}{2}$ to 2 hours, and a *white heat* reached and maintained through a considerable part of this period, the assay may be considered finished and the crucibles removed. When the crucibles are cold they are broken up, the metallic buttons and globules[3] which may adhere to the charcoal lining and slag are separated out with a magnet.

---

[1] Percy's Metallurgy. Fuel. London, 1875, p. 140.

[2] With the same ore the variation between the buttons should not be over a few tenths per cent.

[3] The formation of these globules is prevented by not allowing any fine particles of charcoal or dirt to get into the cavity when it is being charged.

After weighing, the buttons should be fractured and the character of the iron carefully noted. The character of the slag must likewise be carefully observed. If there is too much silica in the flux it has a greenish tint and is transparent; if translucent, blue or grayish enamel, lime, clay, and silica are in the proper proportions; if rough, stony, or crystalline, it is too basic and has a dull lustre. Should the charge be fritted, pulverulent, and the iron diffused in minute particles, the flux has been wrongly proportioned, or the temperature too low. This is apt to happen with ores containing a large proportion of lime and magnesium, as they are always refractory and require the addition of a large proportion of silica and lime. If the slag is but half fused and dark in color, the ore contains an excess of silica or silicates of iron and manganese, which react on the carburetted iron, decomposing it, forming malleable iron and carbonic acid, which escapes through the slag, giving it a spongy appearance. In this case lime should be added. The presence of a small amount of manganese is indicated by the amethystine color of the slag; larger amounts render it olive green, yellowish, or brown.

Titaniferous ores are apt to render the slag copper-colored by the formation of cyano-nitride of titanium. They may also be black and hard, or else bluish and vitreous. Chromic iron ores produce a resinous and dark-colored slag, which is sometimes surrounded with a thin metallic layer. As a general thing the iron buttons are dark-gray or mottled, according to the condition and amount of carbon they contain. If the crucibles are allowed to cool very slowly, the buttons are, as a rule, very graphitic, but if cooled quickly are fine-grained and white, or grayish in color. If the ores are easily reducible, and the proper temperature has been reached and maintained in the furnace, the buttons should be dark, tough, and, if they have not been cooled too quickly, graphitic.

The presence of phosphorus generally renders the buttons white, hard, and brittle; sulphur produces a mottled or reticulated structure; manganese a hard, white, crystalline or close-grained iron; titanium a dull dark-gray button with a crystalline fracture; chromium a well-fused button, with a tin-white, bright crystalline fracture, to a semi-fused white, or light-gray, spongy mass, according to the amount of chromium in the iron.[1]

Proportion of fluxes.—The fluxes should be so adjusted as to obtain a well-fused clean slag, in sufficient quantity to completely

---

[1] Percy's Metallurgy. Iron and Steel. London, 1864. p. 243.

cover the reduced button of metal. Blast furnace slags of the following composition may be taken as types of the slags most desirable:—

$CaO.SiO_2 + Al_2O_3.3SiO_2 =$ lime 30, alumina 14, silica 56 per cent.

$2(2CaO.SiO_2) + 2Al_2O_3.3SiO_2 =$ lime 47, alumina 15, silica 37 per cent.[1]

If the composition of the ore is approximately known, the amount of fluxes necessary to produce a fusible slag can be easily ascertained. When necessary an extra amount of flux must be added in order to produce sufficient slag to cover the button.

As a rule the following fluxes will be found best adapted to ores or metallurgical products, which would be classed under the following several general heads. The proportions given are for 10 grains (0.65 gramme) of ore :[2]—

1. Ores nearly free from gangue, some varieties of magnetite, red and brown hematite, specular iron ore, and micaceous iron ore:—

	Parts.		Parts.		Parts.
China clay	2	Glass	2–2½	Blast-furnace slag	5
Lime	2½	Lime	2½–3	Fluor spar	5
Sand	0–1				

2. Ores, etc., containing silica, varieties of limonite or brown iron ore, refinery slag (tap), and flue cinders:—

	Parts.		Parts.
Glass	1–2¼	China clay	2
Lime	2½–4	Lime	3½–4

3. Ores, containing carbonates of lime, magnesia, protoxide of manganese, etc., calcareous hematites, and spathic ores:—

	Parts.		Parts.
China clay	2	Glass	3–4
Lime	1½	Lime	1½–2
Sand	1		

4. Ores containing silica and alumina, clay iron ores, etc. :—

	Parts.		Parts.
Glass	0–2½	China clay	0–2
Lime	2½–3	Lime	2–3

---

[1] Metallurgy of Iron. H. Bauerman. London, 1882. p. 105.

[2] A universal flux known as Percy's slag, lime 3, silica 2½, and clay 1 part, is sometimes used. School of Mines Quarterly. Vol. II., p. 173.

314 APPENDIX.

5. **Titaniferous ores, or ores containing titanium :**[1]—

	Parts.		Parts.		Parts.
Lime	3	China clay	2	Glass	2½
Glass	2	Lime	2½	Lime	1½
				China clay	1

The ores should be powdered and passed through an 80 mesh sieve (80 meshes per linear inch,—2¼ centimeters). The use of iron mortars should be avoided. When necessary the ore should be dried at about 110° C. Some ores contain hydroscopic as well as combined water.

Fluxes.—*Silica* in the form of white quartz or rock crystal, finely pulverized, is preferred on account of its purity; but white sand used in glass-making forms a good substitute.

Glass, such as plate, crown, or window, finely pulverized, is used. It contains from 60 to 70 per cent. silica, and, being fusible, forms a good substitute for silica or silicates of alumina. Green glass containing oxide of iron, or flint glass containing lead, is objectionable.

China clay, or the hydrated silicate of alumina, forms a very pure and useful flux, as it is practically free from oxide of iron. It is either used in a hydrated or dehydrated state. In the latter case it is pulverized and heated to redness.

Lime.—The common powdered, *unslacked* lime should be used; limestone, chalk, or any varieties of carbonate of lime are good substitutes. Care must be taken that the lime flux is free from sulphates.

Assay in large unlined crucibles.—It is very good practice, and at the same time shows how an iron assay can be made in an unbasqued crucible, to make a series of assays to show the effect of *too little,* the *proper amount,* and *too much carbon* on the result.

The *fluxes and ore* should be pulverized and passed through a 60 mesh sieve. With a series of four, the ore and fluxes may be proportioned as follows:—

	1 and 2.				3 and 4.	
Ore	500 grains	(32 grammes)	Ore	500 grains	(32 grammes)
Glass	250 "	(16 ")	China clay	200 "	(12.8 ")
Lime	300 "	(19.2 ")	Sand	50 "	(3.2 ")
			Lime	250 "	(16 ")

[1] This assay requires the highest possible temperature that can be obtained in an ordinary furnace, and to be subjected to such a temperature a somewhat longer time than usual.

The reducing agent employed is preferably anthracite, but charcoal or coke may be used. In any case it should be pulverized and passed through a 60 mesh sieve. The following amounts are respectively used:—

1. 80 grains (5.12 grammes)
2. 110 " (7.04 ")
3. 120 grains (7.68 grammes)
4. 150 " (9.60 ")

Mix the fluxes and reducing agent thoroughly with the ore, transfer to a crucible about 4 inches high by $1\frac{1}{2}$ inches in diameter (10.12 by 3.79 centimeters). Lute cover on with clay, place on piece of fire-brick and heat in usual manner for an hour or more. In lifting the crucible out of the furnace, care must be taken to shake it as little as possible in order to avoid disseminating globules of iron in the slag. When cold break open the crucible, and collect any small buttons or shots of metals in the slag with a magnet. Weigh, then fracture the largest button and note the character of the iron. If the above scheme has been carefully carried out, using an ordinary hematite or magnetic ore, the results will be something like as follows:—

1. Not being sufficient carbon present the ore is not all reduced; the reduced and unreduced ore being fritted together with slag in an irregular lump.

2. Ore completely reduced and fused into a button. *Iron*, gray. *Slag*, glassy, and transparent, or perhaps dark-grayish, and opaque.

3. Ore completely reduced and fused into a well-melted button. *Iron*, gray to grayish-white. *Slag*, clear, glassy, and transparent, color, gray to greenish by transmitted light.

4. Ore completely reduced, but owing to the excess of carbon present the metal is disseminated through the slag in small shots or globules. *Iron*, gray. *Slag*, glassy, opaque, or translucent.

There should always be present an excess of the amount of carbon necessary to reduce the ore. 100 parts by weight of sesquioxide of iron require $22\frac{1}{2}$ parts of carbon for reduction. Consequently the amount added must be adjusted according to the oxide of iron present.[1]

[1] Percy's Metallurgy. Iron and Steel. p. 243.

THE

METRIC SYSTEM OF WEIGHTS AND MEASURES.

THE United States being the first to introduce the decimal system into the coinage of the country, and to demonstrate its superior utility, it is remarkable that we have hesitated so long in regard to the substitution of the same simple and rational system of weights and measures for the complicated and confused standards in general use.

In May, 1866, the Committee on Coinage, Weights, and Measures presented to the House of Representatives an exhaustive report, accompanied by bills authorizing the introduction of the *metric system* into the various departments of trade, and making all contracts, based on this system of weights and measures, valid before any court in the United States. They said:—

"THE METRIC SYSTEM.

"It is orderly, simple, and perfectly harmonious, having useful relations between all its parts. It is based on the METER, which is the principal and only arbitrary unit. The meter is a measure of length, and was intended to be, and is, very nearly one ten-millionth of the distance on the earth's surface from the equator to the pole. It is 39.37 inches, very nearly.

"The *are* is a surface equal to a square whose side is 10 meters. It is nearly four square rods.

"The *liter* is the unit for measuring capacity, and is equal to the contents of a cube whose edge is a tenth part of the meter. It is a little more than a wine quart.

"The *gramme* is the unit of weight, and is the weight of a cube of water, each edge of the cube being one one-hundredth of the meter. It is equal to 15.432 grains.

"The *stere* is the cubic meter.

"Each of these units is divided decimally, and larger units are formed by multiples of 10, 100, &c. The successive multiples are designated by the prefixes, *deka, hecto, kilo,* and *myria;* the subordinate parts by *deci, centi,* and *milli,* each having its own numerical significance.

"The nomenclature, simple as it is in theory, and designed

from its origin to be universal, can only become familiar by use. Like all strange words, these will become familiar by custom, and obtain popular abbreviations. A system which has incorporated with itself so many different series of weights, and such a nomenclature as 'scruples,' 'pennyweights,' 'avoirdupois,' and with no invariable component word, can hardly protest against a nomenclature whose leading characteristic is a short component word with a prefix signifying number. We are all familiar with *thermometer, barometer, diameter, gasometer,* &c., with *telegram, monogram,* &c., words formed in the same manner.

"After considering every argument for a change of nomenclature, your committee have come to the conclusion that any attempt to conform it to that in present use would lead to confusion of weights and measures, would violate the early learned order and simplicity of metric denomination, and would seriously interfere with that universality of system so essential to international and commercial convenience.

"When it is remembered that of the value of our exports and imports, in the year ending June 30, 1860, in all $762,000,000, the amount of near $700,000,000 was with nations and their dependencies that have now authorized, or taken the preliminary steps to authorize, the metric system, even denominational uniformity for the use of accountants in such vast transactions assumes an important significance. In words of such universal employment, each word should represent the identical thing intended, and no other, and the law of association familiarizes it.

"Your committee unanimously recommend the passage of the bills and joint resolutions appended to this report. . . . . . The metric system is already used in some arts and trades in this country, and is especially adapted to the wants of others. Some of its measures are already manufactured at Bangor, in Maine, to meet an existing demand at home and abroad. The manufacturers of the well-known Fairbanks' scales state: 'For many years we have had a large export demand for our scales with French weights, and the demand and sale are constantly increasing.' Its minute and exact divisions specially adapt it to the use of chemists, apothecaries, the finer operations of the artisan and to all scientific objects. It has always been and is now used in the United States coast survey. Yet in some of the States, owing to the phraseology of their laws, it would be a direct violation of them to use it in the business transactions of the community. It is, therefore, very important to legalize its use, and to give to the people, or that portion of them desiring it, the opportunity for its legal employment, while the knowledge of its characteristics will be thus diffused among men."

WEIGHTS AND MEASURES.
APOTHECARIES' WEIGHT, U. S.

Pound.		Ounces.		Drachms.		Scruples.		Grains.
℔ 1	=	12	=	96	=	288	=	5760
		℥ 1	=	8	=	24	=	480
				ʒ 1	=	3	=	60
						℈ 1	=	gr. 20

The imperial standard Troy weight, at present recognized by the British laws, corresponds with the apothecaries' weight in pounds, ounces, and grains, but differs from it in the division of the ounce, which, according to the former scale, contains twenty pennyweights, each weighing twenty-four grains.

AVOIRDUPOIS WEIGHT.

Pound.		Ounces.		Drachms.		Troy grains.
℔ 1	=	16	=	256	=	7000.
		oz. 1	=	16	=	437.5
				dr. 1	=	27.34375

Relative Value of Troy and Avoirdupois Weights.

Pound.		Pounds.		Pound.	Oz.	Grains.
1 Troy	=	0.822857 Avoirdupois	=	0	13	72.5
1 Avoirdupois	=	1.215277 Troy	=	1	2	280.

WINE MEASURE, U. S.

Gallon.		Pints.		Fluidounces.		Fluidrachms.		Minims.		Cubic inches.
Cong. 1	=	8	=	128	=	1024	=	61440	=	231.
		O 1	=	16	=	128	=	7680	=	28.875
				f℥ 1	=	8	=	480	=	1.8047
						fʒ 1	=	♏ 60	=	0.2256

IMPERIAL MEASURE.
Adopted by all the British College.

Gallon.		Pints.		Fluidounces.		Fluidrachms.		Minims.
1	=	8	=	160	=	1280	=	76800
		1	=	20	=	160	=	9600
				1	=	8	=	480
						1	=	60

Relative Value of Apothecaries' and Imperial Measures.

APOTHECARIES' MEASURE.　　IMPERIAL MEASURE.

	Pints.	Fluidozs.	Fluidrms.	Minims.
1 gallon =	6	13	2	23
1 pint =		16	5	18
1 fluidounce =		1	0	20
1 fluidrachm =			1	2.5
1 minim =				1.04

IMPERIAL MEASURE.　　APOTHECARIES' MEASURE.

	Gallon.	Pints.	Fluidoz.	Fluidrms.	Minims
1 gallon =	1	1	9	5	8
1 pint =		1	3	1	38
1 fluidounce =				7	41
1 fluidrachm =					58
1 minim =					0.96

Relative Value of Weights and Measures in Distilled Water at 60° Fahr.

1. Value of Apothecaries' Weight in Apothecaries' Measure.

		Pints.	Fluidoz.	Fluidr.	Minims.
1 pound	= 0.7900031 pints =	0	12	5	7.2238
1 ounce	= 1.0533376 fluidounces =	0	1	0	25.0020
1 drachm	= 1.0533376 fluidrachms =	0	0	1	3.2002
1 scruple	=	0	0	0	21.0667
1 grain	=	0	0	0	1.0533

2. Value of Apothecaries' Measure in Apothecaries' Weight.

		Pounds.	Oz.	Dr.	Sc.	Gr.	Grains.
1 gallon	= 10.12654270 pounds =	10	1	4	0	8.88 =	58328.886
1 pint	= 1.26581783 pounds =	1	3	1	1	11.11 =	7291.1107
1 fluidounce	= 0.94936332 ounces =	0	0	7	1	15.69 =	455.6944
1 fluidrachm	= 0.94936332 drms. =	0	0	0	2	16.96 =	56.9618
1 minim	= 0.94936332 grains =						1.9493

3. Value of Avoirdupois Weight in Apothecaries' Measure.

		Pints.	Fluidoz.	Fluidrms.	Minims.
1 pound	= 0.9600732 pints =	0	15	2	53.3622
1 ounce	= 0.9600732 fluidounces =	0	0	7	40.8351

4. Value of Apothecaries' Measure in Avoirdupois Weight.

1 gallon   = 8.33269800 pounds.
1 pint   = 1.04158725 pounds.
1 fluidounce = 1.04158725 ounces.

5. Value of Imperial Measure in Apothecaries' and Avoirdupois Weights.

Imperial Measure.	Apothecaries' Weight.	Avoirdupois Weight.	Grains.	Cubic inches.
1 gallon =	12 lb 1 ℥ 6 ʒ 2 ℈ 0 gr. =	10 lb 0 ℥ =	70,000	= 277.27384
1 pint =	1 6 1 2 10 =	1 4 =	8,750	= 34.65923
1 fluidounce =	7 0 17.5 =	1 =	437.5	= 1.73296
1 fluidrachm =	2 14.69 =	=	54.69	= 0.21662
1 minim =		=	0.91	= 0.00361

In converting the weights of liquids heavier or lighter than water into measures, or conversely, a correction must be made for specific gravity. In converting weights into measures, the calculator may proceed as if the liquid was water, and the obtained measure will be the true measure *inversely* as the specific gravity. In the converse operation, of turning measures into weights, the same assumption may be made, and the obtained weight will be the true weight *directly* as the specific gravity.

TABLES

SHOWING THE

RELATIVE VALUES OF FRENCH AND ENGLISH WEIGHTS AND MEASURES, &c.

Measures of Length.

Millimetre	= 0.03937	inch.
Centimetre	= 0.393708	"
Decimetre	= 3.937079	inches.
Metre	= 39.37079	"
"	= 3.2808992	feet.
"	= 1.093633	yard.
Decametre	= 32.808992	feet.
Hectometre	= 328.08992	"
Kilometre	= 3280.8992	"
"	= 1093.633	yards.
Myriametre	= 10936.33	"
"	= 6.2138	miles.
Inch ($\frac{1}{36}$ yard)	= 2.539954	centimetres.
Foot ($\frac{1}{3}$ yard)	= 3.0479449	decimetres.
Yard	= 0.91438348	metre.
Fathom (2 yards)	= 1.82876696	"
Pole or perch (5½ yards)	= 5.029109	metres.
Furlong (220 yards)	= 201.16437	"
Mile (1760 yards)	= 1609.3149	"
Nautical mile	= 1852	"

Superficial Measures.

Square millimetre	=	$\frac{1}{645}$	square inch.	
" "	=	0.00155	" "	
" centimetre	=	0.155006	" "	
" decimetre	=	15.50059	" inches.	
" "	=	0.107643	" foot.	
" metre or centiare	=	1550.05989	" inches.	
" " "	=	10.764299	" feet.	
" " "	=	1.196033	" yard	
Are	=	1076.4299	" feet.	
"	=	119.6033	" yards.	
"	=	0.098845	rood.	
Hectare	=	11960.3326	square yards.	
"	=	2.471143	acres.	
Square inch	=	645.109201	square millimetres.	
" "	=	6.451367	" centimetres	
" foot	=	9.289968	" decimetres.	
" yard	=	0.836097	" metre.	
" rod or perch	=	25.291939	" metres.	
Rood (1210 sq. yards)	=	10.116775	ares.	
Acre (4840 sq. yards)	=	0.404671	hectare.	

Measures of Capacity.

Cubic millimetre	=	0.000061027	cubic inch.
" centimetre or millilitre	=	0.061027	" "
10 " centimetres or centilitre	=	0.61027	" "
100 " " " decilitre	=	6.102705	" inches.
1000 " " " litre	=	61.0270515	" "
" " " " "	=	1.760773	imp'l pint.
" " " " "	=	0.2200967	" gal'n.
Decalitre	=	610.270515	cubic inches.
"	=	2.2009668	imp. gal'ns.
Hectolitre	=	3.531658	cubic feet.
"	=	22.009668	imp. gal'ns.
Cubic metre or stere or kilolitre	=	1.30802	cubic yard.
" " "	=	35.3165807	" feet.
Myrialitre	=	353.165807	" "

WEIGHTS AND MEASURES, ETC.

Cubic inch	= 16.386176	cubic centimetres.
" foot	= 28.315312	" decimetres.
" yard	= 0.764513422	" metre.

American Measures.

Winchester or U.S. gallon (231 cub. in.) = 3.785209 litres.
" " bushel (2150.42 cub. in.) = 35.23719 "
Chaldron (57.25 cubic feet) = 1621.085 "

British Imperial Measures.

Gill	= 0.141983	litre.
Pint (⅛ gallon)	= 0.567932	"
Quart (¼ gallon)	= 1.135864	"
Imperial gallon (277.2738 cub. in.)	= 4.54345797	litres.
Peck (2 gallons)	= 9.0869159	"
Bushel (8 gallons)	= 36.347664	"
Sack (3 bushels)	= 1.09043	hectolitre.
Quarter (8 bushels)	= 2.907813	hectolitres.
Chaldron (12 sacks)	= 13.08516	"

Weights.

Milligramme	=	0.015438395	troy grain.
Centigramme	=	0.15438395	" "
Decigramme	=	1.5438395	" "
Gramme	=	15.438395	" grains.
"	=	0.643	pennyweight.
"	=	0.0321633	oz. troy.
"	=	0.0352889	oz. avoirdupois.
Decagramme	=	154.38395	troy grains.
"	=	5.64	drachms avoirdupois.
Hectogramme	=	3.21633	oz. troy.
"	=	3.52889	oz. avoirdupois.
Kilogramme	=	2.6803	lbs. troy.
"	=	2.205486	lbs. avoirdupois.
Myriagramme	=	26.803	lbs. troy.
"	=	22.05486	lbs. avoirdupois.

Quintal metrique = 100 kilog. = 220.5486 lbs. avoirdupois.
Tonne = 1000 kilog. = 2205.486 " "

Different authors give the following values for the gramme:—

Gramme = 15.44402 troy grains.
" = 15.44242 "
" = 15.4402 "
" = 15.433159 "
" = 15.43234874 "

AVOIRDUPOIS.

Long ton = 20 cwt. = 2240 lbs. = 1015.649 kilogrammes.
Short ton (2000 lbs.) = 906.8296 "
Hundred weight (112 lbs.) = 50.78245 "
Quarter (28 lbs.) = 12.6956144 "
Pound = 16 oz. = 7000 grs. = 453.4148 grammes.
Ounce = 16 dr'ms. = 437.5 grs. = 28.3375 "
Drachm = 27.344 grains = 1.77108 gramme.

TROY (PRECIOUS METALS).

Pound = 12 oz. = 5760 grs. = 373.096 grammes.
Ounce = 20 dwt. = 480 grs. = 31.0913 "
Pennyweight = 24 grs. = 1.55457 gramme.
Grain = 0.064773 "

APOTHECARIES' (PHARMACY).

Ounce = 8 drachms = 480 grs. = 31.0913 gramme.
Drachm = 3 scruples = 60 grs. = 3.8869 "
Scruple = 20 grs. = 1.29546 gramme.

CARAT WEIGHT FOR DIAMONDS.

1 carat = 4 carat grains = 64 carat parts.
" = 3.2 troy grains.
" = 3.273 " "
" = 0.207264 gramme
" = 0.212 "
" = 0.205 "

Great diversity in value.

Proposed Symbols for Abbreviations.

M—myria	— 10000	Mm	Mg	Ml	
K—kilo	— 1000	Km	Kg	Kl	
H—hecto	— 100	Hm	Hg	Hl	Ha
D—deca	— 10	Dm	Dg	Dl	Da
Unit	— 1	metre—m	gramme—g	litre—l	are—a
d—deci	— 0.1	dm	dg	dl	da
c—centi	— 0.01	cm	cg	cl	ca
m—milli	— 0.001	mm	mg	ml	

Km = Kilometre. Hl = Hectolitre. cg = centigramme. c. cm = $\overline{cm}^3$ = cubic centimetre. $\overline{dm}^2$ = sq. dm = square decimetre. Kgm = Kilogrammetre. Kg° = Kilogramme degree.

Celsius or Centigrade.	Fahrenheit.	Réaumur.
— 15°	+ 5°	— 12°
— 10	+ 14	— 8
— 5	+ 23	— 4
0 melting	+ 32	ice 0
+ 5	+ 41	+ 4
+ 10	+ 50	+ 8
+ 15	+ 59	+ 12
+ 20	+ 68	+ 16
+ 25	+ 77	+ 20
+ 30	+ 86	+ 24
+ 35	+ 95	+ 28
+ 40	+104	+ 32
+ 45	+113	+ 36
+ 50	+122	+ 40
+ 55	+131	+ 44
+ 60	+140	+ 48
+ 65	+149	+ 52
+ 70	+158	+ 56
+ 75	+167	+ 60
+ 80	+176	+ 64
+ 85	+185	+ 68
+ 90	+194	+ 72
+ 95	+203	+ 76
+100 boiling	+212	water + 80
+200	+392	+160
+300	+572	+240
+400	+752	+320
+500	+932	+400

$1° C. = 1°.8$ Ft. $= \frac{9}{5}$ Ft. $= 0°.3$ R. $= \frac{4}{5}°$ R.
$1° C. \times \frac{9}{5} = 1°$ Ft. $1°$ Ft. $\times \frac{5}{9} = 1°$ C. $1°$ R. $\times \frac{9}{4} = 1°$ Ft.
$1° C. \times \frac{4}{5} = 1°$ R. $1°$ Ft. $\times \frac{4}{9} = 1°$ R. $1°$ R. $\times \frac{5}{4} = 1°$ C.

Calorie (French) = unit of heat
= kilogramme degree } English.

It is the quantity of heat necessary to raise 1° C. the temperature of 1 kilogramme of distilled water.

Kilogrammetre = Kgm = the power necessary to raise 1 kilogramme, 1 metre high, in one second. It is equal to $\frac{1}{75}$ of a French horse power. An English horse power = 550 foot pounds, while a French horse power = 542.7 foot pounds.

Ready-made Calculations.

No. of units.	Inches to centimetres.	Feet to metres.	Yards to metres.	Miles to Kilometres.	Millimetres to inches.
1	2.53995	0.3047945	0.91438348	1.6093	0.03937079
2	5.0799	0.6095890	1.82876696	3.2186	0.07874158
3	7.6199	0.9143835	2.74315044	4.8279	0.11811237
4	10.1598	1.2197680	3.65753392	6.4373	0.15748316
5	12.6998	1.5239724	4.57191740	8.0466	0.19685395
6	15.2397	1.8287669	5.48630088	9.6559	0.23622474
7	17.7797	2.1335614	6.40068436	11.2652	0.27559553
8	20.3196	2.4383559	7.31506784	12.8745	0.31496632
9	22.8596	2.7431504	8.22945132	14.4838	0.35433711
10	25.3995	3.0479450	9.14383480	16.0930	0.39370790

No. of units.	Centimetres to inches.	Metres to feet.	Metres to yards.	Kilometres to miles.	Square inches to square centimetres.
1	0.3937079	3.2808992	1.093633	0.6213824	6.45136
2	0.7874158	6.5617984	2.187266	1.2427648	12.90272
3	1.1811237	9.8426976	3.280899	1.8641472	19.35408
4	1.5748316	13.1235968	4.374532	2.4855296	25.80544
5	1.9685395	16.4044960	5.468165	3.1069120	32.25680
6	2.3622474	19.6853952	6.561798	3.7282944	38.70816
7	2.7559553	22.9662944	7.655431	4.3496768	45.15952
8	3.1496632	26.2471936	8.749064	4.9710592	51.61088
9	3.5433711	29.5280928	9.842697	5.5924416	58.06224
10	3.9370790	32.8089920	10.936330	6.2138240	64.51360

WEIGHTS AND MEASURES, ETC.

No. of units.	Square feet to sq. metres.	Sq. yards to sq. metres.	Acres to hectares.	Square centimetres to sq. inches.	Sq. metres to sq. feet.
1	0.0929	0.836097	0.404671	0.155	10.7643
2	0.1858	1.672194	0.809342	0.310	21.5286
3	0.2787	2.508291	1.204013	0.465	32.2929
4	0.3716	3.344388	1.618684	0.620	43.0572
5	0.4645	4.180485	2.023355	0.775	53.8215
6	0.5574	5.016582	2.428026	0.930	64.5858
7	0.6503	5.852679	2.832697	1.085	75.3501
8	0.7432	6.688776	3.237368	1.240	86.1144
9	0.8361	7.524873	3.642039	1.395	96.8787
10	0.9290	8.360970	4.046710	1.550	107.6430

No. of units.	Square metres to sq. yards.	Hectares to acres.	Cubic inches to cubic centimetres.	Cubic feet to cubic metres.	Cubic yards to cubic metres.
1	1.196033	2.471143	16.3855	0.02831	0.76451
2	2.392066	4.942286	32.7710	0.05662	1.52902
3	3.588099	7.413429	49.1565	0.08494	2.29354
4	4.784132	9.884572	65.5420	0.11325	3.05805
5	5.980165	12.355715	81.9275	0.14157	3.82257
6	7.176198	14.826858	98.3130	0.16988	4.58708
7	8.372231	17.298001	114.6985	0.19819	5.35159
8	9.568264	19.769144	131.0840	0.22651	6.11611
9	10.764297	22.240287	147.4695	0.25482	6.88062
10	11.960330	24.711430	163.8550	0.28315	7.64513

No. of units.	Cubic centimetres to cubic inches.	Litres to cubic inches.	Hectolitres to cubic feet.	Cubic metres to cubic feet.	Cubic metres to cubic yards.
1	0.06102	61.02705	3.5317	35.31659	1.30802
2	0.12205	122.05410	7.0634	70.63318	2.61604
3	0.18308	183.08115	10.5951	105.94977	3.92406
4	0.24411	244.10820	14.1268	141.26636	5.23208
5	0.30514	305.13525	17.6585	176.58295	6.54010
6	0.36617	366.16230	21.1902	211.89954	7.84812
7	0.42720	427.18935	24.7219	247.21613	9.15614
8	0.48823	488.21640	28.2536	282.53272	10.46416
9	0.54926	549.24345	31.7853	317.84931	11.77218
10	0.61027	610.27050	35.3166	353.16590	13.08020

FRENCH AND ENGLISH WEIGHTS, ETC.

No. of units.	Grains to grammes.	Ounces avoir. to grammes.	Ounces troy to grammes.	Pounds avoir. to kilogrammes.	Pounds troy to kilogrammes.
1	0.064773	28.3375	31.0913	0.4534148	0.373096
2	0.129546	56.6750	62.1826	0.9068296	0.746192
3	0.194319	85.0125	93.2739	1.3602444	1.119288
4	0.259092	113.3500	124.3652	1.8136592	1.492384
5	0.323865	141.6871	155.4565	2.2670740	1.865480
6	0.388638	170.0250	186.5478	2.7204888	2.238576
7	0.453411	198.3625	217.6391	3.1739036	2.611672
8	0.518184	226.7000	248.7304	3.6273184	2.984768
9	0.582957	255.0375	279.8217	4.0807332	3.357864
10	0.647730	283.3750	310.9130	4.5341480	3.730960

No. of units.	Long tons to tonnes of 1000 kilog.	Pounds per square inch to kilogrammes per square centimetre.	Grammes to grains.	Grammes to ounces avoir.	Grammes to ounces troy.
1	1.015649	0.0702774	15.438395	0.0352889	0.0321633
2	2.031298	0.1405548	30.876790	0.0705778	0.0643266
3	3.046947	0.2108322	46.315185	0.1058667	0.0964899
4	4.062596	0.2811096	61.753580	0.1411556	0.1286532
5	5.078245	0.3513870	77.191975	0.1764445	0.1608165
6	6.093894	0.4216644	92.630370	0.2117334	0.1929798
7	7.109543	0.4919418	108.068765	0.2470223	0.2251431
8	8.125192	0.5622192	123.507160	0.2823112	0.2573064
9	9.140841	0.6324966	138.945555	0.3176001	0.2894697
10	10.156490	0.7027740	154.383950	0.3528890	0.3216330

No. of units.	Kilogrammes to pounds avoirdupois.	Kilogrammes to pounds troy.	Metric tonnes of 1000 kilog. to long tons of 2240 pounds.	Kilog. per square millimetre to pounds per square inch.	Kilog. per square centimetre to pounds per square inch.
1	2.205486	2.6803	0.9845919	1422.52	14.22526
2	4.410972	5.3606	1.9691838	2845.05	28.45052
3	6.616458	8.0409	2.9537757	4267.57	42.67578
4	8.821944	10.7212	3.9383676	5690.10	56.90104
5	11.027430	13.4015	4.9229595	7112.63	71.12630
6	13.232916	16.0818	5.9075514	8535.15	85.35156
7	15.438402	18.7621	6.8921433	9957.68	99.57682
8	17.643888	21.4424	7.8767352	11380.20	113.80208
9	19.849374	24.1227	8.8613271	12802.73	128.02734
10	22.054860	26.8030	9.8459190	14225.26	142.25260

HYDROMETERS AND THERMOMETERS.

An areometer is a convenient glass instrument for measuring the density or specific gravity of fluids. Areometer and hydrometer are synonymous terms, the first being derived from the Greek words αραιός, *rare*, and μετρον, *measure;* and the latter from ὕδωρ, *water*, and μετρον, *measure;* hence the same instrument is frequently denominated both hydrometer and areometer. This apparatus is often referred to throughout this work; for instance, in speaking of alcohol, or lye, their strength is stated as being of so many degrees (17° or 36°) Baumé, that is, its force or value is of that specific gravity, corresponding with the degree to which the hydrometer sinks in either the alcohol or alkaline solution. But, for those liquids lighter or rarer than water, viz., alcohol, ethers, etc., the scale is graduated differently than for the heavier or more dense, examples of which are the acids, saline solutions, syrups, and the like. There are several kinds of hydrometers; but that called Baumé's is the most used, and to this our remarks are applied.

They are blown out of a piece of slender glass tubing, and of the form shown by Figs. 124 and 125; A being the stem containing the

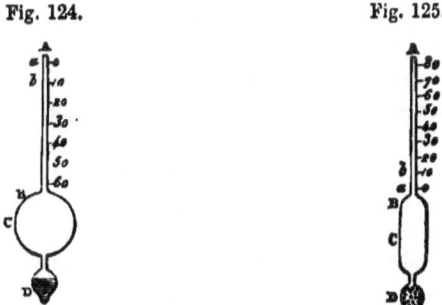

graduated paper scale, B the bulb portion, and D the small globes containing mercury or shot, serving as ballast to maintain the instrument in an upright position, when it is placed in a liquid.

The graduation is accomplished by plunging it into distilled water of 58° F., and weighting the globe with shot or mercury, until the instrument sinks to the line a, which is its zero point. This zero point thus determined is to be marked accurately upon the glass or its accompanying paper scale, and the instrument again plunged into ninety parts of distilled water, holding in solution ten parts of previously dried chloride of sodium or common salt. The point to which it sinks in this liquid, say b, for instance, is then also marked carefully upon the scale, and rated as ten compared with its zero point. The interval between these two points is then spaced off into ten equal divisions, according to which the remainder of the tube is graduated so that each degree is intended to represent a density corresponding to one per cent. of the salt.

The above mode of graduating refers to the hydrometer for liquids denser than water, but that for the liquids rarer than water is a little different from the preceding in form, and necessarily has a modified scale, which is graduated as is shown by Fig. 125. The instrument should be sufficiently heavy in ballast to sink in a saline solution of ten parts of dried chloride of sodium in ninety parts distilled water to the bottom of its stem a, to be marked as the zero of the scale.

Now, when it is again placed in distilled water alone, it floats or sinks to a point somewhere about b, which is to be the ten degree mark. The rest of the stem is then to be accurately divided into as many ten degree intervals as its length will permit, and each subdivision into ten uniform smaller degrees or intervals.

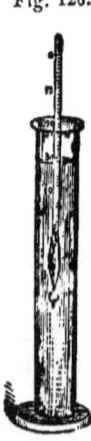

Fig. 126.

As it would be troublesome, and with many impracticable, to estimate the specific gravities of their liquids in a scientific way, these little instruments are a great convenience, for by taking out a portion of the fluid to be tested, and placing it in a glass cylinder, Fig. 126, its degree Baumé may be ascertained by noting the point to which a hydrometer sinks therein, and afterwards its specific gravity, by comparing that with its corresponding degree in the table. For instance, suppose the hydrometer sinks in alcohol to 35°, then its specific gravity is 0.8538, and this again can be translated into its absolute spirit strength by comparison with any accurately calculated alcohol tables. So, also, if a hydrometer for liquids denser than water sinks in lye to 26°, it denotes that the lye has a specific gravity of

1.2268. The presence of foreign matters will cause the hydrometer to give a false indication, and it is, therefore, necessary, when lyes contain impurities, to follow special directions, to ascertain their amount of caustic alkali. When the lye is nearly pure, they answer satisfactorily; and, indeed, under all circumstances, they serve very well for noting a progressive increase or diminution in the strength of lyes or other liquids. The temperature of the liquid should be 58° to 60° F., at the moment of testing it.

Thermometers.—The thermometer is an instrument made of glass exclusively, when intended for practical purposes. Fig. 127 shows one with the scale of Fahrenheit, graduated on the glass, so that, when having been dipped in liquids, it may be easily cleansed. It derives its name from two Greek words, θερμος, *warm*, and μετρον, *measure*, and is, as its title indicates, a measurer of the variation of temperature in bodies. The principle upon which it is constructed, "is the change of volume which takes place in bodies, when their temperature undergoes an alteration, or, in other words, upon their expansion." As it is necessary, in the construction of thermometers, that the material used to measure the change of temperature shall be of uniform expansion, and with a very distant interval between its freezing and boiling point, as fulfilling these requisites better than any other body, metallic mercury is generally used. There are several different thermometrical scales, all constructed upon the same principle, but varying in their graduation; the boiling and freezing points of each, though corresponding in fact, being represented by different numbers. The Fahrenheit scale is most used in this country; that of Celsius, called the Centigrade, in France and the Continent generally, except Spain and Germany, where Reaumur's scale is preferred.

Fig. 127.

The relation between the three scales is shown by Fig. 128. The Fahrenheit scale is most convenient, because of the lesser value of its divisions.

In the graduation of the scale, it is only necessary to have two fixed determinate temperatures, and for these the boiling and freezing points of water are universally chosen. The scales can be extended beyond either of these points, by continuing the graduation. Those degrees below zero or 0° have the minus (—) prefixed, to distinguish them from those above; thus, 55° F. means fifty-five degrees above zero, Fahrenheit's scale, and —9° C., nine

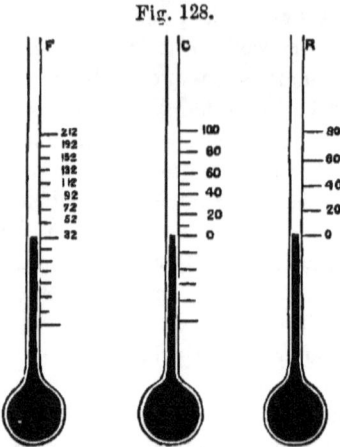

Fig. 128.

degrees below zero, Centigrade scale. The thermometers for general use very seldom, however, extend either way beyond the boiling and freezing points of water, but for manufacturers' use they are graduated sometimes to 400° or 600°.

Centigrade and Fahrenheit.—In the Fahrenheit thermometer the number 0° on the scale corresponds to the temperature of a mixture of salt and ice—the greatest degree of cold that could be artificially produced when the thermometer was originally introduced; 32° (freezing point) corresponds to the temperature of melting ice; and 212° to the temperature of pure boiling water— in both cases, under the ordinary atmospheric pressure of 14.7 pounds per square inch. Each division of the (this) thermometer represents 1° Fah., and between 32° and 212° there are 180°. In the Cent. thermometer, used universally in scientific investigations, 1° corresponds to melting ice, and 100° to boiling water. From the freezing to the boiling point there are 100°.

The accompanying table shows the relation of the Centigrade and Fahrenheit thermometer scales, 5° C. being equal to 9° F., because the interval between the freezing and boiling points of water is divided into 100 and 180 equal parts, and these numbers are respectively multiples of, or 20 times 5 and 9. If the superfluous 32° on the F. side were disposed of, the mutual translation of the scales would be simple, since the two units are to each other inversely as the number of them in any given range.

To reduce F. above melting ice to terms of C., 32° must first be subtracted from the given F. temperature, then multiply the remainder by $\frac{5}{9}$; the product will be the C. term for the given temperature; and conversely divide C. by $\frac{5}{9}$ and add 32 to translate C. into F.; to prove the work, read the terms across the diagram in the table. Below melting ice, the same rules as given above apply, except that where 32 is added above, it should be subtracted here, and *vice versa*.

In the columns at the right hand of each diagram in this table, are found the approximate steam pressures per square inch, due to the adjoining indications of temperature. The pressure is expressed in pounds and in atmospheres.

The high pressures are obtained from the several authors who have deduced and tabulated them from experiments and formulas of Regnault and others; and being hypothetical, accuracy is not claimed for them.

HYDROMETERS AND THERMOMETERS.

COMPARISON OF CENTIGRADE AND FAHRENHEIT SCALES, AND APPROXIMATE STEAM PRESSURE IN POUNDS AND ATMOSPHERES PER SQUARE INCH DUE TO THE TEMPERATURE.

Thermometer		Steam. Non-Condensing Engine.			Thermometer		Steam. Condensing Engine.		
Centi.	Fahr.	Press. per gauge. lbs.	Total Press. Lbs.	Atmos.	Centi.	Fahr.	Press. per gauge.	Back Press. Lbs.	Atmos.
260	500	665	680	46.	100	212	0	14.7	1.
255	491	610	625	42.	95	203	Vacuum, effective. Gauge. Lbs.*	Lbs. 12.	Atmos. 0.85
250	482	560	575	39.	90	194	9½ 4.7	10.	0.7
245	473	515	530	36.	85	185	12¼ 6.2	8.5	0.6
240	464	472	487	33.	80	176	15¼ 7.7	7.	0.5
235	455	430	445	30.	75	167	18¼ 9.1	5.6	0.4
230	446	390	405	27.5	70	158	20½ 10.2	4.5	0.3
225	437	354	369	25.	65	149	22 11.	3.7	
220	428	321	336	23.	60	140	24 11.9	2.8	0.2
215	419	290	305	20.7	55	131	25 12.4	2.3	
210	410	262	277	18.8	50	122	26 12.9	1.3	0.2
205	401	235	250	17.	45	113	26½ 13.3	1.4	
200	392	211	226	15.3	40	104	27¾ 13.6	1.1	0.1
195	383	188	203	13.8	35	95	28¼ 13.8	.9	
190	374	167	182	12.4	30	86	29 13.9	.8	
185	365	148	163	11.1	25	77	*To be added to the pressure indicated by steam gauge to get total pressure on piston.		
180	356	131	146	9.9	20	68			
175	347	115	130	8.8	15	59	M. T. Mines of Brittany, 500 ft., . . . 59 F.		
170	338	100	115	7.8	10	50	Hydr'chloric Ether boils, 52 F.		
165	329	85	100	6.8	5	41	Max. density of water, 39.2 F. −4 C.		
160	320	73	88	6.	0	32	Melting Ice, . 32 F. −0 C.		
155	311	63	78	5.3	−5	23	Blood freezes, . . . 25 F. Castor Oil freezes, . 21 F.		
150	302	55	70	4.7	−10	14	Spirits of turpentine freezes, . . 14 F.		
145	293	45	60	4.1	−15	5			
140	284	37	52	3.5	−20	−4	Brandy freezes, . −7 F.		
135	275	30	45	3.	−25	−13			
130	266	25	40	2.7	−30	−22			
125	257	19	34	2.3	−35	−31			
120	248	14	29	1.9	−40	−40	Mercury freezes, . −40 F.		
115	239	10	25	1.6	−45	−49	Sulphuric Acid (1.641) freezes, . . −45 F.		
110	230	6	21	1.4			Greatest artificial cold, . −166 to −220 F.		
105	221	3	18	1.2			Absolute cold, . $\begin{cases} -459.4\ F. \\ -273.\ C. \end{cases}$		

INDEX.

ABSTRICH, 90
   Acid copper solutions in electrolytic assays of nickel, 196
   solvent agents, 77
Acids for wet assays, 80
Air baths, 24
   excluding fluxes, 79
   ignition with admission of, 31
   exclusion of, 31
Alibegoff on assay of uranium, 256, 257
Allen on examination of New Caledonia nickel ores, 198, 199
Alloys, gold, free from copper, preliminary test for, 168, 169
   of copper, assay, 104
   of gold and silver, with or without copper, 168–181
      assaying, 167–184
      with copper, separated by cupellation, 181
      with silver and copper, quantity of lead required, 169, 170
   of lead, wet assay, 92–95
   of platinum, 186, 187
   electrolytic assay of, 187
   sampling of, 22, 23
   weighing, 29
Almaden, mercury assays at, 242
Amalgam, silver, dry assay of, 143
American assay weights, 70
Ammoniacal nickel solutions, in electrolytic assays of nickel, 196
Ammonium carbonate, 79
Anglesite, 81, 91
Antimonial nickel, 187
   ore, 188
   silver, 128
Antimonium crudum, liquation process for determining, 243
Antimony, 243–250
   assay of, by precipitation, 243, 244
   of lead ore containing, 88

Antimony—
   Becker's method of assaying, 245, 246
   detection of tin in presence of, 232
   determination of, in antimony sulphide, 243
   electrolytic determination of, 249, 250
   fire assays of, 243, 244
   gravimetric assay, 245, 246
   in copper, removal of, 105
   in nickel compounds, removal of, 194
   ores, 243
   oxide, 243
      as a flux, 79
   oxysulphide, 243
   prevents copper from slagging, 99
   roasting and reducing assay of, 244
   sulphide, 243
      inaccuracy of the assays of, 243
   volumetric estimation of, in the presence of tin, 249
   Weil's method of determining, 246–249
   wet assays, 245–250
Apothecary balance, 69
Argentiferous gold, boiling, in nitric acid, 174, 175
   roll assay for, 172–177
   lead, cupellation of, 138–141
Arsenates of copper, 96
Arsenic, 251–255
   acid, reduction of to arsenious acid, 254
   as a flux, 79
   fire assays of, 251, 252
   gravimetric assays of, 252, 253
   in copper, determination of, 127
   in copper, removal of, 105
   in glass, 255
   native, 251

INDEX.

Arsenic—
  ores of, 251
  Pearce's method of determination of, 254, 255
  prevents copper from slagging, 99
  volumetric assays of, 254, 255
  wet assay, 252, 253
  wet assays of, 252-255
  wet method combined with the dry, 253
Arsenical iron in nickel ores, slagging off, 191
Arsenious acid, 251
  estimation of, 254
Arsenizing nickel ores, 188, 190
Assay furnaces, 45-63
  for charcoal, illustrated and described, 49
  liquid, measuring and titration of the, 42-44
  of lead with potassium cyanide in clay crucibles, 84, 85
  of platiniferous ores, 185, 186
  reagents, 74-80
  vessels, 63-69
    for the dry method, 63-68
    for the wet method, 68, 69
Assaying, object of the art of, 17, 18
  various methods employed, 17, 18
Assays of lead in the dry way, 81-92
Atacamite, 96
Atomic weights, 297
Augendrée's process of cupelling gold, 179, 180
Auriferous bismuth, 167
  iron, 168
  lead, 167
    cupellation of, 166, 167
  pyrites in gold tailings, 165
  silver, gold determined in, by Volhard's assay, 181
    grains, separation of, from samples of ores, 181
    pulverulent assay of, 180-184
  steel, 168
Austrian assay weights, 69
Azurite, 96

BALANCE, manipulation of the, 28, 29
Balances and weights, 69, 70
Balling's method of quartation of gold with cadmium, 181, 182

Barium chloride, 80
Bases and salts for wet assays, 80
Basic solvent agents, 77, 78
Becker's method of assaying antimony, 245, 246
  improved by Donath, 245
Beilstein and Jawein on the determination of cadmium, 223
  electrolytic assay of zinc, 210, 211
Belani's method for manganese, 275
Belgium, assay of lead matt in, 84
Berlin School of Mines, furnaces used in the, illustrated and described, 53-55
Berthier on the reducing power of various agents, 75, 76
  method for the dry assay of iron ores, 310
  of determining absolute heating power of fuel, 288-290
Berzelius's method of assaying tungsten, 259
Bischof, assay of lead by, 94
Bismuth, 233-236
  auriferous, 167
  determination of percentage of lead and silver in, 236
  electrolytic precipitation of, by Smith and Knerr, 236
  fire assays of, 234, 235
  glance, 233
  in copper, removal of, 105
  in silver alloy, treatment of, 151
  native, 234
  ochre, 233
  ores, 233
    and compounds free from sulphur, 234
    Joachimsthal process for, 234, 235
    Rose's process for, 235
    sulphurized, 234, 235
    Tamm's process for, 234, 235
    Ullgreen's process for, 236
  oxide, 205
  separation of, from lead, 235, 236
  wet assays of, 235, 236
Bismuthic cupel ash, 234
Bitumen in copper, removal of, 105
Black flux, 74, 75, 77
  and metallic iron, assay of galena with, 85-87

Blast furnaces, 59-61
   lamp, best form of, illustrated and described, 39, 40
Bleiberg, Carinthia, assay of lead matt in, 84
   gravimetric assays of lead in, 92, 93
Blowpipe, refining copper ore with the, 103
   use of, in assaying, 18
Blue vitriol, 96
Bockmann's method of determining sulphur in pyrite's waste, 281
Bock's experiments on gold alloys, 179
Bodewig's method for determining sulphur, 280, 281
Boiling argentiferous gold in nitric acid, 174, 175
Bone ash, making cupels of, 66
Borax glass, 77
   refining copper on the dish with, 99, 100
   use of, in scorification assay of silver, 130
Bournonite, 95, 96
Braunite, 264
Breithauptite, 187
Bromyrite, 128
Bronze, assay of, 119
Brown, Walter L., charge for gold ores containing tellurides, 166
Brunnlechner's modification of Schaffner's assay of zinc, 303-307
Brussels, mint of, assaying gold coins at, 179, 180
Bullion or button balance, 69
Bullion, silver, assays of, 155
Bunsen's method for manganese, 270-272
Bunte's burette for the examination of furnace gases, 294, 295
Burettes, illustrated, 43, 44
Button, weighing the, 29

CADMIUM, 222, 223
   and zinc, separation of, 222
   determined in the same way as zinc, 223
   electrolytic assay of, 222, 223
   in an ammoniacal solution, precipitation of, 222
   in neutral acid solutions, precipitation of, 222

Cadmium—
   in solutions of ammonium or potassium double oxalate, precipitation of, 223
   ores, 222
   quartation of gold with, 181, 182
   sulphide, precipitation of, from an acid solution, 222
Calamine, 207
Calcining vessels, 64
Calcium carbonate, 77
   fluoride, 77, 78
Canby's simplification of Pearce's method of determining arsenic, 255
Carbon, desulphuration with, 78
   in fuels, yield of, 285, 286
Cassiterite, 223
Caustic alkalies, 77
   and carbonates, 78
   carbonates, 78
Centner, assay, 69
Cerusite, 81, 90, 91*
Chamotte, 63
Charcoal and coke furnaces, 49
   and graphite, 79
Charging the sample, 30, 31
Chemical operations, 31-45
   classification of the, 31
Chlorination process, Plattner's, 167
Chrome iron ore, 261
   Calvert's, Britton's, Dittmar's, Hager's, Fels's, and Clarke's methods for decomposing, 263
Chromium, 261-264
   direct assay of, 262
   gravimetric assays of, 262, 263
   indirect assay of, 263
   ores of, 261
   volumetric assay of, 263, 264
   wet assays of, 261-264
Cinnabar, 236
   assay of, 241
   treatment with aqua regia, 241
Clamond's thermo-electric battery, 111
Clark on assay of cadmium, 222
Classen on electrolysis of tin, 233
   on electrolytic assay of silver, 160
   of platinum, 187
   of zinc, 213
   determination of mercury, 242
   on precipitation of cadmium, 223

INDEX.

Classen's method for the electrolytic determination of metals, 114–116
Clay vessels, 63–65
Coal, determination of coking quality of, 286
    of copper in ash of, 286, 287
    of sulphur in, 287
    of volatile products of, 286
Cobalt, 204–207
    and nickel, difficult to separate from copper, 99
    in nickel ores, Donath's assay of, 203, 204
    separation of in nickel ores, 201
    tints produced by, 156
    arsenide in nickel ores, slagging off, 191
    arsenide, slagging off, 191
    assays of, 204–207
    beauty of the colors of, 204
    bloom, 204
    color, assay to determine the quality of, 206
    intensity of, 206, 207
    quality of, 206
    determination of, 204
    of the blue coloring power in, 204
    dry assay, 204
    glace, 204
    in nickel ores, separation of, 203
    in speiss, electrolytic determination of, 199, 200
    metallic oxides in, 205
    object of the smalt assay of, 205
    ores, 204
        containing copper and nickel, roasting of, 205
        entirely pure, roasting of, 205
        impure, roasting, 205
        roasting of, 205, 206
    pyrites, 204
    smalt assay, 204–206
    wet assay, 204
Cobaltine, 204
Cobaltous oxide, 204, 205
Cobenzl's method of assaying tungsten, 260
Coins, mint assay of, 144
    nickel, assay of, 117
        determining the copper in, 201, 202
    samples for producing, 23
    sampling, 22

Coke and charcoal furnaces, 49
    anthracite, and graphite, substitutes for charcoal, 74
Coke cupels, 68
Colorimetric analysis, 17
    assays by, 44, 45
    vessels for, 68, 69
    assays, when employed, 18
    of copper, 104, 125–128
Combustion furnace, illustrated and described, 238
Comparison of Centigrade and Fahrenheit scales and approximate steam pressure in pounds per square inch due to the temperature, 334
Concentrating fluxes, 78, 79
Copper, 95–128
    alloyed with tin, assay of, 119
    alloys of, assay, 104
    and silver in one solution, determination of silver in, 157
    arsenates of, 96
    as recently precipitated sulphide, treatment of, 116
    assay, Cornish, 96, 104
    of, by Haen, 125
    of, with protochloride of tin, 124, 125
    of, with sodium sulphide in an ammoniacal solution, 124
    of, with sulpho-cyanide, 118
    Classen's electrolytic determination of, 115, 116
    colorimetric assays of, 104, 125–128
    determination of arsenic in, 127
    of phosphorus in, 127
    difficult to separate nickel and cobalt from, 99
    dry assays of, 96–104
    electrolytic assays of, 110–116
    Fleitman's assay of, with ferric chloride, 122, 123
    glance, 95
    gravimetric assays, 104–119
    method for the determination of (Genth's), 109, 110
    Heine's assay for, 125, 126
    Herpin's method of testing, 113
    Jaquelin Hubert's assay, for considerable percentages of, 126, 127
    impure (black) precipitated, examination of, 105

Copper—
  indications of a successful assay, 98
  in nickel coins, solution for determining, 201, 202
  in speiss, electrolytic determination of, 199, 200
  in the form of cuprous sulphide, determination of, 116, 117
  lead, and zinc in one solution, determination of, 221
  Lower Harz working assay of, 302
  nickel, 187
  ore, American charge, 98, 99
    assay with lead and borax (Müsen assay), 101, 102
    containing much iron, cobalt, or nickel, treatment of, 101
    Müsen assay of, 101, 102
    not containing lead, antimony, nor zinc, assay of, 101
    refining, 99
      by cupellation, 102, 103
      with the blowpipe, 103
  ores, 95, 96
    charges of, 98, 99
    containing bitumen, treatment of, 105
    differences allowed in assays of, 101
    oxidized, without sulphur, treatment of, 103, 104
    reducing and solvent fusion, 97
    roasting, 97
    very poor, fusing of, in larger quantities, 103
    with antimony or zinc, 95
    with sulphur, antimony, or arsenic, treatment of, 97–103
  oxide as a flux, 79
  Parkes's assay, with potassium cyanide, 120–122
  phosphates of, 96
  poor in silver, dry assay of, 143, 144
  precipitated, correction for iron in, 107
    with iron or zinc, 105
  precipitation of, by the galvanic current, 110, 192, 193
    with iron, 106–108

Copper, precipitation—
  with zinc free from lead and arsenic, 108, 109
  prevention of slagging, 99
  pure precipitated, color of, 107
  pyrites, 95, 276
    treatment of, 97
  red, indication of arsenic, antimony, or selenium in, 112
  reducing with potassium cyanide, 119
  refined, Hampe's method of testing, 113, 114
  refining by itself, without borax and lead, 100, 101
    on the dish with borax, 99, 100
  removal of tin and zinc from, 99
  slag, treatment of, 99
  Steinbeck's method of assaying, 122
  sulphide, 130
  Swedish assay for, 105
  testing for, in the ore residue, 107
  various reducing processes, 119
  volumetric assays of, 104, 119–125
  Weil's method for reducing, 119
  wet assays of, 104–128
Cornish assay of tin, 227
  less accurate than the German, 224
  copper assay, 96, 104
Cornwall, determination of tin in, 223, 224
Crocoisite, 81, 90, 261
Cross method of sampling, 19, 20
Crucible assay for gold ores, 162, 163
  use of, 129
Crucibles for assay of lead, 65, 66
  for dry assay of iron ores, 311
  for lead and copper smelting, illustrated, 65
  for smelting iron, 65
  illustrated, 64, 65
Cupel, loss of silver by absorption of, 146, 147
Cupels, coke, 68
  mould for making, 66
  of bone-ash, 66–68
  simple machine for making, illustrated, 67, 68
Cupellation of the auriferous lead, 166, 167
  preliminary assay of cupriferous alloys by, 169, 170

Cupellation—
  refining copper by, 102, 103
Cupric oxide, 205
  sulphate, 96
    decomposition of, 103
Cupriferous alloys, preliminary assay
  of, by cupellation, 169, 170
  bismuth, 233
  compounds of nickel ores, assays
    of, 191–194
  wet assay for, 192
  rich in copper, assay for, 192
  with small percentage of copper, assay for, 192
  schists, treatment of, 105
  silver, dry assay of, 144–147
Cuprite, 96
Cuprous oxide, 205
  schist, treatment of, 109
  sulphate with calcium and barium sulphates, decomposition of, 103, 104
  sulphide, determination of copper in the form of, 116, 117

DEARSENIZING nickel ores, 191
  Debray's apparatus for precipitation, illustrated and described, 38
Decantation, 39
Decomposing and volatilizing fluxes, 79
Decrepitated common salt, 79
Desiccated mass, pulverizing the, 25
Desiccator, illustrated and described, 40
Desulphuration with carbon, 78
Desulphurizing agents, 78
Determination of moisture, 24
Deutocom's method for determining sulphur, 280
Deville's furnace, 59
D'Hennin's process of separating iridium from gold, 178
Dioptase, 96
Dipping, sampling by, 23
Distillation and sublimation, 35
  furnace, illustrated and described, 61–63
Docimacy, 17
Donath's assay of nickel and cobalt, 203
  improvement of Becker's method of assaying antimony, 245
Draught or wind furnaces, 51–59

Dropping the ore, sampling by, 20, 21
Dross, lead, 91
  silver, 138
Drown's method of determining sulphur in coal, 290, 291
Dry assay, 17
  assays, reagents for, 74–79
  method, working by the, 31–35
Drying disk, illustrated, 24, 25
  precipitates, 39
Ducktown, Tennessee, assaying copper at the copper works of, 122

EARTHY cobalt, 204
  Electrolysis, 17
    separation of gold from platinum by, 183, 184
    assay of nickel, 194–200
      of zinc by Beilstein and Jawein, 210, 211
    assays of manganese, 275, 276
    determination of antimony, 249, 250
      of copper, nickel, and cobalt in speiss, 199, 200
      of mercury, 242
      of metals (Classen's), 114–116
      of silver, 159, 160
      of tin, 232, 233
    processes, assays of lead by, 94, 95
Enargite, 96
England, assay of galena in, 86, 87
  of lead matt in, 83, 84
English assay weights, 69, 70
  commission on errors in assay of coins, 177
Erhard and Schertel's determinations of the fusing points of metals, alloys, etc., 300, 301
Erubescite, 95
Eschka's assay of mercury, 239–241
Escosura on the electrolytic determination of mercury, 242
Evaporation of the solution, 37

FAHLERZ, 95
  Ferric chloride, 80
  Fleitman's assay of copper with, 122, 123
  hydrate, retention of copper by, 126
  oxide, 78, 205

INDEX. 341

Ferrous ammonium sulphate solution, preparation of, 183
   oxide, 205
      prevents copper from slagging, 99
Fikentscher-Nolte method for assaying manganese, 268, 269
Filtering apparatus, illustrated, 38
Filtration, 38
Final reaction, determining the, 40, 41
Fire assays for silver, 129–142
Flasks, measuring, illustrated, 42
Fleitman's assay of copper with ferric chloride, 122, 123
Fletcher's direct-draft crucible furnace, illustrated and described, 57, 58
   injection gas furnace, illustrated and described, 60, 61
Flintshire, assay of lead in, 84
Fluor spar, 77, 78
Fluxes, 78, 79
   for the dry assay of iron ores, 313, 314
   measuring of, 29, 30
   weighing, 29
Fragments, homogeneous, sampling of, 19
   heterogeneous, sampling of, 19–22
Franklinite, 207
Frederickshutte, assay of lead ore in, 84
Freiberg, assay of galena at, 86
   results obtained at, regarding absorption by the cupel, 147
French and English weights and measures, tables of relative value of, 321–328
   Commission on Coinage and Metals, correction table of, for absorption by the cupel, 147
Fresenius on the sources of error in processes for detecting sulphur in pyrites, 280
Fresenius's drying disk, illustrated, 24, 25
Fresenius-Will method for assaying manganese, 267, 268
Fuel, amount of ash in different kinds of, 287
   assays of, 285–291
   determination of absolute heating power, 287
   of ash in, 286
   of heating power of, 287, 288

Fuel, determination—
   of hygroscopic water in, 285
   removal of earthy admixtures from, 284
   yield of carbon from, 285, 286
Fuels, 284–291
   composition of, 284
   physical and chemical behavior of, 291
Furnace, Brown's gas, illustrated and described, 58, 59
   distillation, illustrated and described, 61–63
   Fletcher's direct-draft crucible, illustrated and described, 57, 58
   injector, gas, 60, 61
   for melting with a coke fire, illustrated and described, 57
   gases, examination of, 291–295
   Perrot's gas muffle, illustrated and described, 56, 51
   illustrated and described, 54, 55
   Roessler's, for production of high temperatures, illustrated and described, 56, 57
   sublimation, illustrated and described, 61
   taking the vessels from the, 52, 53
   tools, 70
   Wiessnegg's gas, 56
Furnaces, assay, 45–63
   blast, 59–61
   charcoal and coke, 49
   draught or wind, 51–59
   firing, 52
   for free-burning coal, 55
   for illuminating gas, 55
   for solid, free-burning, flaming fuel, 46–49
   for sublimation and distillation, 61–63
   labor attending, 52
   muffle, 45–51
   oil, of Andouin-Deville, 50
   organic combustion, 63
   used in the Berlin school of mines, illustrated and described, 53–55
Fusing points of metals, alloys, furnace products, rocks, and silicates, according to Erhard and Schertel, 300, 301

Fusing points—
  of metals and furnace products, glowing temperatures, 298, 299
Fusion, 34, 35
  liquating, 35
  mixing, 35
  oxidizing, 34
  precipitating, 35
  purifying, 35
  reducing, 34, 35

GALENA, 81
  assay in an iron pot, 82, 83
  Belgian assay of, 82, 83
  containing large quantities of earths, assay of, 88
  rich, assay of, 82–85
  with more earths, assay of, 85–88
  with zinc blende, pyrites, etc., 88
  etc., without foreign metallic sulphides, 81
Galetti on the effect of manganese-zinc, 219
  titrating solution, 219
  volumetric assay of copper, 119
Garnierite, 188
Gas furnaces, 50, 51
Gay Lussac's apparatus, illustrated and described, 150, 151
  assay of silver bullion, 155
  tables for determining the fineness of silver alloys, 152–154
Genth, Dr., gravimetric method for the determination of copper, 109, 110
Gerlach's method for assaying sulphur earths, 277
German assay of tin, 224
  more accurate than the English or Cornish, 224
  copper assay, 96–104
  smelting works, sampling in, 21
Gersdorffite, 188
Glass, 79
  containing arsenic, 255
Glaucodot, 204
Gold, 161–184
  alloys, Bock's experiments on, 179
  determination by color, 168, 169

Gold, alloys—
  effects produced by various metals in cupelling, 179, 180
  examination of, on the touchstone, 169
  free from copper, preliminary test of, 168, 169
  preparation of the ferrous ammonium sulphate solution, 183
  amalgam, 167, 168
  and silver alloys, with or without copper, 168–181
  sweepings, crucible assay of, 135–138
  with platinum, cupellation of, 186, 187
  as a flux, 79
  brittle, toughening of, 184
  button, laminating the, 173, 174
  coins, standards of, in various countries, 171
  etc., appreciable quantity of the platinum group, metals found in, 180
  cupelling of, to obtain best results, 178, 179
  fire or fusion assays of, 162–167
  Jüptner's volumetric assay of, 182, 183
  loss of, in cupelling, 176
  Lower Harz, working assay of, 303
  mechanical assay by washing, 161, 162
  native, 161
  non-alloys, assay of, 161–167
  ores, 161
    American, assay of, 163
    assay with nothing but litharge as a flux, 163, 164
    containing tellurides, assay of, 165, 166
    crucible assay for, 162, 163
    scorification assay for, 162
    wet assay, 167
    with earths and oxides, 163, 164
    etc., with sulphur, antimony, or arsenite, 164–166
  ore tailings, 165
  pure, preparation of, 171, 172
  pyrites poor in, 164, 165
  quartz, 163

Gold—
 separation of, from platinum by electrolysis, 183, 184
 of iridium from, by fusion, 178
 slag, 163
 smelting with lead, 162–166
 superheating of, 178, 179
 sweepings, 163
 with cadmium, quartation of, 181, 182
 with platinum, cupellation of, 180, 186
Goldsmith's sweepings, sampling, 20, 21
Gorz, results obtained by, in assaying sweepings and other refuse from the manufacture of silverware, 137, 138
Gramme weight, 69
Granulation, sampling by, 23
Graphite and charcoal, 79
 crucibles, 65
Gravimetric analysis, 17
 assay of lead by, 92–94
 assays by, 35–40
 of silver alloys, 158, 159
 vessels for, 68
 assays for copper, 104–119
Greenockite, 222

HAEN'S assay of copper, 125
 Hampe's method for manganese, 274, 275
 of testing refined copper, 113, 114
Hart on the volumetric assay of tin, 231
Hausmannite, 264
Heating power of fuels, determination of, 287, 288
Heat-regulator at United States Assay Office, New York, experiments on, 307–309
Heine's assay for poor copper ores, etc., 125, 126
 colorimetric assay for products poor in copper, 302
Herpin's method of testing copper, 113
Hessite, 128
Heterogeneous fragments, sampling of, 19–22
Homogeneous fragments, sampling of, 19
Horn silver, 128

Hungarian mercurial tetrahedrite, 237, 238
 smelting works, assay of gold at, 165
 speiss assays of copper, 100–101
Hungary, assay of lead in, 88, 89
Hydrogen, apparatus for igniting in a current of, illustrated and described, 117, 118
Hydrometers and thermometers, 329–333
Hydrostatic assay of silver coins, 160
Hygroscopic substances, convenient weighing of, 29
 water in fuels, 285

IDRIA, mercury assay at, 237
 Igniting precipitates, 39, 40
Ihle and Rheinhardt on zinc solution for electrolysis of zinc, 212
Implements and tools, 70–74
Insoluble mercury compounds, analyzing of, 242
Iodide of potassium and starch, assay of silver alloy by, 157, 158
Iodine, determination of tin by means of, 230, 231
Iodyrette, 128
Iridium, action of, on the auriferous silver buttons, 177
 separation of, from gold, by D'Hennin's process of fusion, 178
Iridosmium in gold, 180
Iron, auriferous, 168
 filings, 78
 in nickel ores, separation of, 201
 Lower Harz working assay of, 302
 ores, assay of in large, unlined crucibles, 314
 dry assay of, 310–315
 precipitation of copper with, 106–108
 pyrites, 276, 277
 as a collecting agent, 79

JAQUELIN-HUBERT'S method for considerable percentages of copper, 126, 127
Jüptner's fusion of gold and silver alloys, 181
 volumetric assay of gold, 182, 183

KANDELHARD on the loss of gold in cupelling, 176
Kandelhard's experiments on boiling argentiferous gold in nitric acid, 175
Karmarsch on the hydrostatic assay of coins, 160
Kieselmalachit, 96
Kiliani on electrolytic assay of lead, 94, 95
   on electrolytic assays of silver, 160
Kipp's apparatus for precipitation, illustrated and described, 37
Klausenburg's mining district, differences allowed in assays of lead, 89
Krauss's process of quartation, 182
Küstel on roasting gold ores containing tellurides, 165
Küstel's assay of mercury, 241

LAMINATING the button of gold, 173, 174
Lang's blast furnace, 60
Lead, 81–95
   alloys of, wet assay, 92–95
      and silver, addition of, in gold assays, 170, 171
      assays, combined, 138
   argentiferous, cupellation of, 138–141
   arsenate, 90
   as a flux, 78
   assay by colorimetric process (Bischof), 94
      of, by gravimetric analysis, 92–94
      of, with potassium in clay crucibles, 84–85
      of, with sulphuric acid, 89–90
   assays of, by electrolytic processes, 94, 95
      of, by volumetric processes, 94
      of, in the dry way, 81–92
         inaccuracy of, 81
   auriferous, 167
      cupellation of, 166, 167
   bullion, dry assay of, 143
   carbonate, 81, 90
   chromate, 81, 90, 91
   collecting silver with, 129–138
   copper, and zinc in one solution, determination of, 221
   dross, 91

Lead—
   estimation of sulphur in, 284
   fume, 91
   in tin, determination of, 232
   Lower Harz working assay of, 301, 302
   matt, assaying of, in different countries, 83, 84
   Mohr's process of assay, 93
   molybdate of, 81
   monosulphide with foreign metallic sulphides, 88
   ore containing antimony, assay of, 88
   ores, 81
      examination of, for traces of silver, 141, 142
   oxide, 78, 205
   salts of, 90–92
   oxides free from earths, assaying, 90
      with earths, assaying, 90
   oxidizable nature of, 99
   phosphate, 90, 91
   quantity of, required for alloys of gold with silver and copper, 169, 170
   separation of bismuth from, 235, 236
   silicate (slags), 91, 92
   skimmings, 91
   sulphate, 81, 91
      in copper, removal of, 105
   sulphide, 130
   sweepings, 91
   table of multiples of, in assaying silver alloys, 144
   tailings, 91
   various assays of, 93
Lenssen on the volumetric assay of tin, 231
Leucopyrite, 251
Levol on lead assay, 85
Levol's assay of tin with potassium cyanide, 227, 228
   gas heating apparatus, 175
   method for manganese with iron, 272, 273
Linnaeite, 204
Liquating fusion, 35
Liquation, 35
   process for determining antimonium crudum, 243
Litharge, 76, 90, 138
   assay of gold ore with nothing but, as a flux, 163, 164

Litharge—
    entirely free from silver, preparation of, 76
Litmus tincture, 80
Löwe, assay of lead by, 93
Lower Harz, assay of lead in, 88
    determination of silver, 159
    working assays, 301–303
Luckow on electrolytic assay of zinc, 213
Lunge's method of determining sulphur in pyrites, 279, 280
    sources of error in, 280

MAGNETIC iron pyrites, 276
    pyrites, 188
Malachite, 96
Manganese, 264–276
    Belani's method, 275
    Bunsen's method with iodine, 270–272
    carbide preparation of, from pyrolusite, 264
    electrolytic assays of, 275, 276
    gravimetric assays of, 266–269
    Hampe's method, 274, 275
    injurious influence of, on zinc assay with potassium cyanide, 219
    in nickel ores, treatment of, 203
    Levol's method with iron, 272, 273
    method of Fikentscher-Nolte for, 268, 269
    method of Fresenius-Will for, 267, 268
    methods of Bunsen and Gay-Lussac, 269
    ores, 264
    Volhard's method, 273, 274
    volumetric assays of, 269–275
Manganic oxide, 205
Manganite, 264
Manganous oxide, 205
Manipulations, mechanical, 18–31
Margueritte's method of assaying tungsten, 259
Mariotte's bottle, illustrated, 26, 27
Mascazzini and Parodi on the electrolytic assay of zinc, 211
    process of electrolytic assay of lead, 94, 95
    assay of lead by, 93

Matt, determination of nickel in, 198
Matthey and Johnson's new platinum apparatus, 175
May on the electrolytic assay of lead, 95
Measuring and weighing samples, 28–30
    of fluxes, 29, 30
Mechanical assay of gold by washing, 161, 162
    manipulations, 18–31
Mechernich, assay of lead matt in, 84
Meidinger-Pinkus battery, 111
Mercurial tetrahedrites, 236
Mercury, 236–242
    assays of, at Almaden, 242
    assays yielding free, 237–239
    determination of, in combination with gold, 239–241
    difference allowed in assays of, 240
    electrolytic determination of, 242
    Eschka's assay of, 239–241
    fire assays of, 237–241
    gravimetric assay of, 241
    Idria, Austria, assay of, 240
    in silver alloy, treatment of, 151
    Küstel's assay of, 241
    native, 236
    ores, 236
    Teuber's test for, 240, 241
    volumetric assays of, 241, 242
    wet assays, 241, 242
Metallic sulphides, decomposition of, 78
    method for, 36, 37
    various, the quantities of litharge for their decomposition, 76
Metals for precipitation in wet assays, 80
    fusing points of, 298–300
Metric system, advantages of, in assaying, 70
    of weights and measures, 317, 318
Miargyrite, 128
Microcosmic salt, 77
Millerite, 188
Millot on electrolytic determination of zinc, 213
Mimetene, 90
Minium, 90
Mint cupels, 144, 145
Mints at Brussels and Utrecht, assay of gold coins at, 179, 180

Mispickel, 251
Mixing fusion, 35
  scoops, illustrated, 30, 31
Mohr's assay of lead, 93
  burette, illustrated, 44
  method of estimating arsenious acid, 254
Moisture, determination of, 24
Molybdate of lead, 81
Montana gold assay by washing, 161, 162
Moore on electrolytic assay of zinc, 212
Muffle furnaces, 45–51
  illustrated, 45
Munscheid's gas blast-furnace, 59

NAGYAGITE, 161
Native arsenic, 251
  copper, 95
  silver ores, assay of, 141
  sulphur, 276
Neutral atmosphere, ignition in a, 31
New Caledonia nickel ores, assay of, 198, 199
New York, assay of galena in, 87
Nickel, 187–204
  and cobalt, difficult to separate from copper, 99
    in nickel ores, Donath's assay of, 203, 204
    separation of, in nickel ores, 201
    tints produced by, 156
  arseniate, 188
  coins, assay of, 117, 118
    determining the copper in, 201, 202
  compounds containing antimony, treatment of, 194
    difficult of solution, treatment of, 193, 194
  free from copper, assay of, 188–191
  glance, 188
  in pyrites and matt, determination of, 198
  in speiss, electrolytic determination of, 199, 200
  ores, 187, 188
    arsenizing, 188, 189
    arsenizing and fusion in one operation, 190
    containing metallic sulphides, treatment of, 188

Nickel, ores—
  cupriferous compounds rich in copper, wet assay of, 192
    with small percentage of copper, 192
  dearsenizing, 191
  Donath's assay of nickel and cobalt in, 203, 204
  electrolytic assay of, 94–197
  free from sulphur and rich in arsenic, treatment of, 189
  gravimetric assay of, 194–204
  modifications which may occur in assaying, 189, 190
    which may occur in slagging off the arsenical iron, 191
  New Caledonia, assay of, 198, 199
  reducing and solvent fusion for collecting the metallic arsenides, 189
  separation of cobalt in, 203
    of iron in, 201
    of nickel and cobalt in, 201
  slagging off of the arsenical iron, 190, 191
    the cobalt arsenide, 191
  treatment of, in presence of different metals, 189, 190
    when zinc and manganese are present, 203
    with iron filings, 189
  various assays, 200–204
  volumetric assay of nickel and cobalt (Donath), 203, 204
    with sodium sulphide, 202, 203
  wet assay of, 194–204
    conditions giving inaccurate results, 202
Plattner's fire assay, 188
precipitation of, by zinc, 108
presence of zinc in, effect of, in electrolytic assays, 197
protoxide, 205
silicates, 188

Nickel—
  sulphide, 188
  Winkler's colorimetric assay for, 204
Nickeliferous copper, 188
  iron, 188
  pyrrhotine, 193
  solution from the assay with sulpho-cyanide for determining copper in nickel coins, 201, 202
Nitrate of silver, zinc assay with, 220, 221
Nitric acid, boiling argentiferous gold in, 174–176
Non-alloys, sampling, 18–22

OIL furnaces of Andouin-Deville, 50
  balance, 69
Organic combustion furnaces, 63
Orsat's apparatus for the examination of furnace gases, 291–294
Oxidized copper ores, 96
  ores, charges for, 91
  substances (lead), 90–92
Oxidizing agents, 76
  fusion, 34

PALLADIUM passing into solution from its alloy with gold, 178
Parkes's assay of copper with potassium cyanide, 120–122
Parodi and Mascazzini on the electrolytic assay of zinc, 211
  process of electrolytic assay of lead, 94, 95
Passaic zinc, purity of, 218
Patera's process for separating bismuth from lead, 235, 236
  technical test for uranium, 256
Pearce's method of determining arsenic, 254, 255
  simplified by Canby, 255
Pelouze's volumetric assay of copper, 119
Permanganate, determining the strength of the, 274
  solution, preparation of, 183
Perrot's furnace, illustrated and described, 54, 55
  gas-muffle furnace, illustrated and described, 50, 51
Phosphate of tin, treatment of, 127, 128

Phosphates of copper, 96
Phosphorus in copper, determination of, 127
Pipettes, measuring, illustrated, 42, 43
Pitch blende, 255
Platiniferous ores, assays of, 185, 186
  fire assays, 185
  gold in, 185
  platinum in, 185
  sand in, 185
  wet assay, 185, 186
Platinum, 184–187
  action of, on the auriferous silver button, 177
  alloys of, 186, 187
  electrolytic assay of, 187
  dish, illustrated, 113
  foil, 111
  native, 184, 185
  separation of, from iridium, 187
  of gold from, by electrolysis, 183, 184
  spiral, 111
Plattner's chlorination process for gold, 167
  fire assay of nickel, 188
  method for the determination of temperatures, 299
  muffle furnace, for coal, illustrated and described, 46–48
Polybasite, 128
Potassium carbonate, 77
  Upper Harz assay of lead with, 87, 88
  cyanide, 75, 78, 80
  assay of lead with, in clay crucibles, 84, 85
  Parkes's assay of copper with, 120–122
  preservation of the solution, 122
  standardizing, 121
  ferrocyanide, 75, 78
  zinc assay with, 218, 219
  iodide, 80
  permanganate, 80
  determination of tin by, 231, 232
  sulpho-cyanide, 80
Pourcel's method for chromium, 262
Precipitates, drying, 39
  igniting, 39, 40
Precipitating fusion, 35
  metals by electrolysis, 17
  or desulphurizing agents, 78

Precipitation assay of galena, 81, 82
   Debray's apparatus for, illustrated
     and described, 38
   Kipp's apparatus for, illustrated
     and described, 37
   of the solution, 37
   of the sample, 24–28
Pribram, assay of galena at, 86
Prinsep's principle for the determination of temperatures, 299
Protochloride of tin, assay of copper with, 124, 125
Psilomelane, 264
Pulverized substances, sampling of, 20, 21
Pulverizing the desiccated mass, 25
Pulverulent assay of auriferous silver, 180–184
   sample, weighing a, 29
Pure silver, preparation of, 78, 79
Purifying fusion, 35
Purple copper ore, 95
Pyrargyrite, 128
Pyrites and Matt, determination of nickel in, 198
   determination of sulphur in, 279
   poor in gold, 164, 165
   waste, determination of sulphur in, 281
Pyrolusite, 264
   assays of, 264–276
Pyromorphite, 81, 90, 91
   containing arsenic, 91
Pyrostilbite, 243

QUARTATION, advantages in assaying afforded by, 182
   of gold with cadmium, 181, 182

RASCHETTE'S furnace, 59
   Rammelsbergite, 187, 188
Rammelsberg smelting works, assay at, 88
Rasping, samples by, 19
Raw flux, 75
Reagents, assay, 74–80
   for decomposing, 31, 32
   for dry assays, 74–79
   for volumetric wet assays, 80
   for wet assays, 80
Realgar, assays of, 252
   (red orpiment), 152
Red copper, 96

Red—
   lead ore, 261
   orpiment (realgar), assay of, 252
Reducing agents, 74–76
   fusion, 34, 35
   power, estimation of the, of various agents, 75, 76
   processes for copper, various, 119
Refined slag, 79
Refining dishes, 64
Refractory ores, roasting, 33, 34
Refuse from silver stamping mills, crucible assay of, 136, 137
Remelting, 35
Reodanskite, 188
Rheinhardt and Ihle on zinc solution for electrolysis of zinc, 212
Rheinsand, 163
Rhodium, action of, on the auriferous silver buttons, 177
Riche, assay of lead by, 93
Richter's method of determining the coking quality of coal, 286
Ricketts on lead assay, 85
Roasting, 32, 33
   and reducing assay of galena, 88, 89, 94
   dishes, 32, 33, 64
Robert's process for the preparation of chemically pure gold, 171, 172
Roessler on loss of gold in cupelling, 176, 177
Roessler's small furnace for the production of high temperatures, illustrated and described, 56, 57
Roll assay, drying and annealing of the rolls, 176
   for argentiferous gold, 172–177
   washing the rolls, 175, 176
   weighing of the rolls, 176
Rynoso's assay for phosphorus in phosphor copper, 127

SALT, common, as a flux, 79
   of phosphorus, 77
Saltpetre, 76, 78
Salts of iron, nickel, and lead, 80
   of lead oxide, 90
Sample, charging the, 30, 31
   preparation of the, 24–28
   pulverulent, weighing a, 29
Samples by rasping, 19
   for producing coins, 23
   from the heap, 19

INDEX. 349

Samples—
  sifting, 25-27
  slag, 19
  taken while the ore, etc., is being weighed, 19
  washing, 27, 28
  weighing and measuring, 28-30
Sampling, 18-23
  alloys, 22, 23
  before weighing, 21
  by boring, 23
  by cutting, 22
  by dipping, 23
  by dropping the ore, 20, 21
  by granulation, 23
  by the cross method, 19, 20
  coins, 22
  goldsmith's sweepings, 20, 21
  in some German smelting works, 21
  of alloys, 22, 23
  small ore and pulverized substances, 20, 21
  substances in a state of fusion, 21, 22
    in fragments, 18-20
  while weighing, 20, 21
Saxon assay of tin, 223
Scale, rough, 69
Schaffner's assay of zinc as modified by Brunnlechner, 303-307
  volumetric assay of zinc with sodium sulphide, 214-218
Scheibler's steam apparatus, 24
Schemnitz, differences allowed in assays of lead, 88
Schober's volumetric assay for zinc, 221
Schucht on electrolytic assay of lead, 95
  on electrolytic assay of silver, 160
Schulze's washing apparatus, illustrated, 26, 27
Schwarz's volumetric assay of copper, 119
Scoops, mixing, illustrated, 30, 31
Scorification assay for gold ores of every kind, 162
  use of, 129
  vessels, 64
Sefström's furnace, 59, 60
Selenium in silver, determination of, 158, 159
Sheele's method of assaying tungsten, 259

Sifting samples, 25-27
Silver, 128-160
  alloy, assay of, by iodide of potassium and starch, 157, 158
    Volhard's assay with sulphocyanide, 155, 156
  alloys, assay by gravimetric analysis, 158, 159
  assays of, 143-160
  dry assays of, 143-147
  Gay Lussac's assay of, with sodium chloride, 148-154
  multiples of lead, used in assaying, 144
  preparation of the normal solutions for assays of, 152
  tables for determining the fineness of, 152-154
  volumetric assays of, 148-158
  wet assays for, 147-160
  amalgam, 128
  dry assay of, 143
  and copper in one solution, determination of silver in, 157
  and gold alloys, with or without copper, 168-181
  sweepings, crucible assay of, 135-138
  with platinum, cupellation of, 186, 187
  lead, addition of, in gold assays, 170, 171
  assay, combined, 138
  antimonial, 128
  as a flux, 78, 79
  auriferous, pulverulent assay of, 180-184
  Balling's volumetric assay for, 142, 143
  bromide, 128
  bullion, various assays of, 155
  buttons, action of platinum, rhodium, and iridium on, 177
  charges for crucible assays of, in various countries, 135-138
  for scorification assay in different localities, 133, 134
  chloride in copper, removal of, 105
  crucible assay of, 134-138
  cupriferous, assay of, 144-147
  glance, 128
  in coins, hydrostatic assay of, 160

INDEX.

Silver—
  electrolytic determination of, 159, 160
  examination of lead ores for traces of, 141, 142
  fire assays for, 129–142
  horn, 128
  in nitric acid, dissolving, 175
  iodide, 128
  Lower Harz, working assay of, 303
  mint assay of, 144
  native, 128
  non-alloys of, assays for, 128–143
  number of samples of, for scorification assay, 130
  ore, brittle, 128
  ores, native, assay of, 141
    principal, 128
  preliminary assay of, 144
  pure, preparation of, 78, 79
  quantity of, absorbed by the cupel, 146
  ruby, 128
  scorification assay of, 129–134
  selenium in, determination of, 158, 159
  table of charges for scorification assay, 130–132
  telluride (hessite), 128
  wet assays, 129, 142, 143
  with lead, collecting, 129–138
  with platinum, cupellation of, 186
Silverware, refuse from the manufacture of, assays of, 137, 138
Sire's assaying apparatus, illustrated and described, 149, 150
Skimmings, 90
  and dross, silver, 138
Slag samples, 19
Slags, lead, 91, 92
  silver, crucible assay of, 136
Small ore and pulverized substances, sampling of, 20, 21
Smalt assay of cobalt, 204–206
Smaltine, 204
Smith and Knerr on the electrolytic determination of mercury, 242
Smithsonite, 207
Soapstone crucibles, 65
Sodium carbonate, 77
  chloride, 79, 80
    Gay Lussac's assay of silver alloys with, 148–154
  hyposulphite, 80
  nitroprusside, 80

Sodium—
  sulphide, 80
    in an ammoniacal solution, assay of copper with, 124
    test of strength of, 124
    volumetric assay of nickel ore with, 202, 203
    of zinc with, 214–218
Solution, empirical, 41
  evaporation of, 37
  precipitation of the, 37
  preparation of the standard, 41, 42
Solutions, decinormal, 41
  normal, 41
Solvent agents, 77, 78
Speiss, electrolytic determination of copper, nickel, and cobalt in, 199, 200
Starch and iodide of potassium, use of, in assaying silver, 157, 158
  paste, 80
Steel, auriferous, 168
Steinbeck's method of assaying copper, 122
Stephanite, 128
Stibnite, 243
Stohmann's siphon pipette, 43
Storer, assay of lead by, 93, 94
Stromeyerite, 128
Sublimation and distillation, 35
  furnaces for, 61–63
  furnace, illustrated and described, 61
Substances in a state of fusion, sampling, 21, 22
  in fragments, sampling, 18–20
  lead, oxidized, 90–92
Sulphide of iron, 130
  of zinc, 130
Sulphides, metallic, method for, 36, 37
Sulphocyanide, assay of copper with, 118
  assay of silver alloy with, 155, 156
Sulphur, 276–284
  amount of, in coal or its ash, 287
  assays by distillation, 277
  for the determination of the quantity of, in a substance, 277–284
  Bodewig's method for, 280, 281
  determination of, in pyrites, pyritous ores, etc., 279

INDEX.   351

Sulphur, determination of, in—
    waste, 281
  Deutocom's method for, 280
  dry assay, Hungarian method, 278
    (raw matt assay) of, 277, 278
  earths, 277
  estimation of in metallic lead, 284
  gravimetric assays, 279–281
    determination of, in pyrites, according to Lunge, 279, 280
    Lower Harz working assay of, 303
    native, 276
    ores, 276
    process for determining small quantities of, in materials and products of the iron works at Creuzot, 279
    volumetric assay, Wildenstein's method, 282, 283
    assays, 282–284
    determination of, in ores, 283
  wet assays, 279–284
  working test for determining residue of, in roasting charge in Silesian zinc works, 281
Sulphuretted copper ores, 95, 96
Sulphuric acid, accuracy of assay of lead with, 89, 90
  assay of lead with, 89, 90
Sulphurized nickel and cobalt ores, 130
  substances, 81–90
Sulphurizing agents, 78
Swedish assay, 37
  for copper, 105
Sweepings and other refuse from manufactures of silverware, assaying of, 137, 138
  goldsmiths', samples of, 20, 21
Sylvanite, 161

TABLE for determining the fineness of silver alloys, 152–154
Tabular synopses, 297–301
Talbott's method of separating tungsten from tin, 260
Tarnowitz, assay of lead matt in, 83
Taylor, Charles, on heat regulator, U. S. Assay Office, N. Y., 309.

Tellerium, white, 161
Tellurides in gold ores, 165, 166
Tenny on the electrolytic assay of lead, 95
Tetradymite, 233
Tetrahedrite, 95
Teuber's test for mercury, 240, 241
Thermostats, 24
Tin, 223–233
  assays, accurate results obtained, 231
  comparative accuracy of German and English or Cornish assays of, 224
  Cornish assay of, 227
  determination of, by means of potassium permanganate, 231, 232
  of, by iodine, 230, 231
  of lead in, 232
  electrolytic determination of, 233, 234
  fire assays for, 224–229
  of object of, 224
  German assay of, 224
  gravimetric assays of, 228–239
  in copper, removal of, 105
  separation of, 99
  in presence of antimony, detection of, 232
  in separate grains, treatment of, 226, 227
  in silver alloy, treatment of, 151
  in tin slags, determination of, 229, 230
  Levol's assay of, with potassium cyanide, 227, 228
  losses of, by German and Cornish methods, 224
  ore, slagging off of, 226
  ores, 223
    with foreign metallic sulphides, arsenides, and antimonides, treatment of, 225, 226
    with many earthy admixtures, assay of, 225
  oxide, reducing, 224
  Saxon assay of, 223
  slags, treatment of, 229, 230
  volumetric assays, 230–232
  wet assay of, 228–233
Tincture of Brazil wood, 80
Tinstone, 223
  and metals in combination with, specific gravities of, 225

Tinstone—
    determination of, by washing, 223, 224
    digestion of, with aqua regia, 228, 229
Tookey, use of platinum tube by, 175
Tools and implements, 70–74
Torry and Eaton on effect of other metals in assay of copper, 120
    H. G., on heat regulator, U. S. Assay Office, N. Y., 307–309
Touchstone, examination of gold alloys on the, 169
Toughening of brittle gold, 184
Tungsten, 258–261
    Berzelius's method of assaying, 259
    Cobenzl's method of assaying, 260
    fire assay of, 259
    gravimetric assays, 259, 260
    Margueritte's method of assaying, 259
    preparation of the titrating solution for, 261
    separation from tin, 260
    Sheele's method of assaying, 259
    volumetric assay of, 261
    wet assays, 259–261
    Zettner's method of volumetric assay of, 261

ULLMANITE, 188
    Upper Harz, assay of galena in, 81, 82, 87
        with potassium carbonate in, 87, 88
Uranium, 255–258
    analytical process for, 255
    determination of with potassium dichromate and iodine, 258
    difficulty of determining oxysalts of, 257
    gravimetric assays of, 255–257
    ores, 255
    Patera's technical test, 256
    precipitation of, in presence of alkaline earths, 256, 257
    separation from calcium, 256
    volumetric assay of, 257
    wet assays of, 255–258
Utrecht, mint of, assaying gold coins at, 179, 180

VALENTINITE, 243
    Vanning, art of, 28
        shovel, illustrated, 27
Van Riemsdijk on cupelling pure gold, 178, 179
Varvicite, 264
Vessels, assay, 63–69
    for the dry method, 63–68
    for the wet method, 68, 69
    of bone-ash, 66–68
Vitriol, blue, 96
Volatilizing fluxes, 79
Volhard's method for manganese, 273, 274
Volhard's volumetric assay of copper, 119
Volumetric analysis, 17
    assays by, 40–45
    vessels for, 68
    assay of gold, 182, 183
    assays, advantages of, 18
    for copper, 104
    wet assays, reagents for, 80

WAD, 264
    Wait's machine for making cupels, illustrated, 67, 68
Wales, assay of lead matt in, 84
Wash-bottle, illustrated, 38
Washing apparatus, Schulze's, 26, 27
    samples, 27, 28
Water-bath, illustrated, 24
Weighing alloys, 29
    and measuring samples, 28–30
    a pulverulent sample, 29
    fluxes, 29
    hygroscopic substances, 29
    the button, 29
Weights and balances, 69, 70
    measures (English), 319, 320
Weil's method of volumetric determination of sulphur in ores, 283
Welch's furnace, 59
Welter's law, 288
Wernicke on electrolytic assay of lead, 95
Wet assay, 17
    assays of copper, 104–128
        reagents for, 80
    method, operations by the, 35–45
White flux, 75
    tellurium, 161
Whittell, Dr. A. P., humid assay for silver, 155

INDEX.

Wiessnegg's gas furnace, 56
Wildenstein's volumetric assay of sulphur, 282, 283
Willemite, 207
Wind furnaces for free burning coal, 55
    for illuminating gas, 55
Winkler's proposed colorimetric assay for nickel, 204
Wolfram, 258
Wrought-iron vessels, 65, 66
Wulfenite, 81, 90, 91

YELLOW lead ore, 90, 91
    orpiment, artificial, 252
        assays of, 252
        native, 252
Yver, on separation of zinc and cadmium, 222

ZETTNER'S method of the volumetric assay of tungsten, 261
Zimmermann's method of determining uranium, 258
Zinc, 207, 221
    and cadmium, separation of, 222
    and manganese in nickel ores, 203
    assay, admixtures having a disturbing effect, to be removed, 217
        coloration of the hydrated ferric oxide in, 216
        preparation of the titrating solutions, 220
        quantity of fluid to be used in Schaffner's assay, 216
        quantity of hydrated ferric acid in, 216
        Schaffner's, points to be observed in, 216
        uniform light requisite during the titration, 218
        with potassium ferro-cyanide, 218, 219
    assays, indicators for final reaction, 215, 216
    as zinc oxide, determination of, 210
    blende, 207, 208
    bloom, 207
    carbonate, 207
    combined with sulphur, assay for, 220, 221
    commercial, purification of, 218

Zinc—
    copper, and lead in one solution, determination of, 221
    determination of, as zinc sulphide, 209, 210
        by decomposing the sulphide of zinc with chloride of silver, and determining the zinc from the equivalent content of chlorine, 219, 220
        from its combinations with sulphur by decomposition with nitrate of silver, 220, 221
    dissolving of, with hydrochloric acid, 108
    distillation assay, 207, 208
    dry assays, 207–209
    electrolytic assay of, according to Beilstein and Jawein, 210, 211
    for fixing the standard solution, 218
    free from lead and arsenic, precipitation of copper with, 107, 108
    granulated, 108
    granulation of, 80
    gravimetric assays, 209–214
    in a slightly acid solution in the presence of a strong mineral acid, determination of, 213, 214
    in copper, separation of, 99
    indirect assay, 208, 209
    in ores, electrolytic method of determining, 213
    in presence of copper, precipitation of, 214, 215
    of metals soluble in ammonia, assays for, 214, 215
    Lower Harz working assay of, 302
    ore, electrolytic assay, 210–214
        removal of various metals from, 217, 218
        when copper is present, assay for, 211
    ores, 207
    oxide, determination of, 210
    precipitation of, by various processes, 212, 213
        from its sulphate solution, 211
    Schaffner's volumetric assay with sodium sulphide, 214–218

Zinc—
   Schober's volumetric assay for, 221
   silicate, 207
   sulphate solution, precipitation of zinc from, 211, 212

Zinc—
   sulphide, determination of, 209, 210
   volumetric assays, 214–221
   wet assays, 209–221

Zinkite, 207

CATALOGUE
OF
Practical and Scientific Books
PUBLISHED BY
HENRY CAREY BAIRD & CO.
INDUSTRIAL PUBLISHERS, BOOKSELLERS AND IMPORTERS.
810 Walnut Street, Philadelphia.

☞ Any of the Books comprised in this Catalogue will be sent by mail, free of postage, to any address in the world, at the publication prices.

☞ A Descriptive Catalogue, 90 pages, 8vo., will be sent free and free of postage, to any one in any part of the world, who will furnish his address.

☞ Where not otherwise stated, all of the Books in this Catalogue are bound in muslin.

AMATEUR MECHANICS' WORKSHOP:
A treatise containing plain and concise directions for the manipulation of Wood and Metals, including Casting, Forging, Brazing, Soldering and Carpentry. By the author of the "Lathe and Its Uses." Seventh edition. Illustrated. 8vo. . . . $2.50

ANDRES.—A Practical Treatise on the Fabrication of Volatile and Fat Varnishes, Lacquers, Siccatives and Sealing Waxes.
From the German of ERWIN ANDRES, Manufacturer of Varnishes and Lacquers. With additions on the Manufacture and Application of Varnishes, Stains for Wood, Horn, Ivory, Bone and Leather. From the German of DR. EMIL WINCKLER and LOUIS E. ANDES. The whole translated and edited by WILLIAM T. BRANNT. With 11 illustrations. 12mo. $2.50

ARLOT.—A Complete Guide for Coach Painters:
Translated from the French of M. ARLOT, Coach Painter, for eleven years Foreman of Painting to M. Eherler, Coach Maker, Paris. By A. A. FESQUET, Chemist and Engineer. To which is added an Appendix, containing Information respecting the Materials and the Practice of Coach and Car Painting and Varnishing in the United States and Great Britain. 12mo. . . . $1.25

(1)

ARMENGAUD, AMOROUX, AND JOHNSON.—The Practical Draughtsman's Book of Industrial Design, and Machinist's and Engineer's Drawing Companion:
Forming a Complete Course of Mechanical Engineering and Architectural Drawing. From the French of M. Armengaud the elder, Prof. of Design in the Conservatoire of Arts and Industry, Paris, and MM. Armengaud the younger, and Amoroux, Civil Engineers. Re-written and arranged with additional matter and plates, selections from and examples of the most useful and generally employed mechanism of the day. By WILLIAM JOHNSON, Assoc. Inst. C. E. Illustrated by fifty folio steel plates, and fifty wood-cuts. A new edition, 4to., half morocco $7.50

ARMSTRONG.—The Construction and Management of Steam Boilers:
By R. ARMSTRONG, C. E. With an Appendix by ROBERT MALLET, C. E., F. R. S. Seventh Edition. Illustrated. 1 vol. 12mo. 75

ARROWSMITH.—Paper-Hanger's Companion:
A Treatise in which the Practical Operations of the Trade are Systematically laid down: with Copious Directions Preparatory to Papering; Preventives against the Effect of Damp on Walls; the various Cements and Pastes Adapted to the Several Purposes of the Trade; Observations and Directions for the Panelling and Ornamenting of Rooms, etc. By JAMES ARROWSMITH. 12mo., cloth $1.00

ASHTON.—The Theory and Practice of the Art of Designing Fancy Cotton and Woollen Cloths from Sample:
Giving full instructions for reducing drafts, as well as the methods of spooling and making out harness for cross drafts and finding any required reed; with calculations and tables of yarn. By FREDERIC T. ASHTON, Designer, West Pittsfield, Mass. With fifty-two illustrations. One vol. folio $6.00

ASKINSON.—Perfumes and their Preparation:
A Comprehensive Treatise on Perfumery, containing Complete Directions for Making Handkerchief Perfumes, Smelling-Salts, Sachets, Fumigating Pastils; Preparations for the Care of the Skin, the Mouth, the Hair; Cosmetics, Hair Dyes, and other Toilet Articles. By G.W. ASKINSON. Translated from the German by ISIDOR FURST. Revised by CHARLES RICE. 32 Illustrations. 8vo. $3.00

BAIRD.—Miscellaneous Papers on Economic Questions.
By Henry Carey Baird. (*In preparation.*)

BAIRD.—The American Cotton Spinner, and Manager's and Carder's Guide:
A Practical Treatise on Cotton Spinning; giving the Dimensions and Speed of Machinery, Draught and Twist Calculations, etc.; with notices of recent Improvements: together with Rules and Examples for making changes in the sizes and numbers of Roving and Yarn. Compiled from the papers of the late ROBERT H. BAIRD. 12mo. $1.50

BAIRD.—Standard Wages Computing Tables:
An Improvement in all former Methods of Computation, so arranged that wages for days, hours, or fractions of hours, at a specified rate per day or hour, may be ascertained at a glance. By T. Spangler Baird. Oblong folio $5.00

BAKER.—Long-Span Railway Bridges:
Comprising Investigations of the Comparative Theoretical and Practical Advantages of the various Adopted or Proposed Type Systems of Construction; with numerous Formulæ and Tables. By B. Baker. 12mo. $1.50

BAKER.—The Mathematical Theory of the Steam-Engine:
With Rules at length, and Examples worked out for the use of Practical Men. By T. Baker, C. E., with numerous Diagrams. Sixth Edition, Revised by Prof. J. R. Young. 12mo. . 75

BARLOW.—The History and Principles of Weaving, by Hand and by Power:
Reprinted, with Considerable Additions, from "Engineering," with a chapter on Lace-making Machinery, reprinted from the Journal of the "Society of Arts." By Alfred Barlow. With several hundred illustrations. 8vo., 443 pages $10.00

BARR.—A Practical Treatise on the Combustion of Coal:
Including descriptions of various mechanical devices for the Economic Generation of Heat by the Combustion of Fuel, whether solid, liquid or gaseous. 8vo. $2.50

BARR.—A Practical Treatise on High Pressure Steam Boilers:
Including Results of Recent Experimental Tests of Boiler Materials, together with a Description of Approved Safety Apparatus, Steam Pumps, Injectors and Economizers in actual use. By Wm. M. Barr. 204 Illustrations. 8vo. $3.00

BAUERMAN.—A Treatise on the Metallurgy of Iron:
Containing Outlines of the History of Iron Manufacture, Methods of Assay, and Analysis of Iron Ores, Processes of Manufacture of Iron and Steel, etc., etc. By H. Bauerman, F. G. S., Associate of the Royal School of Mines. Fifth Edition, Revised and Enlarged. Illustrated with numerous Wood Engravings from Drawings by J. B. Jordan. 12mo. $2.00

BAYLES.—House Drainage and Water Service:
In Cities, Villages and Rural Neighborhoods. With Incidental Consideration of Certain Causes Affecting the Healthfulness of Dwellings. By James C. Bayles, Editor of "The Iron Age" and "The Metal Worker." With numerous illustrations. 8vo. cloth,

BEANS.—A Treatise on Railway Curves and Location of Railroads:
By E. W. Beans, C. E. Illustrated. 12mo. Tucks . $1.50

BECKETT.—A Rudimentary Treatise on Clocks, and Watches and Bells:
By Sir Edmund Beckett, Bart., LL. D., Q. C. F. R. A. S. With numerous illustrations. Seventh Edition, Revised and Enlarged. 12mo. $2.25

BELL.—Carpentry Made Easy:
Or, The Science and Art of Framing on a New and Improved System. With Specific Instructions for Building Balloon Frames, Barn Frames, Mill Frames, Warehouses, Church Spires, etc. Comprising also a System of Bridge Building, with Bills, Estimates of Cost, and valuable Tables. Illustrated by forty-four plates, comprising nearly 200 figures. By WILLIAM E. BELL, Architect and Practical Builder. 8vo. $5.00

BEMROSE.—Fret-Cutting and Perforated Carving:
With fifty-three practical illustrations. By W. BEMROSE, JR. 1 vol. quarto $2.50

BEMROSE.—Manual of Buhl-work and Marquetry:
With Practical Instructions for Learners, and ninety colored designs. By W. BEMROSE, JR. 1 vol. quarto . . . $3.00

BEMROSE.—Manual of Wood Carving:
With Practical Illustrations for Learners of the Art, and Original and Selected Designs. By WILLIAM BEMROSE, JR. With an Introduction by LLEWELLYN JEWITT, F. S. A., etc. With 128 illustrations, 4to. $2.50

BILLINGS.—Tobacco:
Its History, Variety, Culture, Manufacture, Commerce, and Various Modes of Use. By E. R. BILLINGS. Illustrated by nearly 200 engravings. 8vo. $3.00

BIRD.—The American Practical Dyers' Companion:
Comprising a Description of the Principal Dye-Stuffs and Chemicals used in Dyeing, their Natures and Uses; Mordants, and How Made; with the best American, English, French and German processes for Bleaching and Dyeing Silk, Wool, Cotton, Linen, Flannel, Felt, Dress Goods, Mixed and Hosiery Yarns, Feathers, Grass, Felt, Fur, Wool, and Straw Hats, Jute Yarn, Vegetable Ivory, Mats, Skins, Furs, Leather, etc., etc. By Wood, Aniline, and other Processes, together with Remarks on Finishing Agents, and Instructions in the Finishing of Fabrics, Substitutes for Indigo, Water-Proofing of Materials, Tests and Purification of Water, Manufacture of Aniline and other New Dye Wares, Harmonizing Colors, etc., etc.; embracing in all over 800 Receipts for Colors and Shades, *accompanied by 170 Dyed Samples of Raw Materials and Fabrics.* By F. J. BIRD, Practical Dyer, Author of "The Dyers' Hand-Book." 8vo. $10.00

BLINN.—A Practical Workshop Companion for Tin, Sheet-Iron, and Copper-plate Workers:
Containing Rules for describing various kinds of Patterns used by Tin, Sheet-Iron and Copper plate Workers; Practical Geometry; Mensuration of Surfaces and Solids; Tables of the Weights of Metals, Lead-pipe, etc.; Tables of Areas and Circumferences of Circles; Japan, Varnishes, Lackers, Cements, Compositions, etc., etc. By LEROY J. BLINN, Master Mechanic. With One Hundred and Seventy Illustrations. 12mo. $2.50

BOOTH.—Marble Worker's Manual:
Containing Practical Information respecting Marbles in general, their Cutting, Working and Polishing; Veneering of Marble; Mosaics; Composition and Use of Artificial Marble, Stuccos, Cements, Receipts, Secrets, etc., etc. Translated from the French by M. L. BOOTH. With an Appendix concerning American Marbles. 12mo., cloth $1.50

BOOTH and MORFIT.—The Encyclopædia of Chemistry, Practical and Theoretical:
Embracing its application to the Arts, Metallurgy, Mineralogy, Geology, Medicine and Pharmacy. By JAMES C. BOOTH, Melter and Refiner in the United States Mint, Professor of Applied Chemistry in the Franklin Institute, etc., assisted by CAMPBELL MORFIT, author of "Chemical Manipulations," etc. Seventh Edition. Complete in one volume, royal 8vo., 978 pages, with numerous wood-cuts and other illustrations $3.50

BRAMWELL.—The Wool Carder's Vade-Mecum:
A Complete Manual of the Art of Carding Textile Fabrics. By W. C. BRAMWELL. Third Edition, revised and enlarged. Illustrated. Pp. 400. 12mo. $2.50

BRANNT.—A Practical Treatise on Animal and Vegetable Fats and Oils:
Comprising both Fixed and Volatile Oils, their Physical and Chemical Properties and Uses, the Manner of Extracting and Refining them, and Practical Rules for Testing them; as well as the Manufacture of Artificial Butter, Lubricants, including Mineral Lubricating Oils, etc., and on Ozokerite. Edited chiefly from the German of DRS. KARL SCHAEDLER, G. W. ASKINSON, and RICHARD BRUNNER, with Additions and Lists of American Patents relating to the Extraction, Rendering, Refining, Decomposing, and Bleaching of Fats and Oils. By WILLIAM T. BRANNT. Illustrated by 244 engravings. 739 pages. 8vo. $12.50

BRANNT.—A Practical Treatise on the Manufacture of Soap and Candles:
Based upon the most Recent Experiences in the Practice and Science; comprising the Chemistry, Raw Materials, Machinery, and Utensils and Various Processes of Manufacture, including a great variety of formulas. Edited chiefly from the German of Dr. C. Deite, A. Engelhardt, Dr. C. Schaedler and others; with additions and lists of American Patents relating to these subjects. By WM. T. BRANNT. Illustrated by 163 engravings. 677 pages. 8vo. . . $7.50

BRANNT.—A Practical Treatise on the Raw Materials and the Distillation and Rectification of Alcohol, and the Preparation of Alcoholic Liquors, Liqueurs, Cordials, Bitters, etc.:
Edited chiefly from the German of Dr. K. Stammer, Dr. F. Elsner, and E. Schubert. By WM. T. BRANNT. Illustrated by thirty-one engravings. 12mo. $2.50

BRANNT—WAHL.—The Techno-Chemical Receipt Book:
Containing several thousand Receipts covering the latest, most important, and most useful discoveries in Chemical Technology, and their Practical Application in the Arts and the Industries. Edited chiefly from the German of Drs. Winckler, Elsner, Heintze, Mierzinski, Jacobsen, Koller, and Heinzerling, with additions by WM. T. BRANNT and WM. H. WAHL, PH. D. Illustrated by 78 engravings. 12mo. 495 pages $2.00

BROWN.—Five Hundred and Seven Mechanical Movements:
Embracing all those which are most important in Dynamics, Hydraulics, Hydrostatics, Pneumatics, Steam-Engines, Mill and other Gearing, Presses, Horology and Miscellaneous Machinery; and including many movements never before published, and several of which have only recently come into use. By HENRY T. BROWN 12mo. $1.00

BUCKMASTER.—The Elements of Mechanical Physics:
By J. C. BUCKMASTER. Illustrated with numerous engravings. 12mo. $1.00

BULLOCK.—The American Cottage Builder:
A Series of Designs, Plans and Specifications, from $200 to $20,000, for Homes for the People; together with Warming, Ventilation, Drainage, Painting and Landscape Gardening. By JOHN BULLOCK, Architect and Editor of "The Rudiments of Architecture and Building," etc., etc. Illustrated by 75 engravings. 8vo. $3.00

BULLOCK.—The Rudiments of Architecture and Building:
For the use of Architects, Builders, Draughtsmen, Machinists, Engineers and Mechanics. Edited by JOHN BULLOCK, author of "The American Cottage Builder." Illustrated by 250 Engravings. 8vo. $3.00

BURGH.—Practical Rules for the Proportions of Modern Engines and Boilers for Land and Marine Purposes.
By N. P. BURGH, Engineer. 12mo. $1.50

BYLES.—Sophisms of Free Trade and Popular Political Economy Examined.
By a BARRISTER (SIR JOHN BARNARD BYLES, Judge of Common Pleas). From the Ninth English Edition, as published by the Manchester Reciprocity Association. 12mo. . . . $1.25

BOWMAN.—The Structure of the Wool Fibre in its Relation to the Use of Wool for Technical Purposes:
Being the substance, with additions, of Five Lectures, delivered at the request of the Council, to the members of the Bradford Technical College, and the Society of Dyers and Colorists. By F. H. BOWMAN, D. Sc., F. R. S. E., F. L. S. Illustrated by 32 engravings. 8vo. $6.50

BYRNE.—Hand-Book for the Artisan, Mechanic, and Engineer:
Comprising the Grinding and Sharpening of Cutting Tools, Abrasive Processes, Lapidary Work, Gem and Glass Engraving, Varnishing and Lackering, Apparatus, Materials and Processes for Grinding and

Polishing, etc. By OLIVER BYRNE. Illustrated by 185 wood engravings. 8vo. $5.00

BYRNE.—Pocket-Book for Railroad and Civil Engineers:
Containing New, Exact and Concise Methods for Laying out Railroad Curves, Switches, Frog Angles and Crossings; the Staking out of work; Levelling; the Calculation of Cuttings; Embankments; Earthwork, etc. By OLIVER BYRNE. 18mo., full bound, pocket-book form $1.75

BYRNE.—The Practical Metal-Worker's Assistant:
Comprising Metallurgic Chemistry; the Arts of Working all Metals and Alloys; Forging of Iron and Steel; Hardening and Tempering; Melting and Mixing; Casting and Founding; Works in Sheet Metal; the Processes Dependent on the Ductility of the Metals; Soldering; and the most Improved Processes and Tools employed by Metal-Workers. With the Application of the Art of Electro-Metallurgy to Manufacturing Processes; collected from Original Sources, and from the works of Holtzapffel, Bergeron, Leupold, Plumier, Napier, Scoffern, Clay, Fairbairn and others. By OLIVER BYRNE. A new, revised and improved edition, to which is added an Appendix, containing The Manufacture of Russian Sheet-Iron. By JOHN PERCY, M. D., F. R. S. The Manufacture of Malleable Iron Castings, and Improvements in Bessemer Steel. By A. A. FESQUET, Chemist and Engineer. With over Six Hundred Engravings, Illustrating every Branch of the Subject. 8vo. $5.00

BYRNE.—The Practical Model Calculator:
For the Engineer, Mechanic, Manufacturer of Engine Work, Naval Architect, Miner and Millwright. By OLIVER BYRNE. 8vo., nearly 600 pages $3.00

CABINET MAKER'S ALBUM OF FURNITURE:
Comprising a Collection of Designs for various Styles of Furniture. Illustrated by Forty-eight Large and Beautifully Engraved Plates. Oblong, 8vo. $2.00

CALLINGHAM.—Sign Writing and Glass Embossing:
A Complete Practical Illustrated Manual of the Art. By JAMES CALLINGHAM. 12mo. $1.50

CAMPIN.—A Practical Treatise on Mechanical Engineering:
Comprising Metallurgy, Moulding, Casting, Forging, Tools, Workshop Machinery, Mechanical Manipulation, Manufacture of Steam-Engines, etc. With an Appendix on the Analysis of Iron and Iron Ores. By FRANCIS CAMPIN, C. E. To which are added, Observations on the Construction of Steam Boilers, and Remarks upon Furnaces used for Smoke Prevention; with a Chapter on Explosions. By R. ARMSTRONG, C. E., and JOHN BOURNE. Rules for Calculating the Change Wheels for Screws on a Turning Lathe, and for a Wheel, cutting Machine. By J. LA NICCA. Management of Steel, Including Forging, Hardening, Tempering, Annealing, Shrinking and Expansion; and the Case-hardening of Iron. By G. EDE. 8vo. Illustrated with twenty-nine plates and 100 wood engravings $5.00

CAREY.—A Memoir of Henry C. Carey.
By Dr. Wm. Elder. With a portrait. 8vo., cloth . . 75

CAREY.—The Works of Henry C. Carey:
Harmony of Interests: Agricultural, Manufacturing and Commercial. 8vo. $1.25
Manual of Social Science. Condensed from Carey's "Principles of Social Science." By Kate McKean. 1 vol. 12mo. . $2.00
Miscellaneous Works. With a Portrait. 2 vols. 8vo. $10.00
Past, Present and Future. 8vo. $2.50
Principles of Social Science. 3 volumes, 8vo. . . $7.50
The Slave-Trade, Domestic and Foreign; Why it Exists, and How it may be Extinguished (1853). 8vo. . . $2.00
The Unity of Law: As Exhibited in the Relations of Physical, Social, Mental and Moral Science (1872). 8vo. . . $2.50

CLARK.—Tramways, their Construction and Working:
Embracing a Comprehensive History of the System. With an exhaustive analysis of the various modes of traction, including horse-power, steam, heated water and compressed air; a description of the varieties of Rolling stock, and ample details of cost and working expenses. By D. Kinnear Clark. Illustrated by over 200 wood engravings, and thirteen folding plates. 1 vol. 8vo. . $9.00

COLBURN.—The Locomotive Engine:
Including a Description of its Structure, Rules for Estimating its Capabilities, and Practical Observations on its Construction and Management. By Zerah Colburn. Illustrated. 12mo. . $1.00

COLLENS.—The Eden of Labor; or, the Christian Utopia.
By T. Wharton Collens, author of "Humanics," "The History of Charity," etc. 12mo. Paper cover, $1.00; Cloth . $1.25

COOLEY.—A Complete Practical Treatise on Perfumery:
Being a Hand-book of Perfumes, Cosmetics and other Toilet Articles. With a Comprehensive Collection of Formulæ. By Arnold J. Cooley. 12mo. $1.50

COOPER.—A Treatise on the use of Belting for the Transmission of Power.
With numerous illustrations of approved and actual methods of arranging Main Driving and Quarter Twist Belts, and of Belt Fastenings. Examples and Rules in great number for exhibiting and calculating the size and driving power of Belts. Plain, Particular and Practical Directions for the Treatment, Care and Management of Belts. Descriptions of many varieties of Beltings, together with chapters on the Transmission of Power by Ropes; by Iron and Wood Frictional Gearing; on the Strength of Belting Leather; and on the Experimental Investigations of Morin, Briggs, and others. By John H. Cooper, M. E. 8vo. $3.50

CRAIK.—The Practical American Millwright and Miller.
By David Craik, Millwright. Illustrated by numerous wood engravings and two folding plates. 8vo. . . . $3.50

CROSS.—The Cotton Yarn Spinner:
Showing how the Preparation should be arranged for Different Counts of Yarns by a System more uniform than has hitherto been practiced; by having a Standard Schedule from which we make all our Changes. By RICHARD CROSS. 122 pp. 12mo. . 75

CRISTIANI.—A Technical Treatise on Soap and Candles:
With a Glance at the Industry of Fats and Oils. By R. S. CRISTIANI, Chemist. Author of "Perfumery and Kindred Arts." Illustrated by 176 engravings. 581 pages, 8vo. . . . $15.00

COAL AND METAL MINERS' POCKET BOOK:
Of Principles, Rules, Formulæ, and Tables, Specially Compiled and Prepared for the Convenient Use of Mine Officials, Mining Engineers, and Students preparing themselves for Certificates of Competency as Mine Inspectors or Mine Foremen. Revised and Enlarged edition. Illustrated, 565 pages, small 12mo., cloth. . $2.00
Pocket book form, flexible leather with flap . . $2.75

DAVIDSON.—A Practical Manual of House Painting, Graining, Marbling, and Sign-Writing:
Containing full information on the processes of House Painting in Oil and Distemper, the Formation of Letters and Practice of Sign-Writing, the Principles of Decorative Art, a Course of Elementary Drawing for House Painters, Writers, etc., and a Collection of Useful Receipts. With nine colored illustrations of Woods and Marbles, and numerous wood engravings. By ELLIS A. DAVIDSON. 12mo.
$3 00

DAVIES.—A Treatise on Earthy and Other Minerals and Mining:
By D. C. DAVIES, F. G. S., Mining Engineer, etc. Illustrated by 76 Engravings. 12mo. $5 00

DAVIES.—A Treatise on Metalliferous Minerals and Mining:
By D. C. DAVIES, F. G. S., Mining Engineer, Examiner of Mines, Quarries and Collieries. Illustrated by 148 engravings of Geological Formations, Mining Operations and Machinery, drawn from the practice of all parts of the world. Fifth Edition, thoroughly Revised and much Enlarged by his son, E. Henry Davies. 12mo., 524 pages $5.00

DAVIES.—A Treatise on Slate and Slate Quarrying:
Scientific, Practical and Commercial. By D. C. DAVIES, F. G S., Mining Engineer, etc. With numerous illustrations and folding plates. 12mo. $2.00

DAVIS.—A Practical Treatise on the Manufacture of Brick, Tiles and Terra-Cotta:
Including Stiff Clay, Dry Clay, Hand Made, Pressed or Front, and Roadway Paving Brick, Enamelled Brick, with Glazes and Colors, Fire Brick and Blocks, Silica Brick, Carbon Brick, Glass Pots, Re-

torts, Architectural Terra-Cotta, Sewer Pipe, Drain Tile, Glazed and Unglazed Roofing Tile, Art Tile, Mosaics, and Imitation of Intarsia or Inlaid Surfaces. Comprising every product of Clay employed in Architecture, Engineering, and the Blast Furnace. With a Detailed Description of the Different Clays employed, the Most Modern Machinery, Tools, and Kilns used, and the Processes for Handling, Disintegrating, Tempering, and Moulding the Clay into Shape, Drying, Setting, and Burning. By Charles Thomas Davis. Third Edition. Revised and in great part rewritten. Illustrated by 261 engravings. 662 pages $5.00

DAVIS.—A Treatise on Steam-Boiler Incrustation and Methods for Preventing Corrosion and the Formation of Scale:
By CHARLES T. DAVIS. Illustrated by 65 engravings. 8vo. $1.50

DAVIS.—The Manufacture of Paper:
Being a Description of the various Processes for the Fabrication, Coloring and Finishing of every kind of Paper, Including the Different Raw Materials and the Methods for Determining their Values, the Tools, Machines and Practical Details connected with an intelligent and a profitable prosecution of the art, with special reference to the best American Practice. To which are added a History of Paper, complete Lists of Paper-Making Materials, List of American Machines, Tools and Processes used in treating the Raw Materials, and in Making, Coloring and Finishing Paper. By CHARLES T. DAVIS. Illustrated by 156 engravings. 608 pages, 8vo. $6.00

DAVIS.—The Manufacture of Leather:
Being a description of all of th Processes for the Tanning, Tawing, Currying, Finishing and Dyeing of every kind of Leather; including the various Raw Materials and the Methods for Determining their Values; the Tools, Machines, and all Details of Importance connected with an Intelligent and Profitable Prosecution of the Art, with Special Reference to the Best American Practice. To which are added Complete Lists of all American Patents for Materials, Processes, Tools, and Machines for Tanning, Currying, etc. By CHARLES THOMAS DAVIS. Illustrated by 302 engravings and 12 Samples of Dyed Leathers. One vol., 8vo., 824 pages . . . $25.00

DAWIDOWSKY—BRANNT.—A Practical Treatise on the Raw Materials and Fabrication of Glue, Gelatine, Gelatine Veneers and Foils, Isinglass, Cements, Pastes, Mucilages, etc.;
Based upon Actual Experience. By F. DAWIDOWSKY, Technical Chemist. Translated from the German, with extensive additions, including a description of the most Recent American Processes, by WILLIAM T. BRANNT, Graduate of the Royal Agricultural College of Eldena, Prussia. 35 Engravings. 12mo. . . . $2.50

DE GRAFF.—The Geometrical Stair-Builders' Guide:
Being a Plain Practical System of Hand-Railing, embracing all its necessary Details, and Geometrically Illustrated by twenty-two Steel Engravings; together with the use of the most approved principles of Practical Geometry. By SIMON DE GRAFF, Architect. 4to. $2.50

DE KONINCK—DIETZ.—A Practical Manual of Chemical Analysis and Assaying:
As applied to the Manufacture of Iron from its Ores, and to Cast Iron, Wrought Iron, and Steel, as found in Commerce. By L. L. DE KONINCK, Dr. Sc., and E. DIETZ, Engineer. Edited with Notes, by ROBERT MALLET, F. R. S., F. S. G., M. I. C. E., etc. American Edition, Edited with Notes and an Appendix on Iron Ores, by A. A. FESQUET, Chemist and Engineer. 12mo. . . . $1.50

DUNCAN.—Practical Surveyor's Guide:
Containing the necessary information to make any person of common capacity, a finished land surveyor without the aid of a teacher By ANDREW DUNCAN. Revised. 72 engravings, 214 pp. 12mo. $1.50

DUPLAIS.—A Treatise on the Manufacture and Distillation of Alcoholic Liquors:
Comprising Accurate and Complete Details in Regard to Alcohol from Wine, Molasses, Beets, Grain, Rice, Potatoes, Sorghum, Asphodel, Fruits, etc.; with the Distillation and Rectification of Brandy, Whiskey, Rum, Gin, Swiss Absinthe, etc., the Preparation of Aromatic Waters, Volatile Oils or Essences, Sugars, Syrups, Aromatic Tinctures, Liqueurs, Cordial Wines, Effervescing Wines, etc., the Ageing of Brandy and the improvement of Spirits, with Copious Directions and Tables for Testing and Reducing Spirituous Liquors, etc., etc. Translated and Edited from the French of MM. DUPLAIS, Ainé et Jeune. By M. MCKENNIE, M. D. To which are added the United States Internal Revenue Regulations for the Assessment and Collection of Taxes on Distilled Spirits. Illustrated by fourteen folding plates and several wood engravings. 743 pp. 8vo. $10 00

DUSSAUCE.—Practical Treatise on the Fabrication of Matches, Gun Cotton, and Fulminating Powder.
By Professor H. DUSSAUCE. 12mo. $3 00

DYER AND COLOR-MAKER'S COMPANION:
Containing upwards of two hundred Receipts for making Colors, on the most approved principles, for all the various styles and fabrics now in existence; with the Scouring Process, and plain Directions for Preparing, Washing-off, and Finishing the Goods. 12mo. $1.00

EDWARDS.—A Catechism of the Marine Steam-Engine,
For the use of Engineers, Firemen, and Mechanics. A Practical Work for Practical Men. By EMORY EDWARDS, Mechanical Engineer. Illustrated by sixty-three Engravings, including examples of the most modern Engines. Third edition, thoroughly revised, with much additional matter. 12mo. 414 pages . . $2 00

EDWARDS.—Modern American Locomotive Engines,
Their Design, Construction and Management. By EMORY EDWARDS. Illustrated 12mo. $2.00

EDWARDS.—The American Steam Engineer:
Theoretical and Practical, with examples of the latest and most approved American practice in the design and construction of Steam Engines and Boilers. For the use of engineers, machinists, boilermakers, and engineering students. By EMORY EDWARDS. Fully illustrated, 419 pages. 12mo. . . . $2.50

EDWARDS.—Modern American Marine Engines, Boilers, and Screw Propellers,
Their Design and Construction. Showing the Present Practice of the most Eminent Engineers and Marine Engine Builders in the United States. Illustrated by 30 large and elaborate plates. 4to. $5.00

EDWARDS.—The Practical Steam Engineer's Guide
In the Design, Construction, and Management of American Stationary, Portable, and Steam Fire-Engines, Steam Pumps, Boilers, Injectors, Governors, Indicators, Pistons and Rings, Safety Valves and Steam Gauges. For the use of Engineers, Firemen, and Steam Users. By EMORY EDWARDS. Illustrated by 119 engravings. 420 pages. 12mo. $2 50

EISSLER.—The Metallurgy of Gold:
A Practical Treatise on the Metallurgical Treatment of Gold-Bearing Ores, including the Processes of Concentration and Chlorination, and the Assaying, Melting, and Refining of Gold. By M. EISSLER. With 132 Illustrations. 12mo. $5.00

EISSLER.—The Metallurgy of Silver:
A Practical Treatise on the Amalgamation, Roasting, and Lixiviation of Silver Ores, including the Assaying, Melting, and Refining of Silver Bullion. By M. EISSLER. 124 Illustrations. 336 pp. 12mo. $4.25

ELDER.—Conversations on the Principal Subjects of Political Economy.
By DR. WILLIAM ELDER. 8vo. $2 50

ELDER.—Questions of the Day,
Economic and Social. By DR. WILLIAM ELDER. 8vo. . $3.00

ERNI.—Mineralogy Simplified.
Easy Methods of Determining and Classifying Minerals, including Ores, by means of the Blowpipe, and by Humid Chemical Analysis, based on Professor von Kobell's Tables for the Determination of Minerals, with an Introduction to Modern Chemistry. By HENRY ERNI, A.M., M.D., Professor of Chemistry. Second Edition, rewritten, enlarged and improved. 12mo. $3 00

FAIRBAIRN.—The Principles of Mechanism and Machinery of Transmission
Comprising the Principles of Mechanism, Wheels, and Pulleys, Strength and Proportions of Shafts, Coupling of Shafts, and Engaging and Disengaging Gear. By SIR WILLIAM FAIRBAIRN, Bart C. E. Beautifully illustrated by over 150 wood-cuts. In one volume, 12mo $2.50

FLEMING.—Narrow Gauge Railways in America.
A Sketch of their Rise, Progress, and Success. Valuable Statistics as to Grades, Curves, Weight of Rail, Locomotives, Cars, etc. By HOWARD FLEMING. Illustrated, 8vo. $1 00

FORSYTH.—Book of Designs for Headstones, Mural, and other Monuments:
Containing 78 Designs. By JAMES FORSYTH. With an Introduction by CHARLES BOUTELL, M. A. 4to., cloth . . $5 00

FRANKEL—HUTTER.—A Practical Treatise on the Manufacture of Starch, Glucose, Starch-Sugar, and Dextrine:
Based on the German of LADISLAUS VON WAGNER, Professor in the Royal Technical High School, Buda-Pest, Hungary, and other authorities. By JULIUS FRANKEL, Graduate of the Polytechnic School of Hanover. Edited by ROBERT HUTTER, Chemist, Practical Manufacturer of Starch-Sugar. Illustrated by 58 engravings, covering every branch of the subject, including examples of the most Recent and Best American Machinery. 8vo., 344 pp. . $3.50

GARDNER.—The Painter's Encyclopædia:
Containing Definitions of all Important Words in the Art of Plain and Artistic Painting, with Details of Practice in Coach, Carriage, Railway Car, House, Sign, and Ornamental Painting, including Graining, Marbling, Staining, Varnishing, Polishing, Lettering, Stenciling, Gilding, Bronzing, etc. By FRANKLIN B. GARDNER. 158 Illustrations. 12mo. 427 pp. $2.00

GARDNER.—Everybody's Paint Book:
A Complete Guide to the Art of Outdoor and Indoor Painting, Designed for the Special Use of those who wish to do their own work, and consisting of Practical Lessons in Plain Painting, Varnishing, Polishing, Staining, Paper Hanging, Kalsomining, etc., as well as Directions for Renovating Furniture, and Hints on Artistic Work for Home Decoration. 38 Illustrations. 12mo., 183 pp. . $1.00

GEE.—The Goldsmith's Handbook:
Containing full instructions for the Alloying and Working of Gold, including the Art of Alloying, Melting, Reducing, Coloring, Collecting, and Refining; the Processes of Manipulation, Recovery of Waste; Chemical and Physical Properties of Gold; with a New System of Mixing its Alloys; Solders, Enamels, and other Useful Rules and Recipes. By GEORGE E. GEE. 12mo. . . $1.75

GEE.—The Silversmith's Handbook:
Containing full instructions for the Alloying and Working of Silver, including the different modes of Refining and Melting the Metal; its Solders; the Preparation of Imitation Alloys; Methods of Manipulation; Prevention of Waste; Instructions for Improving and Finishing the Surface of the Work; together with other Useful Information and Memoranda. By GEORGE E. GEE. Illustrated. 12mo. $1.75

GOTHIC ALBUM FOR CABINET-MAKERS:
Designs for Gothic Furniture. Twenty-three plates. Oblong $2.00

GRANT.—A Handbook on the Teeth of Gears:
Their Curves, Properties, and Practical Construction. By GEORGE B. GRANT. Illustrated. Third Edition, enlarged. 8vo. $1.00

GREENWOOD.—Steel and Iron:
Comprising the Practice and Theory of the Several Methods Pursued in their Manufacture, and of their Treatment in the Rolling-Mills, the Forge, and the Foundry. By WILLIAM HENRY GREENWOOD, F. C. S. With 97 Diagrams, 536 pages. 12mo. $2.00

GREGORY.—Mathematics for Practical Men:
Adapted to the Pursuits of Surveyors, Architects, Mechanics, and Civil Engineers. By OLINTHUS GREGORY. 8vo., plates $3.00

GRISWOLD.—Railroad Engineer's Pocket Companion for the Field:
Comprising Rules for Calculating Deflection Distances and Angles, Tangential Distances and Angles, and all Necessary Tables for Engineers; also the Art of Levelling from Preliminary Survey to the Construction of Railroads, intended Expressly for the Young Engineer, together with Numerous Valuable Rules and Examples. By W. GRISWOLD. 12mo., tucks $1.75

GRUNER.—Studies of Blast Furnace Phenomena:
By M. L. GRUNER, President of the General Council of Mines of France, and lately Professor of Metallurgy at the Ecole des Mines. Translated, with the author's sanction, with an Appendix, by L. D. B. GORDON, F. R. S. E., F. G. S. 8vo. . . . $2.50

Hand-Book of Useful Tables for the Lumberman, Farmer and Mechanic:
Containing Accurate Tables of Logs Reduced to Inch Board Measure, Plank, Scantling and Timber Measure; Wages and Rent, by Week or Month; Capacity of Granaries, Bins and Cisterns; Land Measure, Interest Tables, with Directions for Finding the Interest on any sum at 4, 5, 6, 7 and 8 per cent., and many other Useful Tables. 32 mo., boards. 186 pages 25

HASERICK.—The Secrets of the Art of Dyeing Wool, Cotton, and Linen,
Including Bleaching and Coloring Wool and Cotton Hosiery and Random Yarns. A Treatise based on Economy and Practice. By E. C. HASERICK. *Illustrated by 323 Dyed Patterns of the Yarns or Fabrics.* 8vo. $7.50

HATS AND FELTING:
A Practical Treatise on their Manufacture. By a Practical Hatter. Illustrated by Drawings of Machinery, etc. 8vo. . . $1.25

HOFFER.—A Practical Treatise on Caoutchouc and Gutta Percha,
Comprising the Properties of the Raw Materials, and the manner of Mixing and Working them; with the Fabrication of Vulcanized and Hard Rubbers, Caoutchouc and Gutta Percha Compositions, Waterproof Substances, Elastic Tissues, the Utilization of Waste, etc., etc. From the German of RAIMUND HOFFER. By W. T. BRANNT. Illustrated 12mo. $2.50

HAUPT.—Street Railway Motors:
With Descriptions and Cost of Plants and Operation of the Various Systems now in Use. 12mo. $1.75

HAUPT—RHAWN.—A Move for Better Roads:
Essays on Road-making and Maintenance and Road Laws, for which Prizes or Honorable Mention were Awarded through the University of Pennsylvania by a Committee of Citizens of Philadelphia, with a Synopsis of other Contributions and a Review by the Secretary, LEWIS M. HAUPT, A. M., C. E.; also an Introduction by WILLIAM H. RHAWN, Chairman of the Committee. 319 pages. 8vo. $2.00

HUGHES.—American Miller and Millwright's Assistant:
By WILLIAM CARTER HUGHES. 12mo. . . . $1.50

HULME.—Worked Examination Questions in Plane Geometrical Drawing:
For the Use of Candidates for the Royal Military Academy, Woolwich; the Royal Military College, Sandhurst; the Indian Civil Engineering College, Cooper's Hill; Indian Public Works and Telegraph Departments; Royal Marine Light Infantry; the Oxford and Cambridge Local Examinations, etc. By F. EDWARD HULME, F. L. S., F. S. A., Art-Master Marlborough College. Illustrated by 300 examples. Small quarto $2.50

JERVIS.—Railroad Property:
A Treatise on the Construction and Management of Railways; designed to afford useful knowledge, in the popular style, to the holders of this class of property; as well as Railway Managers, Officers, and Agents. By JOHN B. JERVIS, late Civil Engineer of the Hudson River Railroad, Croton Aqueduct, etc. 12mo., cloth $2.00

KEENE.—A Hand-Book of Practical Gauging:
For the Use of Beginners, to which is added a Chapter on Distillation, describing the process in operation at the Custom-House for ascertaining the Strength of Wines. By JAMES B. KEENE, of H. M. Customs. 8vo. $1.25

KELLEY.—Speeches, Addresses, and Letters on Industrial and Financial Questions:
By HON. WILLIAM D. KELLEY, M. C. 544 pages, 8vo. . $2.50

KELLOGG.—A New Monetary System:
The only means of Securing the respective Rights of Labor and Property, and of Protecting the Public from Financial Revulsions. By EDWARD KELLOGG. Revised from his work on "Labor and other Capital." With numerous additions from his manuscript. Edited by MARY KELLOGG PUTNAM. Fifth edition. To which is added a Biographical Sketch of the Author. One volume, 12mo. Paper cover $1.00
Bound in cloth 1.25

KEMLO.—Watch-Repairer's Hand-Book:
Being a Complete Guide to the Young Beginner, in Taking Apart, Putting Together, and Thoroughly Cleaning the English Lever and other Foreign Watches, and all American Watches. By F. KEMLO, Practical Watchmaker. With Illustrations. 12mo. . $1.25

KENTISH.—A Treatise on a Box of Instruments,
And the Slide Rule; with the Theory of Trigonometry and Logarithms, including Practical Geometry, Surveying, Measuring of Timber, Cask and Malt Gauging, Heights, and Distances. By THOMAS KENTISH. In one volume. 12mo. $1.25

KERL.—The Assayer's Manual:
An Abridged Treatise on the Docimastic Examination of Ores, and Furnace and other Artificial Products. By BRUNO KERL, Professor in the Royal School of Mines. Translated from the German by WILLIAM T. BRANNT. Second American edition, edited with Extensive Additions by F. LYNWOOD GARRISON, Member of the American Institute of Mining Engineers, etc. Illustrated by 87 engravings. 8vo. $3.00

KICK.—Flour Manufacture.
A Treatise on Milling Science and Practice. By FREDERICK KICK Imperial Regierungsrath, Professor of Mechanical Technology in the Imperial German Polytechnic Institute, Prague. Translated from the second enlarged and revised edition with supplement by H. H. P. POWLES, Assoc. Memb Institution of Civil Engineers. Illustrated with 28 Plates, and 167 Wood-cuts. 367 pages. 8vo. . $10.00

KINGZETT.—The History, Products, and Processes of the Alkali Trade:
Including the most Recent Improvements. By CHARLES THOMAS KINGZETT, Consulting Chemist. With 23 illustrations. 8vo. $2.50

KIRK.—The Founding of Metals:
A Practical Treatise on the Melting of Iron, with a Description of the Founding of Alloys; also, of all the Metals and Mineral Substances used in the Art of Founding. Collected from original sources. By EDWARD KIRK, Practical Foundryman and Chemist. Illustrated. Third edition. 8vo. $2.50

LANDRIN.—A Treatise on Steel:
Comprising its Theory, Metallurgy, Properties, Practical Working, and Use. By M. H. C. LANDRIN, JR., Civil Engineer. Translated from the French, with Notes, by A. A. FESQUET, Chemist and Engineer. With an Appendix on the Bessemer and the Martin Processes for Manufacturing Steel, from the Report of Abram S. Hewitt United States Commissioner to the Universal Exposition, Paris, 1867. 12mo. $3.00

LANGBEIN.—A Complete Treatise on the Electro-Deposition of Metals:
Translated from the German, with Additions, by WM. T. BRANNT. 125 illustrations. 8vo. $4.00

LARDNER.—The Steam-Engine:
For the Use of Beginners. Illustrated. 12mo. . . . 75

LEHNER.—The Manufacture of Ink:
Comprising the Raw Materials, and the Preparation of Writing, Copying and Hektograph Inks, Safety Inks, Ink Extracts and Powders, etc. Translated from the German of SIGMUND LEHNER, with additions by WILLIAM T. BRANNT. Illustrated. 12mo. $2.00

LARKIN.—The Practical Brass and Iron Founder's Guide:
A Concise Treatise on Brass Founding, Moulding, the Metals and their Alloys, etc.; to which are added Recent Improvements in the Manufacture of Iron, Steel by the Bessemer Process, etc., etc. By JAMES LARKIN, late Conductor of the Brass Foundry Department in Reany, Neafie & Co.'s Penn Works, Philadelphia. New edition, revised, with extensive additions. 12mo. . . . $2.50

LEROUX.—A Practical Treatise on the Manufacture of Worsteds and Carded Yarns:
Comprising Practical Mechanics, with Rules and Calculations applied to Spinning; Sorting, Cleaning, and Scouring Wools; the English and French Methods of Combing, Drawing, and Spinning Worsteds, and Manufacturing Carded Yarns. Translated from the French of CHARLES LEROUX, Mechanical Engineer and Superintendent of a Spinning-Mill, by HORATIO PAINE, M. D., and A. A. FESQUET, Chemist and Engineer. Illustrated by twelve large Plates. To which is added an Appendix, containing Extracts from the Reports of the International Jury, and of the Artisans selected by the Committee appointed by the Council of the Society of Arts, London, on Woolen and Worsted Machinery and Fabrics, as exhibited in the Paris Universal Exposition, 1867. 8vo. $5.00

LEFFEL.—The Construction of Mill-Dams:
Comprising also the Building of Race and Reservoir Embankments and Head-Gates, the Measurement of Streams, Gauging of Water Supply, etc. By JAMES LEFFEL & Co. Illustrated by 58 engravings. 8vo. $2.50

LESLIE.—Complete Cookery:
Directions for Cookery in its Various Branches. By MISS LESLIE. Sixtieth thousand. Thoroughly revised, with the addition of New Receipts. 12mo. $1.50

LE VAN.—The Steam Engine and the Indicator:
Their Origin and Progressive Development; including the Most Recent Examples of Steam and Gas Motors, together with the Indicator, its Principles, its Utility, and its Application. By WILLIAM BARNET LE VAN. Illustrated by 205 Engravings, chiefly of Indicator-Cards. 469 pp. 8vo. $4.00

LIEBER.—Assayer's Guide:
Or, Practical Directions to Assayers, Miners, and Smelters, for the Tests and Assays, by Heat and by Wet Processes, for the Ores of all the principal Metals, of Gold and Silver Coins and Alloys, and of Coal, etc. By OSCAR M. LIEBER. Revised. 283 pp. 12mo. $1.50

Lockwood's Dictionary of Terms:
Used in the Practice of Mechanical Engineering, embracing those Current in the Drawing Office, Pattern Shop, Foundry, Fitting, Turning, Smith's and Boiler Shops, etc., etc., comprising upwards of Six Thousand Definitions. Edited by a Foreman Pattern Maker, author of "Pattern Making." 417 pp. 12mo. . . . $3.00

LUKIN.—Amongst Machines:
Embracing Descriptions of the various Mechanical Appliances used in the Manufacture of Wood, Metal, and other Substances. 12mo.
$1.75

LUKIN.—The Boy Engineers:
What They Did, and How They Did It. With 30 plates. 18mo.
$1.75

LUKIN.—The Young Mechanic:
Practical Carpentry. Containing Directions for the Use of all kinds of Tools, and for Construction of Steam-Engines and Mechanical Models, including the Art of Turning in Wood and Metal. By JOHN LUKIN, Author of "The Lathe and Its Uses," etc. Illustrated. 12mo. $1.75

MAIN and BROWN.—Questions on Subjects Connected with the Marine Steam-Engine:
And Examination Papers; with Hints for their Solution. By THOMAS J. MAIN, Professor of Mathematics, Royal Naval College, and THOMAS BROWN, Chief Engineer, R. N. 12mo., cloth. $1.00

MAIN and BROWN.—The Indicator and Dynamometer:
With their Practical Applications to the Steam-Engine. By THOMAS J. MAIN, M. A. F. R., Ass't S. Professor Royal Naval College, Portsmouth, and THOMAS BROWN, Assoc. Inst. C. E., Chief Engineer R. N., attached to the R. N. College. Illustrated. 8vo. . $1.00

MAIN and BROWN.—The Marine Steam-Engine.
By THOMAS J. MAIN, F. R. Ass't S. Mathematical Professor at the Royal Naval College, Portsmouth, and THOMAS BROWN, Assoc. Inst. C. E., Chief Engineer R. N. Attached to the Royal Naval College. With numerous illustrations. 8vo. .

MAKINS.—A Manual of Metallurgy:
By GEORGE HOGARTH MAKINS. 100 engravings. Second edition rewritten and much enlarged. 12mo., 592 pages . . $3.00

MARTIN.—Screw-Cutting Tables, for the Use of Mechanical Engineers:
Showing the Proper Arrangement of Wheels for Cutting the Threads of Screws of any Required Pitch; with a Table for Making the Universal Gas-Pipe Thread and Taps. By W. A. MARTIN, Engineer. 8vo. . 50

MICHELL.—Mine Drainage:
Being a Complete and Practical Treatise on Direct-Acting Underground Steam Pumping Machinery. With a Description of a large number of the best known Engines, their General Utility and the Special Sphere of their Action, the Mode of their Application, and their Merits compared with other Pumping Machinery. By STEPHEN MICHELL. Illustrated by 137 engravings. 8vo., 277 pages . $6.00

MOLESWORTH.—Pocket-Book of Useful Formulæ and Memoranda for Civil and Mechanical Engineers.
By GUILFORD L. MOLESWORTH, Member of the Institution of Civil Engineers, Chief Resident Engineer of the Ceylon Railway. Full-bound in Pocket-book form $1.00

MOORE.—The Universal Assistant and the Complete Mechanic:
Containing over one million Industrial Facts, Calculations, Receipts, Processes, Trades Secrets, Rules, Business Forms, Legal Items, Etc., in every occupation, from the Household to the Manufactory. By R. Moore. Illustrated by 500 Engravings. 12mo. . $2.50

MORRIS.—Easy Rules for the Measurement of Earthworks:
By means of the Prismoidal Formula. Illustrated with Numerous Wood-Cuts, Problems, and Examples, and concluded by an Extensive Table for finding the Solidity in cubic yards from Mean Areas. The whole being adapted for convenient use by Engineers, Surveyors, Contractors, and others needing Correct Measurements of Earthwork. By Elwood Morris, C. E. 8vo. $1.50

MAUCHLINE.—The Mine Foreman's Hand-Book
Of Practical and Theoretical Information on the Opening, Ventilating, and Working of Collieries. Questions and Answers on Practical and Theoretical Coal Mining. Designed to Assist Students and Others in Passing Examinations for Mine Foremanships. By Robert Mauchline, Ex-Inspector of Mines. A New, Revised and Enlarged Edition. Illustrated by 114 engravings. 8vo. 337 pages $3.75

NAPIER.—A System of Chemistry Applied to Dyeing.
By James Napier, F. C. S. A New and Thoroughly Revised Edition. Completely brought up to the present state of the Science, including the Chemistry of Coal Tar Colors, by A. A. Fesquet, Chemist and Engineer. With an Appendix on Dyeing and Calico Printing, as shown at the Universal Exposition, Paris, 1867. Illustrated. 8vo. 422 pages $3.50

NEVILLE.—Hydraulic Tables, Coefficients, and Formulæ, for finding the Discharge of Water from Orifices, Notches, Weirs, Pipes, and Rivers:
Third Edition, with Additions, consisting of New Formulæ for the Discharge from Tidal and Flood Sluices and Siphons; general information on Rainfall, Catchment-Basins, Drainage, Sewerage, Water Supply for Towns and Mill Power. By John Neville. C. E. M R I. A.; Fellow of the Royal Geological Society of Ireland. Thick, 12mo. $5.50

NEWBERY.—Gleanings from Ornamental Art of every style:
Drawn from Examples in the British, South Kensington, Indian, Crystal Palace, and other Museums, the Exhibitions of 1851 and 1862, and the best English and Foreign works. In a series of 100 exquisitely drawn Plates, containing many hundred examples. By Robert Newbery. 4to. $12.50

NICHOLLS.—The Theoretical and Practical Boiler-Maker and Engineer's Reference Book:
Containing a variety of Useful Information for Employers of Labor Foremen and Working Boiler-Makers. Iron, Copper, and Tinsmiths

Draughtsmen, Engineers, the General Steam-using Public, and for the Use of Science Schools and Classes. By SAMUEL NICHOLLS. Illustrated by sixteen plates, 12mo. $2.50

NICHOLSON.—A Manual of the Art of Bookbinding:
Containing full instructions in the different Branches of Forwarding, Gilding, and Finishing. Also, the Art of Marbling Book-edges and Paper. By JAMES B. NICHOLSON. Illustrated. 12mo., cloth $2.25

NICOLLS.—The Railway Builder:
A Hand-Book for Estimating the Probable Cost of American Railway Construction and Equipment. By WILLIAM J. NICOLLS, Civil Engineer. Illustrated, full bound, pocket-book form . $2.00

NORMANDY.—The Commercial Handbook of Chemical Analysis:
Or Practical Instructions for the Determination of the Intrinsic or Commercial Value of Substances used in Manufactures, in Trades, and in the Arts. By A. NORMANDY. New Edition, Enlarged, and to a great extent rewritten. By HENRY M. NOAD, Ph.D., F.R.S., thick 12mo. $5.00

NORRIS.—A Handbook for Locomotive Engineers and Machinists:
Comprising the Proportions and Calculations for Constructing Locomotives; Manner of Setting Valves; Tables of Squares, Cubes, Areas, etc., etc. By SEPTIMUS NORRIS, M. E. New edition. Illustrated, 12mo. $1.50

NYSTROM.—A New Treatise on Elements of Mechanics:
Establishing Strict Precision in the Meaning of Dynamical Terms; accompanied with an Appendix on Duodenal Arithmetic and Metrology. By JOHN W. NYSTROM, C. E. Illustrated. 8vo. $2.00

NYSTROM.—On Technological Education and the Construction of Ships and Screw Propellers:
For Naval and Marine Engineers. By JOHN W. NYSTROM, late Acting Chief Engineer, U.S.N. Second edition, revised, with additional matter. Illustrated by seven engravings. 12mo. . $1.50

O'NEILL.—A Dictionary of Dyeing and Calico Printing:
Containing a brief account of all the Substances and Processes in use in the Art of Dyeing and Printing Textile Fabrics; with Practical Receipts and Scientific Information. By CHARLES O'NEILL, Analytical Chemist. To which is added an Essay on Coal Tar Colors and their application to Dyeing and Calico Printing. By A. A. FESQUET, Chemist and Engineer. With an appendix on Dyeing and Calico Printing, as shown at the Universal Exposition, Paris, 1867. 8vo., 491 pages $3.50

ORTON.—Underground Treasures:
How and Where to Find Them. A Key for the Ready Determination of all the Useful Minerals within the United States. By JAMES ORTON, A.M., Late Professor of Natural History in Vassar College, N. Y.; Cor. Mem. of the Academy of Natural Sciences, Philadelphia, and of the Lyceum of Natural History, New York; author of the "Andes and the Amazon," etc. A New Edition, with Additions. Illustrated $1.50

OSBORN.—The Prospector's Field Book and Guide:
In the Search for and the Easy Determination of Ores and Other Useful Minerals. By Prof. H. S. OSBORN, LL. D., Author of "The Metallurgy of Iron and Steel;" "A Practical Manual of Minerals, Mines, and Mining." Illustrated by 44 Engravings. 12mo. $1.50

OSBORN.—A Practical Manual of Minerals, Mines and Mining:
Comprising the Physical Properties, Geologic Positions, Local Occurrence and Associations of the Useful Minerals; their Methods of Chemical Analysis and Assay: together with Various Systems of Excavating and Timbering, Brick and Masonry Work, during Driving, Lining, Bracing and other Operations, etc. By Prof. H. S. OSBORN, LL. D., Author of the "Metallurgy of Iron and Steel." Illustrated by 171 engravings from original drawings. 8vo. $4.50

OVERMAN.—The Manufacture of Steel:
Containing the Practice and Principles of Working and Making Steel. A Handbook for Blacksmiths and Workers in Steel and Iron, Wagon Makers, Die Sinkers, Cutlers, and Manufacturers of Files and Hardware, of Steel and Iron, and for Men of Science and Art. By FREDERICK OVERMAN, Mining Engineer, Author of the "Manufacture of Iron," etc. A new, enlarged, and revised Edition. By A. A. FESQUET, Chemist and Engineer. 12mo. . . $1.50

OVERMAN.—The Moulder's and Founder's Pocket Guide:
A Treatise on Moulding and Founding in Green-sand, Dry-sand, Loam, and Cement; the Moulding of Machine Frames, Mill-gear, Hollow-ware, Ornaments, Trinkets, Bells, and Statues; Description of Moulds for Iron, Bronze, Brass, and other Metals; Plaster of Paris, Sulphur, Wax, etc.; the Construction of Melting Furnaces, the Melting and Founding of Metals; the Composition of Alloys and their Nature, etc., etc. By FREDERICK OVERMAN, M. E. A new Edition, to which is added a Supplement on Statuary and Ornamental Moulding, Ordnance, Malleable Iron Castings, etc. By A. A. FESQUET, Chemist and Engineer. Illustrated by 44 engravings. 12mo. . $2.00

PAINTER, GILDER, AND VARNISHER'S COMPANION.
Containing Rules and Regulations in everything relating to the Arts of Painting, Gilding, Varnishing, Glass-Staining, Graining, Marbling, Sign-Writing, Gilding on Glass, and Coach Painting and Varnishing; Tests for the Detection of Adulterations in Oils, Colors, etc.; and a Statement of the Diseases to which Painters are peculiarly liable, with the Simplest and Best Remedies. Sixteenth Edition. Revised, with an Appendix. Containing Colors and Coloring—Theoretical and Practical. Comprising descriptions of a great variety of Additional Pigments, their Qualities and Uses, to which are added, Dryers, and Modes and Operations of Painting, etc. Together with Chevreul's Principles of Harmony and Contrast of Colors. 12mo. Cloth $1.50

PALLETT.—The Miller's, Millwright's, and Engineer's Guide.
By HENRY PALLETT. Illustrated. 12mo. . . . $2.00

PERCY.—The Manufacture of Russian Sheet-Iron.
By JOHN PERCY, M. D., F. R. S., Lecturer on Metallurgy at the Royal School of Mines, and to The Advance Class of Artillery Officers at the Royal Artillery Institution, Woolwich; Author of "Metallurgy." With Illustrations. 8vo., paper . . 50 cts.

PERKINS.—Gas and Ventilation:
Practical Treatise on Gas and Ventilation. With Special Relation to Illuminating, Heating, and Cooking by Gas. Including Scientific Helps to Engineer-students and others. With Illustrated Diagrams. By E. E. PERKINS. 12mo., cloth $1.25

PERKINS AND STOWE.—A New Guide to the Sheet-iron and Boiler Plate Roller:
Containing a Series of Tables showing the Weight of Slabs and Piles to Produce Boiler Plates, and of the Weight of Piles and the Sizes of Bars to produce Sheet-iron; the Thickness of the Bar Gauge in decimals; the Weight per foot, and the Thickness on the Bar or Wire Gauge of the fractional parts of an inch; the Weight per sheet, and the Thickness on the Wire Gauge of Sheet-iron of various dimensions to weigh 112 lbs. per bundle; and the conversion of Short Weight into Long Weight, and Long Weight into Short. Estimated and collected by G. H. PERKINS and J. G. STOWE. $2.50

POWELL—CHANCE—HARRIS.—The Principles of Glass Making.
By HARRY J. POWELL, B. A. Together with Treatises on Crown and Sheet Glass; by HENRY CHANCE, M. A. And Plate Glass, by H. G. HARRIS, Asso. M. Inst. C. E. Illustrated 18mo. . $1.50

PROCTOR.—A Pocket-Book of Useful Tables and Formulæ for Marine Engineers:
By FRANK PROCTOR. Second Edition, Revised and Enlarged. Full-bound pocket-book form $1.50

REGNAULT.—Elements of Chemistry:
By M. V. REGNAULT. Translated from the French by T. FORREST BETTON, M. D., and edited, with Notes, by JAMES C. BOOTH, Melter and Refiner U. S. Mint, and WILLIAM L. FABER, Metallurgist and Mining Engineer. Illustrated by nearly 700 wood-engravings. Comprising nearly 1,500 pages. In two volumes, 8vo., cloth . $7.50

RICHARDS.—Aluminium:
Its History, Occurrence, Properties, Metallurgy and Applications, including its Alloys. By JOSEPH W. RICHARDS, A. C., Chemist and Practical Metallurgist, Member of the Deutsche Chemische Gesellschaft. Illust. Third edition, enlarged and revised (1895) . $6.00

RIFFAULT, VERGNAUD, and TOUSSAINT.—A Practical Treatise on the Manufacture of Colors for Painting:
Comprising the Origin, Definition, and Classification of Colors; the Treatment of the Raw Materials; the best Formulæ and the Newest Processes for the Preparation of every description of Pigment, and the Necessary Apparatus and Directions for its Use; Dryers; the Testing, Application, and Qualities of Paints, etc., etc. By MM. RIFFAULT, VERGNAUD, and TOUSSAINT. Revised and Edited by M.

F. MALEPEYRE. Translated from the French, by A. A. FESQUET, Chemist and Engineer. Illustrated by Eighty engravings. In one vol., 8vo., 659 pages $5.00

ROPER.—A Catechism of High-Pressure, or Non-Condensing Steam-Engines:
Including the Modelling, Constructing, and Management of Steam-Engines and Steam Boilers. With valuable illustrations. By STEPHEN ROPER, Engineer. Sixteenth edition, revised and enlarged. 18mo., tucks, gilt edge $2.00

ROPER.—Engineer's Handy-Book:
Containing a full Explanation of the Steam-Engine Indicator, and its Use and Advantages to Engineers and Steam Users. With Formulæ for Estimating the Power of all Classes of Steam-Engines; also, Facts, Figures, Questions, and Tables for Engineers who wish to qualify themselves for the United States Navy, the Revenue Service, the Mercantile Marine, or to take charge of the Better Class of Stationary Steam-Engines. Sixth edition. 16mo., 690 pages, tucks, gilt edge $3.50

ROPER.—Hand-Book of Land and Marine Engines:
Including the Modelling, Construction, Running, and Management of Land and Marine Engines and Boilers. With illustrations. By STEPHEN ROPER, Engineer. Sixth edition. 12mo., tucks, gilt edge. $3.50

ROPER.—Hand-Book of the Locomotive:
Including the Construction of Engines and Boilers, and the Construction, Management, and Running of Locomotives. By STEPHEN ROPER. Eleventh edition. 18mo., tucks, gilt edge . $2.50

ROPER.—Hand-Book of Modern Steam Fire-Engines.
With illustrations. By STEPHEN ROPER, Engineer. Fourth edition, 12mo., tucks, gilt edge $3.50

ROPER.—Questions and Answers for Engineers.
This little book contains all the Questions that Engineers will be asked when undergoing an Examination for the purpose of procuring Licenses, and they are so plain that any Engineer or Fireman of ordinary intelligence may commit them to memory in a short time. By STEPHEN ROPER, Engineer. Third edition . . $3.00

ROPER.—Use and Abuse of the Steam Boiler.
By STEPHEN ROPER, Engineer. Eighth edition, with illustrations. 18mo., tucks, gilt edge $2.00

ROSE.—The Complete Practical Machinist:
Embracing Lathe Work, Vise Work, Drills and Drilling, Taps and Dies, Hardening and Tempering, the Making and Use of Tools, Tool Grinding, Marking out Work, etc. By JOSHUA ROSE. Illustrated by 356 engravings. Thirteenth edition, thoroughly revised and in great part rewritten. In one vol., 12mo., 439 pages $2.50

ROSE.—Mechanical Drawing Self-Taught:
Comprising Instructions in the Selection and Preparation of Drawing Instruments, Elementary Instruction in Practical Mechanical Draw

ing, together with Examples in Simple Geometry and Elementary Mechanism, including Screw Threads, Gear Wheels, Mechanical Motions, Engines and Boilers. By JOSHUA ROSE, M. E. Illustrated by 330 engravings. 8vo, 313 pages $4.00

ROSE.—The Slide-Valve Practically Explained:
Embracing simple and complete Practical Demonstrations of the operation of each element in a Slide-valve Movement, and illustrating the effects of Variations in their Proportions by examples carefully selected from the most recent and successful practice. By JOSHUA ROSE, M. E. Illustrated by 35 engravings . $1.00

ROSS.—The Blowpipe in Chemistry, Mineralogy and Geology:
Containing all Known Methods of Anhydrous Analysis, many Working Examples, and Instructions for Making Apparatus. By LIEUT.-COLONEL W. A. ROSS, R. A., F. G. S. With 120 Illustrations. 12mo. $2.00

SHAW.—Civil Architecture:
Being a Complete Theoretical and Practical System of Building, containing the Fundamental Principles of the Art. By EDWARD SHAW, Architect. To which is added a Treatise on Gothic Architecture, etc. By THOMAS W. SILLOWAY and GEORGE M. HARDING, Architects. The whole illustrated by 102 quarto plates finely engraved on copper. Eleventh edition. 4to. $7.50

SHUNK.—A Practical Treatise on Railway Curves and Location, for Young Engineers.
By W. F. SHUNK, C. E. 12mo. Full bound pocket-book form $2.00

SLATER.—The Manual of Colors and Dye Wares.
By J. W. SLATER. 12mo. $3.00

SLOAN.—American Houses:
A variety of Original Designs for Rural Buildings. Illustrated by 26 colored engravings, with descriptive references. By SAMUEL SLOAN, Architect. 8vo. $1.50

SLOAN.—Homestead Architecture:
Containing Forty Designs for Villas, Cottages, and Farm-houses, with Essays on Style, Construction, Landscape Gardening, Furniture, etc., etc. Illustrated by upwards of 200 engravings. By SAMUEL SLOAN, Architect. 8vo. $3.50

SLOANE.—Home Experiments in Science.
By T. O'CONOR SLOANE, E. M., A. M., Ph. D. Illustrated by 91 engravings. 12mo. $1.50

SMEATON.—Builder's Pocket-Companion:
Containing the Elements of Building, Surveying, and Architecture; with Practical Rules and Instructions connected with the subject. By A. C. SMEATON, Civil Engineer, etc. 12mo. . . $1.50

SMITH.—A Manual of Political Economy.
By E. PESHINE SMITH. A New Edition, to which is added a full Index. 12mo. $1.25

SMITH.—Parks and Pleasure-Grounds:
Or Practical Notes on Country Residences, Villas, Public Parks, and Gardens. By CHARLES H. J. SMITH, Landscape Gardener and Garden Architect, etc., etc. 12mo. $2.00

SMITH.—The Dyer's Instructor:
Comprising Practical Instructions in the Art of Dyeing Silk, Cotton, Wool, and Worsted, and Woolen Goods; containing nearly 800 Receipts. To which is added a Treatise on the Art of Padding; and the Printing of Silk Warps, Skeins, and Handkerchiefs, and the various Mordants and Colors for the different styles of such work. By DAVID SMITH, Pattern Dyer. 12mo. . . . $2.00

SMYTH.—A Rudimentary Treatise on Coal and Coal-Mining.
By WARRINGTON W. SMYTH, M. A., F. R. G., President R. G. S. of Cornwall. Fifth edition, revised and corrected. With numerous illustrations. 12mo. $1.75

SNIVELY.—Tables for Systematic Qualitative Chemical Analysis.
By JOHN H. SNIVELY, Phr. D. 8vo. $1.00

SNIVELY.—The Elements of Systematic Qualitative Chemical Analysis:
A Hand-book for Beginners. By JOHN H. SNIVELY, Phr. D. 16mo. $2.00

STOKES.—The Cabinet-Maker and Upholsterer's Companion:
Comprising the Art of Drawing, as applicable to Cabinet Work; Veneering, Inlaying, and Buhl-Work; the Art of Dyeing and Staining Wood, Ivory, Bone, Tortoise-Shell, etc. Directions for Lackering, Japanning, and Varnishing; to make French Polish, Glues, Cements, and Compositions; with numerous Receipts, useful to workmen generally. By STOKES. Illustrated. A New Edition, with an Appendix upon French Polishing, Staining, Imitating, Varnishing, etc., etc. 12mo $1.25

STRENGTH AND OTHER PROPERTIES OF METALS:
Reports of Experiments on the Strength and other Properties of Metals for Cannon. With a Description of the Machines for Testing Metals, and of the Classification of Cannon in service. By Officers of the Ordnance Department, U. S. Army. By authority of the Secretary of War. Illustrated by 25 large steel plates. Quarto . $10.00

SULLIVAN.—Protection to Native Industry.
By Sir EDWARD SULLIVAN, Baronet, author of "Ten Chapters on Social Reforms." 8vo. $1.00

SULZ.—A Treatise on Beverages:
Or the Complete Practical Bottler. Full instructions for Laboratory Work, with Original Practical Recipes for all kinds of Carbonated Drinks, Mineral Waters, Flavorings, Extracts, Syrups, etc. By CHAS. HERMAN SULZ, Technical Chemist and Practical Bottler Illustrated by 428 Engravings. 816 pp. 8vo . . $10.00

SYME.—Outlines of an Industrial Science.
By David Syme. 12mo. $2.00

TABLES SHOWING THE WEIGHT OF ROUND, SQUARE, AND FLAT BAR IRON, STEEL, ETC.,
By Measurement. Cloth 63

TAYLOR.—Statistics of Coal:
Including Mineral Bituminous Substances employed in Arts and Manufactures; with their Geographical, Geological, and Commercial Distribution and Amount of Production and Consumption on the American Continent. With Incidental Statistics of the Iron Manufacture. By R. C. Taylor. Second edition, revised by S. S. Haldeman. Illustrated by five Maps and many wood engravings. 8vo., cloth $6.00

TEMPLETON.—The Practical Examiner on Steam and the Steam-Engine:
With Instructive References relative thereto, arranged for the Use of Engineers, Students, and others. By William Templeton, Engineer. 12mo. $1.00

THAUSING.—The Theory and Practice of the Preparation of Malt and the Fabrication of Beer:
With especial reference to the Vienna Process of Brewing. Elaborated from personal experience by Julius E. Thausing, Professor at the School for Brewers, and at the Agricultural Institute, Mödling, near Vienna. Translated from the German by William T. Brannt, Thoroughly and elaborately edited, with much American matter, and according to the latest and most Scientific Practice, by A. Schwarz and Dr. A. H. Bauer. Illustrated by 140 Engravings. 8vo., 815 pages $10.00

THOMAS.—The Modern Practice of Photography:
By R. W. Thomas, F. C. S. 8vo. . . . 25

THOMPSON.—Political Economy. With Especial Reference to the Industrial History of Nations:
By Robert E. Thompson, M. A., Professor of Social Science in the University of Pennsylvania. 12mo. $1.50

THOMSON.—Freight Charges Calculator:
By Andrew Thomson, Freight Agent. 24mo. . . $1.25

TURNER'S (THE) COMPANION:
Containing Instructions in Concentric, Elliptic, and Eccentric Turning; also various Plates of Chucks, Tools, and Instruments; and Directions for using the Eccentric Cutter, Drill, Vertical Cutter, and Circular Rest; with Patterns and Instructions for working them. 12mo. $1.25

TURNING: Specimens of Fancy Turning Executed on the Hand or Foot-Lathe:
With Geometric, Oval, and Eccentric Chucks, and Elliptical Cutting Frame. By an Amateur. Illustrated by 30 exquisite Photographs. 4to. $3.00

VAILE.—Galvanized-Iron Cornice-Worker's Manual:
Containing Instructions in Laying out the Different Mitres, and Making Patterns for all kinds of Plain and Circular Work. Also, Tables of Weights, Areas and Circumferences of Circles, and other Matter calculated to Benefit the Trade. By CHARLES A. VAILE. Illustrated by twenty-one plates. 4to. $5.00

VILLE.—On Artificial Manures:
Their Chemical Selection and Scientific Application to Agriculture. A series of Lectures given at the Experimental Farm at Vincennes, during 1867 and 1874-75. By M. GEORGES VILLE. Translated and Edited by WILLIAM CROOKES, F. R. S. Illustrated by thirty-one engravings. 8vo., 450 pages $6.00

VILLE.—The School of Chemical Manures:
Or, Elementary Principles in the Use of Fertilizing Agents. From the French of M. GEO. VILLE, by A. A. FESQUET, Chemist and Engineer. With Illustrations. 12mo. $1.25

VOGDES.—The Architect's and Builder's Pocket-Companion and Price-Book:
Consisting of a Short but Comprehensive Epitome of Decimals, Duodecimals, Geometry and Mensuration; with Tables of United States Measures, Sizes, Weights, Strengths, etc., of Iron, Wood, Stone, Brick, Cement and Concretes, Quantities of Materials in given Sizes and Dimensions of Wood, Brick and Stone; and full and complete Bills of Prices for Carpenter's Work and Painting; also, Rules for Computing and Valuing Brick and Brick Work, Stone Work, Painting, Plastering, with a Vocabulary of Technical Terms, etc. By FRANK W. VOGDES, Architect, Indianapolis, Ind. Enlarged, revised, and corrected. In one volume, 368 pages, full-bound, pocket-book form, gilt edges $2.00
Cloth 1.50

VAN CLEVE.—The English and American Mechanic:
Comprising a Collection of Over Three Thousand Receipts, Rules, and Tables, designed for the Use of every Mechanic and Manufacturer. By B. FRANK VAN CLEVE. Illustrated. 500 pp. 12mo. $2.00

WAHNSCHAFFE.—A Guide to the Scientific Examination of Soils:
Comprising Select Methods of Mechanical and Chemical Analysis and Physical Investigation. Translated from the German of Dr. F. WAHNSCHAFFE. With additions by WILLIAM T. BRANNT. Illustrated by 25 engravings. 12mo. 177 pages . . . $1.50

WALL.—Practical Graining:
With Descriptions of Colors Employed and Tools Used. Illustrated by 47 Colored Plates, Representing the Various Woods Used in Interior Finishing. By WILLIAM E. WALL. 8vo. . $2.50

WALTON.—Coal-Mining Described and Illustrated:
By THOMAS H. WALTON, Mining Engineer. Illustrated by 24 large and elaborate Plates, after Actual Workings and Apparatus. $5.00

WARE.—The Sugar Beet.
Including a History of the Beet Sugar Industry in Europe, Varieties of the Sugar Beet, Examination, Soils, Tillage, Seeds and Sowing, Yield and Cost of Cultivation, Harvesting, Transportation, Conservation, Feeding Qualities of the Beet and of the Pulp, etc. By LEWIS S. WARE, C. E., M. E. Illustrated by ninety engravings. 8vo.
$4.00

WARN.—The Sheet-Metal Worker's Instructor:
For Zinc, Sheet-Iron, Copper, and Tin-Plate Workers, etc. Containing a selection of Geometrical Problems; also, Practical and Simple Rules for Describing the various Patterns required in the different branches of the above Trades. By REUBEN H. WARN, Practical Tin-Plate Worker. To which is added an Appendix, containing Instructions for Boiler-Making, Mensuration of Surfaces and Solids, Rules for Calculating the Weights of different Figures of Iron and Steel, Tables of the Weights of Iron, Steel, etc. Illustrated by thirty-two Plates and thirty-seven Wood Engravings. 8vo. . $3.00

WARNER.—New Theorems, Tables, and Diagrams, for the Computation of Earth-work:
Designed for the use of Engineers in Preliminary and Final Estimates, of Students in Engineering, and of Contractors and other non-professional Computers. In two parts, with an Appendix. Part I. A Practical Treatise; Part II. A Theoretical Treatise, and the Appendix, Containing Notes to the Rules and Examples of Part I.; Explanations of the Construction of Scales, Tables, and Diagrams, and a Treatise upon Equivalent Square Bases and Equivalent Level Heights. The whole illustrated by numerous original engravings, comprising explanatory cuts for Definitions and Problems, Stereometric Scales and Diagrams, and a series of Lithographic Drawings from Models; Showing all the Combinations of Solid Forms which occur in Railroad Excavations and Embankments. By JOHN WARNER, A. M., Mining and Mechanical Engineer. Illustrated by 14 Plates. A new, revised and improved edition. 8vo. $4.00

WATSON.—A Manual of the Hand-Lathe:
Comprising Concise Directions for Working Metals of all kinds, Ivory, Bone and Precious Woods; Dyeing, Coloring, and French Polishing; Inlaying by Veneers, and various methods practised to produce Elaborate work with Dispatch, and at Small Expense. By EGBERT P. WATSON, Author of "The Modern Practice of American Machinists and Engineers." Illustrated by 78 engravings. $1.50

WATSON.—The Modern Practice of American Machinists and Engineers
Including the Construction, Application, and Use of Drills, Lathe Tools, Cutters for Boring Cylinders, and Hollow-work generally, with the most Economical Speed for the same; the Results verified by Actual Practice at the Lathe, the Vise, and on the Floor. Together

with Workshop Management, Economy of Manufacture, the Steam Engine, Boilers, Gears, Belting, etc., etc. By EGBERT P. WATSON. Illustrated by eighty-six engravings. 12mo. . . . $2.50

WATSON.—The Theory and Practice of the Art of Weaving by Hand and Power:
With Calculations and Tables for the Use of those connected with the Trade. By JOHN WATSON, Manufacturer and Practical Machine-Maker. Illustrated by large Drawings of the best Power Looms. 8vo. $6.00

WATT.—The Art of Soap Making:
A Practical Hand-book of the Manufacture of Hard and Soft Soaps, Toilet Soaps, etc., including many New Processes, and a Chapter on the Recovery of Glycerine from Waste Leys. By ALEXANDER WATT. Ill. 12mo. $3.00

WEATHERLY.—Treatise on the Art of Boiling Sugar, Crystallizing, Lozenge-making, Comfits, Gum Goods,
And other processes for Confectionery, etc., in which are explained, in an easy and familiar manner, the various Methods of Manufacturing every Description of Raw and Refined Sugar Goods, as sold by Confectioners and others. 12mo. $1.50

WIGHTWICK.—Hints to Young Architects:
Comprising Advice to those who, while yet at school, are destined to the Profession; to such as, having passed their pupilage, are about to travel; and to those who, having completed their education, are about to practise. Together with a Model Specification involving a great variety of instructive and suggestive matter. By GEORGE WIGHTWICK, Architect. A new edition, revised and considerably enlarged; comprising Treatises on the Principles of Construction and Design. By G. HUSKISSON GUILLAUME, Architect. Numerous Illustrations. One vol. 12mo. $2.00

WILL.—Tables of Qualitative Chemical Analysis.
With an Introductory Chapter on the Course of Analysis. By Professor HEINRICH WILL, of Giessen, Germany. Third American, from the eleventh German edition. Edited by CHARLES F. HIMES Ph. D., Professor of Natural Science, Dickinson College, Carlisle, Pa 8vo. $1.50

WILLIAMS.—On Heat and Steam:
Embracing New Views of Vaporization, Condensation, and Explosion. By CHARLES WYE WILLIAMS, A. I. C. E. Illustrated 8vo. $2.50

WILSON.—A Treatise on Steam Boilers:
Their Strength, Construction, and Economical Working. By ROBERT WILSON. Illustrated 12mo. $2.50

WILSON.—First Principles of Political Economy:
With Reference to Statesmanship and the Progress of Civilization. By Professor W. D. WILSON, of the Cornell University. A new and revised edition. 12mo. $1.50

WOHLER.—A Hand-Book of Mineral Analysis:
By F. WÖHLER, Professor of Chemistry in the University of Göttingen. Edited by HENRY B. NASON, Professor of Chemistry in the Rensselaer Polytechnic Institute, Troy, New York. Illustrated. 12mo. $2.50

WORSSAM.—On Mechanical Saws:
From the Transactions of the Society of Engineers, 1869. By S. W. WORSSAM, JR. Illustrated by eighteen large plates. 8vo. $2.50

RECENT ADDITIONS.

BRANNT.—Varnishes, Lacquers, Printing Inks and Sealing-Waxes:
Their Raw Materials and their Manufacture, to which is added the Art of Varnishing and Lacquering, including the Preparation of Putties and of Stains for Wood, Ivory, Bone, Horn, and Leather. By WILLIAM T. BRANNT. Illustrated by 39 Engravings, 338 pages. 12mo. $3.00

BRANNT—The Practical Scourer and Garment Dyer:
Comprising Dry or Chemical Cleaning; the Art of Removing Stains; Fine Washing; Bleaching and Dyeing of Straw Hats, Gloves, and Feathers of all kinds; Dyeing of Worn Clothes of all fabrics, including Mixed Goods, by One Dip; and the Manufacture of Soaps and Fluids for Cleansing Purposes. Edited by WILLIAM T. BRANNT, Editor of "The Techno-Chemical Receipt Book." Illustrated. 203 pages. 12mo. $2.00

BRANNT.—Petroleum.
Its History, Origin, Occurrence, Production, Physical and Chemical Constitution, Technology, Examination and Uses; Together with the Occurrence and Uses of Natural Gas. Edited chiefly from the German of Prof. Hans Hoefer and Dr. Alexander Veith, by WM. T. BRANNT. Illustrated by 3 Plates and 284 Engravings. 743 pp. 8vo. $7.50

BRANNT.—A Practical Treatise on the Manufacture of Vinegar and Acetates, Cider, and Fruit-Wines:
Preservation of Fruits and Vegetables by Canning and Evaporation; Preparation of Fruit-Butters, Jellies, Marmalades, Catchups, Pickles, Mustards, etc. Edited from various sources. By WILLIAM T. BRANNT. Illustrated by 79 Engravings. 479 pp. 8vo. $5.00

BRANNT.—The Metal Worker's Handy-Book of Receipts and Processes:
Being a Collection of Chemical Formulas and Practical Manipulations for the working of all Metals; including the Decoration and Beautifying of Articles Manufactured therefrom, as well as their Preservation. Edited from various sources. By WILLIAM T. BRANNT. Illustrated. 12mo. $2.50

DEITE.—A Practical Treatise on the Manufacture of Perfumery:
Comprising directions for making all kinds of Perfumes, Sachet Powders, Fumigating Materials, Dentifrices, Cosmetics, etc., with a full account of the Volatile Oils, Balsams, Resins, and other Natural and Artificial Perfume-substances, including the Manufacture of Fruit Ethers, and tests of their purity. By Dr. C. DEITE, assisted by L. BORCHERT, F. EICHBAUM, E. KUGLER, H. TOEFFNER, and other experts. From the German, by WM. T. BRANNT. 28 Engravings. 358 pages. 8vo. $3.00

EDWARDS.—American Marine Engineer, Theoretical and Practical:
With Examples of the latest and most approved American Practice. By EMORY EDWARDS. 85 illustrations. 12mo. . . $2.50

EDWARDS.—900 Examination Questions and Answers:
For Engineers and Firemen (Land and Marine) who desire to obtain a United States Government or State License. Pocket-book form, gilt edge $1.50

POSSELT.—Technology of Textile Design:
Being a Practical Treatise on the Construction and Application of Weaves for all Textile Fabrics, with minute reference to the latest Inventions for Weaving. Containing also an Appendix, showing the Analysis and giving the Calculations necessary for the Manufacture of the various Textile Fabrics. By E. A. POSSELT, Head Master Textile Department, Pennsylvania Museum and School of Industrial Art, Philadelphia, with over 1000 illustrations. 292 pages. 4to. $5.00

POSSELT.—The Jacquard Machine Analysed and Explained:
With an Appendix on the Preparation of Jacquard Cards, and Practical Hints to Learners of Jacquard Designing. By E. A. POSSELT. With 230 illustrations and numerous diagrams, 127 pp. 4to. $3.00

POSSELT.—The Structure of Fibres, Yarns and Fabrics:
Being a Practical Treatise for the Use of all Persons Employed in the Manufacture of Textile Fabrics, containing a Description of the Growth and Manipulation of Cotton, Wool, Worsted, Silk, Flax, Jute, Ramie, China Grass and Hemp, and Dealing with all Manufacturers' Calculations for Every Class of Material, also Giving Minute Details for the Structure of all kinds of Textile Fabrics, and an Appendix of Arithmetic, specially adapted for Textile Purposes. By E. A. POSSELT. Over 400 Illustrations. quarto. . $10.00

RICH.—Artistic Horse-Shoeing:
A Practical and Scientific Treatise, giving Improved Methods of Shoeing, with Special Directions for Shaping Shoes to Cure Different Diseases of the Foot, and for the Correction of Faulty Action in Trotters. By GEORGE E. RICH. 62 Illustrations. 153 pages. 12mo. $1.00

RICHARDSON.—Practical Blacksmithing:
A Collection of Articles Contributed at Different Times by Skilled Workmen to the columns of "The Blacksmith and Wheelwright," and Covering nearly the Whole Range of Blacksmithing, from the Simplest Job of Work to some of the Most Complex Forgings. Compiled and Edited by M. T. RICHARDSON.
Vol. I.   210 Illustrations.   224 pages.   12mo.   .   . $1.00
Vol. II. 230 Illustrations.   262 pages.   12mo.   .   . $1.00
Vol. III. 390 Illustrations.   307 pages.   12mo.   .   . $1.00
Vol. IV. 276 Illustrations.   276 pages.   12mo.   .   . $1.00

RICHARDSON —The Practical Horseshoer:
Being a Collection of Articles on Horseshoeing in all its Branches which have appeared from time to time in the columns of "The Blacksmith and Wheelwright," etc. Compiled and edited by M. T. RICHARDSON.   174 illustrations. $1.00

ROPER.—Instructions and Suggestions for Engineers and Firemen:
By STEPHEN ROPER, Engineer. 18mo. Morocco   . $2.00

ROPER.—The Steam Boiler: Its Care and Management:
By STEPHEN ROPER, Engineer. 12mo., tuck, gilt edges. $2.00

ROPER.—The Young Engineer's Own Book:
Containing an Explanation of the Principle and Theories on which the Steam Engine as a Prime Mover is Based. By STEPHEN ROPER, Engineer. 160 illustrations, 363 pages. 18mo., tuck . $3.00

ROSE.—Modern Steam-Engines:
An Elementary Treatise upon the Steam-Engine, written in Plain language; for Use in the Workshop as well as in the Drawing Office. Giving Full Explanations of the Construction of Modern Steam-Engines: Including Diagrams showing their Actual operation. Together with Complete but Simple Explanations of the operations of Various Kinds of Valves, Valve Motions, and Link Motions, etc., thereby Enabling the Ordinary Engineer to clearly Understand the Principles Involved in their Construction and Use, and to Plot out their Movements upon the Drawing Board. By JOSHUA ROSE, M. E. Illustrated by 422 engravings. Revised. 358 pp. .   . $6.00

ROSE.—Steam Boilers:
A Practical Treatise on Boiler Construction and Examination, for the Use of Practical Boiler Makers, Boiler Users, and Inspectors; and embracing in plain figures all the calculations necessary in Designing or Classifying Steam Boilers. By JOSHUA ROSE, M. E. Illustrated by 73 engravings. 250 pages. 8vo. $2.50

SCHRIBER.—The Complete Carriage and Wagon Painter:
A Concise Compendium of the Art of Painting Carriages, Wagons, and Sleighs, embracing Full Directions in all the Various Branches, including Lettering, Scrolling, Ornamenting, Striping, Varnishing, and Coloring, with numerous Recipes for Mixing Colors. 73 Illustrations. 177 pp. 12mo. $1.00

www.ingramcontent.com/pod-product-compliance
Lightning Source LLC
Chambersburg PA
CBHW031413230426
43668CB00007B/299